# Van Nostrand Reinhold Soil Science Series

**Editor: Charles W. Finkl, Jnr., Florida Atlantic University**

SOIL CLASSIFICATION / *Charles W. Finkl, Jnr.*
CHEMISTRY OF IRRIGATED SOILS / *Rachel Levy*
SOIL SALINITY: Two Decades of Research in Irrigated Agriculture /
      *H. Frenkel and A. Meiri*
ANDOSOLS / *Kim H. Tan*
PODZOLS / *Peter Buurman*
SOIL NUTRIENT AVAILABILITY: Chemistry and Concepts / *Y. K. Soon*
ADSORPTION PHENOMENA / *Robert D. Harter*
SOIL EROSION AND ITS CONTROL / *R. P. C. Morgan*
SOIL MICROMORPHOLOGY / *Georges Stoops and Hari Eswaran*
LAND EVALUATIONS / *Donald A. Davidson*
SOIL MINERAL WEATHERING / *J. A. Kittrick*
CHEMISTRY OF SOIL SOLUTIONS / *Adel M. Elprince*


Related Titles of Interest
THE ENCYCLOPEDIA OF SOIL SCIENCE, PART I / *Rhodes W. Fairbridge
      and Charles W. Finkl, Jnr.*
SOIL MECHANICS, 3rd edition / *R. F. Craig*
SURFICIAL DEPOSITS OF THE UNITED STATES / *Charles B. Hunt*
EROSION AND SEDIMENT YIELD / *Jonathan N. Laronne and
      M. Paul Mosley*
MINED-LAND REHABILITATION / *Dennis L. Law*

# SOIL EROSION AND ITS CONTROL

Edited by

**R. P. C. MORGAN**
**Silsoe College**
**Silsoe, Bedford, England**

A Hutchinson Ross Publication

VNR VAN NOSTRAND REINHOLD COMPANY
New York

Manufactured in the United States of America.

Published by Van Nostrand Reinhold Company Inc.
135 West 50th Street
New York, New York 10020

Van Nostrand Reinhold Company Limited
Molly Millars Lane
Wokingham, Berkshire RG11 2PY, England

Van Nostrand Reinhold
480 Latrobe Street
Melbourne, Victoria 3000, Australia

Macmillan of Canada
Division of Gage Publishing Limited
164 Commander Boulevard
Agincourt, Ontario MIS 3C7, Canada

15   14   13   12   11   10   9   8   7   6   5   4   3   2   1

**Library of Congress Cataloging in Publication Data**
Main entry under title:
Soil erosion and its control.
    (Van Nostrand Reinhold soil science series)
    "A Hutchinson Ross publication."
    Includes index.
    1. Soil erosion—Addresses, essays, lectures.   2. Soil conservation—
Addresses, essays, lectures.   I. Morgan, R. P. C. (Royston Philip Charles),
1942-   II. Series.
S623.S574   1986      631.4'5        85-11083
ISBN 0-442-26441-0

# CONTENTS

**Series Editor's Foreword**                                                vii
**Preface**                                                                  xi
**Contents by Author**                                                      xiii

**Introduction**                                                             1

## PART I: AWARENESS AND ATTITUDES

**Editor's Comments on Papers 1, 2, and 3**                                   4

1   **LOWDERMILK, W. C.:**   Man-made Deserts                                 9
        *Pacific Affairs* **8(4):**409–419 (1935)

*Ind* 2   **JACKS, G. V.:**   Soil Erosion                                   20
        *The Rape of the Earth. A Survey of Soil Erosion,* G. V. Jacks and R. O. Whyte,
            Faber, London, 1939, pp. 17–28

3   **ECKHOLM, E. P.:**   Two Costly Lessons: The Dust Bowl and the Virgin
        Lands                                                                32
        *Losing Ground,* E. P. Eckholm, Pergamon, Oxford, 1976,
            pp. 46–57, 190–191

## PART II: MEASUREMENT AND INVENTORY

**Editor's Comments on Papers 4, 5, and 6**                                  46

4   **HUDSON, N. W.:**   The Design of Field Experiments on Soil Erosion     52
        *Jour. Agric. Eng. Research* **2:**56–65 (1957)

5   **DUNNE, T.:**   Studying Patterns of Soil Erosion in Kenya             62
        *FAO Soils Bull.* **33:**109–122 (1977)

*Ind* 6   **RAPP, A.:**   Soil Erosion and Sedimentation in Tanzania and Lesotho    76
        *Ambio* **4:**154–163 (1975)

## PART III: ASSESSMENT AND EVALUATION

**Editor's Comments on Papers 7 Through 13**                                 88

7   **GRANT, K. E.:**   Erosion in 1973–74: The Record and the Challenge    95
        *Jour. Soil Water Conserv.* **30:**29–32 (1975)

8   **JIANG, D., L. QI and J. TAN:**   Soil Erosion and Conservation in the
        Wuding River Valley, China                                          99
        *Soil Conservation: Problems and Prospects,* R. P. C. Morgan, ed., Wiley,
            Chichester, 1981, pp. 461–479

Contents

**9** **DAS, D. C.:** Soil Conservation Practices and Erosion Control in India—
A Case Study 118
*FAO Soils Bull.* **33:**11-31, 39-50 (1977)

**10** **RICHTER, G.:** On the Soil Erosion Problem in the Temperate Humid
Area of Central Europe 151
*GeoJournal* **4:**279-287 (1980)

**11** **MORGAN, R. P. C.:** Soil Erosion and Conservation in Britain 160
*Prog. Phys. Geog.* **4:**24-47 (1980)

**12** **TEMPLE, P. H.:** Measurements of Runoff and Soil Erosion at an
Erosion Plot Scale with Particular Reference to Tanzania 184
*Geog. Annaler* **54-A:**203-220 (1972)

**13** **MAENE, L. M., and W. SULAIMAN:** Status of Soil Conservation Research
in Peninsular Malaysia and Its Future Development 202
*Soil Science and Agricultural Development in Malaysia Conf. Proc.,*
Malaysian Society of Soil Science, Kuala Lumpur, 1980, pp. 307-324

### PART IV: MODELING

**Editor's Comments on Papers 14 Through 17** 222

**14** **MEYER, L. D., and W. H. WISCHMEIER:** Mathematical Simulation of the
Process of Soil Erosion by Water 227
*Am. Soc. Agric. Eng. Trans.* **12:**754-758 (1969)

**15** **BENNETT, J. P.:** Concepts of Mathematical Modeling of Sediment
Yield 233
*Water Resources Research* **10:**485-492 (1974)

**16** **FOSTER, G. R., L. J. LANE, J. D. NOWLIN, J. M. LAFLEN and R. A. YOUNG:**
Estimating Erosion and Sediment Yield on Field-Sized Areas 241
*Am. Soc. Agric. Eng. Trans.* **24:**1253-1262 (1981)

**17** **MORGAN, R. P. C., D. D. V. MORGAN and H. J. FINNEY:** A Predictive
Model for the Assessment of Soil Erosion Risk 251
*Jour. Agric. Eng. Research* **30:**245-253 (1984)

### PART V: SOCIAL AND ECONOMIC ASPECTS

**Editor's Comments on Papers 18, 19, and 20** 262

**18** **HUDSON, N. W.:** Non Technical Constraints on Soil Conservation 268
*South-East Asian Regional Symposium on Problems of Soil Erosion and
Sedimentation,* T. Tingsanchali and H. Eggers eds., Asian Institute of
Technology, Bangkok, 1981, pp. 15-26

**19** **TEMPLE, P. H.:** Soil and Water Conservation Policies in the Uluguru
Mountains, Tanzania 280
*Geog. Annaler* **54-A:**110-123 (1972)

**20** **RUNOLFSSON, S.:** Soil Conservation in Iceland 293
*The Breakdown and Restoration of Ecosystems,* M. W. Holdgate and
M. J. Woodman, eds., Plenum, New York, 1978, pp. 231-238

**Author Citation Index** 301
**Subject Index** 307
**About the Editor** 311

# SERIES EDITOR'S FOREWORD

The Van Nostrand Reinhold Soil Science Series attempts to provide cogent summaries of the field by reproducing classical and modern papers, ones that provide keys to understanding of critical turning points in the development of the discipline. Scientific literature today is so vast and widely dispersed, especially in a multifaceted discipline like soil science, that much valuable information becomes ignored by default. Many pioneering works are now coveted by libraries, and retrieval from the archives is not easy. In fact, many important papers published in the ephemeral literature are no longer available to serious or committed researchers through interlibrary loan. Other professionals devoted to teaching or burdened with administrative duties must be hard pressed to keep up with comprehensive arrays of technical literature spread through scores of journals. Most of us can, at best, skim only a few select journals to make copies of tables of contents, abstracts and summaries, and reviews in order to remain abreast of specialized and often limited aspects of the robust field of soil science as a whole.

This series in soil science, developed as a practical solution to this problem, reprints key papers and investigative landmarks that relate to a common theme. The papers are reproduced in facsimile, either in their entirety or in significant part, so readers can follow major original events in the field, not peruse paraphrased or abbreviated versions of others. Some foreign works have been especially translated for use in the series. Occasionally short, foreign language articles are reproduced from French or German journals.

Essays by the volume editor provide running commentaries that introduce readers to highlights in the field, provide critical evaluation of the significance of the various papers, and discuss the development of selected topics or subject areas. It is hoped that the volume editor's comments will ease the transition for the seasoned investigator who wishes to step into a new field of research as well as provide students and professors with a compact working library of most important scientific advances in soil science.

Areas of specialization in soil science are divided by the International Society of Soil Science into seven divisions or "commissions." The first six commissions cover soil physics, chemistry, mineralogy, biology, fertility, and technology. Because the scope of the field is so great, we concentrate initially on topics traditionally devoted to the seventh commission: soil morphology, genesis, classification, and geography. The series thus begins with volumes dealing with the major soils of the world: their recognition, characteristics, formation, distribution, and classification. Other volumes concentrate on topics in agronomy, soil-plant relationships, soil engineering topics, or melds of pure science with soil systems. The Van Nostrand Reinhold Soil Science Series plows deeply through the field, picking significant but timely topics on an eclectic basis.

Each volume in the series is edited by a specialist or authority in the area covered by the book. The volume editor's efforts reflect a concerted worldwide search, review, selection, and distillation of the primary literature contained in journals and monographs and in industrial and governmental reports. Individual volumes thus represent an information-selection and repackaging program of value to libraries, students, and professionals.

The books contain a preface, introduction, and highlight commentaries by the volume editor. Many volumes contain rare papers that are hard to locate and obtain, as well as landmark papers published in English for the first time. All volumes contain author citation and subject indexes of the contained papers, usually twenty to fifty key papers in a given subject area.

The papers reprinted in this volume deal with different aspects of soil erosion, one of a series of problems confronting modern agriculturalists. Along with drought, salinization, and deteriorating quality of irrigation water, erosion is one of the most important agricultural problems of the world. Erosion of topsoil is a primary contributor to nonpoint pollution of surface waters and is a major factor in the loss of valuable plant nutrients from croplands. Estimates of erosion losses in the United States during the 1970s ranged as high as 4 billion Mg (metric tons) annually. At this rate of soil loss, it will take only about 100 years to wash away every single hectare of cropland now being farmed in the United States. Representing about a 30 percent increase over soil loss in the 1930s, the infamous dust bowl years of the Great Plains, much of the problem ironically can be traced to new techniques of farming. Designed to substitute capital, petroleum, or technology for labor, mechanization seems to have coaxed many land managers into assuming that basic conservation techniques are now obsolete or no longer required in modern farming operations. Such a notion could hardly be farther from the truth. Many prior civilizations have noted, often after the fact, that when soil resources are exploited without regard for environmental stability, agricultural ecosystems eventually fail catastrophically.

Soil erosion may be attributed to natural or geological processes and to losses that are due, in one way or another, to human activity. Geological erosion, on the average tends, to be slower than soil formation so that weathered materials accumulate in the landscape providing a suitable medium for plant growth. The balance between rates of soil development and soil-eroding processes is, however, increasingly upset by human use of the land. Many agriculturalists and land developers, for example, regard soil in economic terms merely as a means to turning a quick profit. They do not consider themselves as custodians of the soil resource, the greatest of all assests in many nations. Human erosion accelerates the erosive forces of wind and water by removing the resisting forces (such as vegetation cover) that retard erosion. The lack of effective conservation techniques that employ windbreaks, contour cropping, grass waterways, hydraulic channels, pasture furrows, terracing, tile drains, stubble mulching, and cover crops, to name but a few, has contributed to serious soil loss in many regions.

The risks of soil erosion will eventually challenge the resources of producers (farmers, foresters, and ranchers) and the skill and resourcefulness of researchers, educators, and politicians as they attempt to achieve the impossible dream—produce more food at lower costs and protect the soil resource base at the same time. As losses from soil erosion increase due to the misuse of croplands in industrialized societies, from the clearing of tropical rainforests in the Amazon Basin, and from the expanding desertification of semiarid regions (e.g. the Sahel region of Africa), the need to learn from prior successes in erosion control will become pregnant. The papers reprinted in this volume provide a historical perspective to the pervasive nature of soil erosion, recognition of the erosion problem, and attempts to reduce soil losses by conservation. Some critical turning points and benchmark studies of soil erosion are reprinted in this volume as part of an effort to highlight important stages in our understanding of the problem.

CHARLES W. FINKL, JNR.

# PREFACE

An appreciation of our current understanding of soil erosion and its control can only be gained by bringing together studies from the fields of science, including social science, technology, and economics. Important contributions have been made by agriculturalists, agronomists, foresters, rangeland specialists, and land resource planners as well as by civil and agricultural engineers in their search for practical solutions to erosion problems. The need to design and implement soil conservation measures in the field has meant that the subject has developed mainly at the applied end of the research spectrum through a combination of trial and error and engineering experience. It is, therefore, essentially an empirical science. Only in the last two decades has a more fundamental base been provided through studies by geomorphologists and agricultural engineers on the mechanics of erosion. This development has been accompanied by a greater realization by soil scientists of the role of recent erosion and sedimentation in pedogenesis. Another trend has been the greater importance attached to social and economic aspects of soil conservation in the search for explanations of the failure of many erosion control schemes.

This background has produced a richness of literature but a multiplicity of approaches. While the interdisciplinary nature of soil erosion and conservation is widely recognized, research results are scattered over a large number of journals, and scientists often do not have a comprehensive view. This problem is compounded by important bodies of literature being published in French, German, Russian, and Chinese, in addition to English. Since much work is carried out by consultancy and advisory services, there is also a large volume of material in the form of reports, handbooks, and pamphlets with limited circulation.

In selecting items for this Soil Science volume I have focused on the more practical aspects of the subject and avoided topics like erosion processes and soil formation. This prevents overlap with other similar collections, notably in geomorphology and soil science, in which key contributions have already appeared. Although the editorial comments represent a personal view, readers have the opportunity to examine the material selected for themselves. Whether readers' views are in sympathy with the opinions expressed is not important. What matters is that, through agreement or disagreement, a stimulus is provided to question the way our thoughts are progressing. Hopefully this will result in more considered policies for research and for the implementation of soil conservation measures in the future. At present we are wasting resources by continually repeating research that has already been done elsewhere. This repetition occurs because scientists are unaware of what goes on in other disciplines and of what is published in other languages. It also occurs because insufficient thought is given to whether the questions that the research is answering are still appropriate. Thus we continue to collect data from erosion plots in the field with little idea of how accurate the measurements are. More important, we have even less idea of how accurate we need them to be.

A historical approach is adopted only for Parts I and IV. For Parts II, III, and V, papers are selected over a rather narrow time range but from a variety of disciplines. The opportunity has been taken to combine work in scientific journals and books with that in bulletins, reports, and conference proceedings, thereby bringing the nonjournal sources to a wider readership.

The intention is to present a historical and spatial perspective on the state of the art with a view to provoking discussion on what we should be doing to improve soil conservation practice in the future. The volume is aimed at all who are concerned with soil erosion and its control.

R. P. C. MORGAN

# CONTENTS BY AUTHOR

Bennett, J. P., 233
Das, D. C., 118
Dunne, T., 62
Eckholm, E. P., 32
Finney, H. J., 251
Foster, G. R., 241
Grant, K. E., 95
Hudson, N. W., 52, 268
Jacks, G. V., 20
Jiang, D., 99
Laflen, J. M., 241
Lane, L. J., 241
Lowdermilk, W. C., 9
Maene, L. M., 202

Meyer, L. D., 227
Morgan, D. D. V., 251
Morgan, R. P. C., 160, 251
Nowlin, J. D., 241
Qi, L., 99
Rapp, A., 76
Richter, G., 151
Runolfsson, S., 293
Sulaiman, W., 202
Tan, J., 99
Temple, P. H., 184, 280
Wischmeier, W. H., 227
Young, R. A., 241

# SOIL EROSION AND ITS CONTROL

# INTRODUCTION

The purpose of this set of papers is to discuss, through the medium of an anthology, the significance of soil erosion today. Part I is concerned with changes in our awareness of and attitudes towards soil erosion. Assessment of the severity and spatial extent of the problem, however, requires reliable data on rates of erosion. Many local and global-scale evaluations have been attempted, but are they comparable? Is there a need for a standardized data-collecting system? In Part II we examine the techniques of data collection from small-scale hillslope erosion plots to regional scale assessments using reservoir surveys and remote sensing. Particular attention is paid to the accuracy of the information obtained.

Part III presents seven case studies of soil erosion ranging from countries where it is recognized as a major problem, namely the United States, China and India, to those where erosion is of lower magnitude, as in the United Kingdom and West Germany. From the evidence contained in these studies, some of the questions raised in Part I are reexamined. Do the semi-humid regions really have the worst problem? Is there really no erosion problem in western Europe? Part IV concentrates on the development of predictive models of soil erosion for use as design tools in planning conservation measures. The papers reviewed relate to the new generation of models intended to replace the Universal Soil Loss Equation.

Part V examines the need for awareness of the cultural environment when designing soil conservation strategies. Since the technology of erosion control is reasonably well established, it is becoming fashionable to argue that it is the social and economic constraints that prevent its successful implementation. This argument presupposes that in developing the technology the right questions have been answered and that the technology is, therefore, appropriate. With a better understanding of erosion mechanics, can refinements be made to the technology to make it more acceptable? Can a comparison of successful and unsuccessful implementation of erosion control measures allow guidelines to be developed for future conservation schemes? Will political ideologies render these guidelines unacceptable?

Part I

# AWARENESS AND ATTITUDES

# Editor's Comments
# on Papers 1, 2, and 3

**1    LOWDERMILK**
*Man-made Deserts*

**2    JACKS**
Excerpt from *Soil Erosion*

**3    ECKHOLM**
*Two Costly Lessons: The Dust Bowl and the Virgin Lands*

It is increasingly difficult to find new ways of introducing texts on soil erosion. Themes typically relate to man's continuous struggle with the problem of erosion from the earliest days of settled agriculture, or to the impending crisis resulting from the need to feed an increasing world population against a background of land going out of agricultural production through erosion. The common message of the three papers reviewed here is how erosion develops when the land or, perhaps more strictly, the soil is exploited without regard for long-term or even medium-term environmental stability. Failure to understand the ecological sensitivity of the environment when natural ecosystems based on forests or grasslands are replaced by agricultural ecosystems has catastrophic effects. The dust clouds over the United States in 1934, described by Eckholm in Paper 3, constitute a short-term example. More important, however, are the medium-term effects of desertification and human suffering, such as occurred following the 1930s Dust Bowl. In the long term, erosion may be a major contributory factor in the collapse of civilizations. Examples are described by Lowdermilk (Paper 1) in Mesopotamia and the Mediterranean. A similar situation may exist today in the Sahel region of Africa.

In Paper 2, Jacks states that the semi-humid regions of the world are the most prone to erosion, and both Lowdermilk and Eckholm provide circumstantial evidence to support this (Papers 1 and 3). Studies of sediment yields in rivers by Fournier (1960) and Walling and Webb (1982) have also since confirmed this. Since these regions are ecologically the most sensitive, disturbance by man results in extensive deterioration of the land in association with desertification. This situation contrasts with the more localized land degradation that occurs with the misuse of more humid regions. Small changes in the vegetation cover in semi-arid and semi-humid regions bring about higher proportionate increases in rates of erosion than is the case in other environments (Kirkby, 1980). The relatively high natural rates of erosion in these areas, however, make it difficult to distinguish the effects of man-induced changes in the vegetation from natural changes. A long-term progressive climatic deterioration may not be readily identifiable against the background of short-term fluctuations in rainfall, which happen to coincide with a period of greater pressure on and mismanagement of the fragile ecosystems. Hare (1977) concluded that it was the coincidence of drought and misuse of the land which brought about the 1968–1973 desertification of the Sahel.

In contrast with the situation in the semi-humid regions of the world, there is, according to Jacks, no problem of erosion in western Europe. He clearly states that the agricultural practices adopted there have led to improvements in soil fertility and have promoted an agricultural ecosystem that is well adapted to its environment. Recent research (Richter, 1983; Morgan, 1985) suggests that this view is ill founded and that rates of soil loss, particularly for sandy and sandy loam soils low in organic matter, are too high to allow sustained agricultural use in western Europe far into the next century.

The main concern of the authors whose works are reviewed here is the effect of erosion on agricultural productivity. It was against this background that research scientists in the 1960s attempted to determine what the maximum acceptable rate of erosion or soil loss tolerance should be. All soils in the United States were assigned tolerance values from 4.5 to 11.2 t ha$^{-1}$ y$^{-1}$, depending on their depth, previous erosion history, and fertility (Mannering, 1981). By the mid-1970s it was recognized that these values needed revising. On the one hand, there are some soils in parts of Iowa and Illinois with topsoil depths of 30 cm or more where the loss of the topmost 5 cm through erosion would have little effect on productivity. On the other hand, there has been greater attention paid in recent years to the pollution of water bodies. Sediment, either in its own right or through the chemicals adsorbed to it, is a major pollutant. Since the rates of erosion from hillsides that create pollution problems are much lower than those that were accepted as critical for agricultural productivity, the dictates of water quality mean that tolerance values need to be lowered (Moldenhauer and Onstad, 1975). Erosion is thus potentially a severer problem than is indicated in these introductory papers.

The period since the 1930s has seen the emergence of many texts on soil conservation describing the technology available for erosion control. Along with this, enormous advances have been made in our understanding of erosion in terms of the mechanics of the processes involved. Yet, despite this research, has any real progress been made? The types of problems described by Eckholm in 1976 are the same as those discussed by Jacks in 1939 and, based on Lowdermilk's research, the same as those which must have prevailed as Mesopotamia declined.

In addition to discussing the breakdown of civilization in the Middle East, Lowdermilk draws attention to overgrazing in Turkestan, the desiccation of large areas of the Sahara, and gullying in the loess area of northwest China. A common theme, he points out, was the failure to employ any conservation measures as agriculture extended into the steeper slopes of the uplands. Still, throughout early civilization, some aspects of conservation were known. Terraces, now in disrepair, occur on the hillslopes around Petra in Jordan, an important center from 200 B.C. to 200 A.D. They are also found in Judea and around Tyr (Tyre) and Saida (Sidon), those of the latter being built by the Phoenicians and representing one of the earliest conservation structures in the world. The most extensive terrace systems were those built by the Incas, in which each terrace is comprised of a 1- to 5-m high stone wall backfilled with rocks and gravel to within 1 m of its surface and then covered with topsoil carried from the lowlands. The inhabitants of Carthage understood the principles of dry farming and water spreading. The Romans gave much attention to agriculture; the use of legumes in rotations was widely recommended and the maintenance of soil fertility was considered a virtue. Terracing and contouring were widely practiced (Lowdermilk, 1953; Troeh, Hobbs, and Donahue, 1982).

The principles of agronomic and mechanical measures for soil conservation, along with soil management, have been restated many times throughout history, each time with some development, refinement, and increase in understanding. The work of Jethro Tull and Charles Townsend was concerned with sustaining soil fertility, especially on sandy soils. Their ideas were repeated in early American agricultural history by Jared Eliot and Samuel Deane (Stallings, 1957). Up until the early 1800s, research was carried out by innovative farmers and enthusiastic amateurs, but later in the nineteenth century, with the founding of organizations like Rothamsted Experimental Station and the Dokuchayev Soil Institute, proper research trials on soil fertility were established and the results disseminated in scientific journals.

Generally, it seems that whenever erosion was recognized as a problem, the techniques for controlling it were also known but were not widely used. Thus, although erosion has become more serious as agriculture has extended onto more marginal land and although our knowledge of soil conservation measures has improved, the overall perception of the problem has not changed. Through case studies of the Dust Bowl in the 1930s and the plowing-up of the Virgin Lands of the Soviet Union in the 1950s, Eckholm shows how previous lessons are rarely learned. Information on erosion control does not diffuse well in time or space, and scientific work is frequently ignored— especially by governments—until a catastrophe occurs.

There is some evidence, however, to show that the situation is not always bad. Jacks describes the Huang He (Yellow River) in China as an outstanding and eternal symbol of the mortality of civilization. With an annual sediment load of 2,500 million tons of which 1,600 million comes from the loess plateau, the erosion rate is one of the highest in the world. Soil loss on hillslopes varies from 50 to 200 t ha$^{-1}$ y$^{-1}$. Yet amelioration is possible. Bench terracing has locally reduced soil loss by 95%, afforestation by 75–90%, and grass-growing by 60–80% (Gong and Jiang, 1977; Gong and Mou, 1984). The extent of these improvements is examined more fully in a later section of this volume.

Another example also considered in detail later is Iceland, where, at the time of settlement in 875 A.D., about 40,000 km$^2$ was covered with the climax vegetation of grassland and birch forest. By 1200 A.D. this resource carried sufficient livestock to support a population of 80,000. Overgrazing, however, led to severe erosion by the end of the Middle Ages. Erosion, combined with loss of grazing land through burial beneath deposits of ash laid down in periodic volcanic eruptions and a hostile climate, caused the population to decline to 50,000 by 1700 and the grassland area to fall to 20,000 km$^2$. Since that time, a more efficient use of pasture, improved management of lowland meadows, and drainage of the marshland have increased stocking capacities, and the population has risen to 220,000. Although erosion is still severe, an effective soil conservation service is gradually reducing its impact (Friðriksson, 1972).

## REFERENCES

Fournier, F., 1960, *Climat et érosion: la relation entre l'érosion du sol par l'eau et les précipitations atmosphériques,* Presses Universitaires, Paris, 201p.

Friðriksson, S., 1972, Grass and grass utilization in Iceland, *Ecology* **53:**785-796.

Gong, S., and D. Jiang, 1977, Soil erosion and its control in small gully water-sheds in the rolling loess area on the middle reaches of the Yellow River, *Paris Symposium on Erosion and Solid Matter Transported in Inland Waters,* Preprint. Also published in *Scientia Sinica* **22**(6), 1978.

Gong, S., and J. Mou, 1984, Utilization of water and sediment resources of the Wuding River, in *River Basin Strategy, International Seminar on the Relevance of River Basin Approach for Coordinated Land and Water Conservation and Management,* University of Linköping, pp. 38–47.

Hare, F. K., 1977, Climate and desertification, in *Desertification: Its Causes and Consequences,* Pergamon, Oxford, pp. 63–167.

Kirkby, M. J., 1980, The problem, in *Soil Erosion,* M. J. Kirkby and R.P.C. Morgan, eds., Wiley, Chichester, pp. 1–16.

Lowdermilk, W. C., 1953, Conquest of the land through seven thousand years, *USDA Soil Conservation Service, Agricultural Information Bull. No. 99,* 30p.

Mannering, J. V., 1981, The use of soil loss tolerances as a strategy for soil conservation, in *Soil Conservation: Problems and Prospects,* R. P. C. Morgan, ed., Wiley, Chichester, pp. 337–349.

Moldenhauer, W. C., and C. A. Onstad, 1975, Achieving specified soil loss levels, *Jour. Soil and Water Conserv.* **30:**166–168.

Morgan, R. P. C., 1985, Soil erosion measurement and soil conservation research in cultivated areas of the UK, *Geog. Jour.* **151:**11-20.

Richter, G., 1983, Aspects and problems of soil erosion hazard in the EEC countries, in *Soil Erosion,* A. G. Prendergast, ed., Abridged Proceedings, Workshop on Soil Erosion and Conservation: Assessment of the Problems and State of the Art in EEC Countries, Commission of the European Communities, Luxembourg, pp. 9–15.

Stallings, J. H., 1957, *Soil Conservation,* Prentice-Hall, Englewood Cliffs, N.J.

Troeh, F. R., J. A. Hobbs, and R. L. Donahue, 1982, *Soil and Water Conservation for Productivity and Environmental Protection,* Prentice-Hall, Englewood Cliffs, N.J.

Walling, D. E., and B. W. Webb, 1982, Patterns of sediment yield, in *Background to Palaeohydrology,* K. J. Gregory, ed., Wiley, Chichester, pp. 69–100.

**8**

# 1

# MAN-MADE DESERTS

## W. C. LOWDERMILK

THE history of civilizations is a record of struggles against the progressive desiccation of civilized lands. The more ancient the civilization, the drier and more wasted, usually, is the supporting country. In fact, so devastating seems the occupation of man that, with a few striking exceptions, a desert or near-desert condition is often associated with his long habitation of a region. Two major factors are believed to account for the growth of man-made deserts. In the first place, semi-arid to semi-humid regions proved the most favorable sites for the early development of human culture. Such areas, however, stand in a condition of delicate ecological balance between humid and true desert climates. Comparatively slight disturbances of the coverage of vegetation and soils, such as are brought about by human occupation for grazing and cultivation, are sufficient to extend the borders of the desert far beyond the natural true desert into more humid climates.

In the second place, processes of soil erosion are accelerated by the exposure of soil surfaces hitherto protected by complete mantles of vegetation, whether grass or forest, by heavy grazing and cultivation. It is only within the past decade that experimental studies of these processes have been made. So enormous have been the differences in soil wastage and superficial run-off of rain waters from bared sloping lands, as compared with similar surfaces protected by a complete coverage of vegetation, that new light is thrown on the problem of the decadence of former civilizations. Aside from other important factors the history of civilizations may be interpreted in terms of soil erosion, so direct is the relation between the productive condition of soils and the prosperity of a people. The operation of mankind's exploitative and destructive activities is often decisive in zones of delicate balance between soil formation and destruction, between rain absorption and rapid run-off.

Recently the archaeologists have turned back the pages of history, not merely centuries, but thousands of years. Their post-mortems on buried civilizations suggest that it has been the hand of man, more than climatic change, which has reduced once rich and populous regions to desolation and poverty. After a long struggle, a civilization either died or its people migrated to more productive regions. Many ancient civilizations, once revelling in a golden age of prosperity, are crumbling in ruins or lie buried in sands and debris, largely caused by the destructive treatment of the lands on which they were dependent for sustenance. If modern peoples are to escape a similar fate by man-induced impoverishment and the desiccation of their lands, it would seem well to take a measure of these destructive processes and forces, and by intelligent land planning and land use provide for the sustained productivity of agricultural lands and the protection of grass lands and forests for food, textiles, raw materials and continued water supply.

It is evident that climatic changes have occurred in the past and are still in progress. Such changes follow the pace of land movements and are comparatively slow in terms of human history. Superimposed upon them there may be a rapid growth of human populations and their activity, as well as that of their herds, which can produce increased desiccation equivalent in effect to changes of climate. It becomes important to discover how far human occupation is rendering the earth less inhabitable and at the same time to discover means by which such processes of deterioration may be held in check and productivity sustained. It is possible for man and his animals to render regions uninhabitable, especially in zones of delicate ecological balance between humid and true desert climates. Man-made deserts may extend from semi-arid climates to humid climates, under certain conditions. In the light of this conception, of man-induced desiccation, it is in place to examine what is now known about the results of human occupation, in the way of increasing aridity due to destruction of vegetative cover, and how these desert conditions are rapidly being brought about in various areas throughout the world.

New knowledge concerning desert conditions resulting from overgrazing of domestic herds, especially in periods of drought, has interesting implications. Aerial moving pictures taken by Mr. and Mrs. Martin Johnson in Africa show wild herds, numbering scores of thousands, blackening the landscape as they trot across the plains, in clouds of dust, to water holes. The landscape shows every evidence of destruction of vegetation and breakdown of surface soils, resembling the effects of over-grazing by domestic herds in America. The Westover and Enlow Expedition in Russian Turkistan during 1934 reported that water holes were usually marked by active sand dunes, due to the complete destruction of surrounding vegetation by converging herds. This suggests the possibility that in the remote past enormous wild grazing herds, during drought periods, may have so utterly destroyed vegetation as to set in motion desert-producing processes, not unlike those induced by domestic herds in the past and present.

Many students have attributed desiccation and the consequent drying up of streams to the removal of forests. That is only part of the story. The great enemy of the human race is soil erosion, which has been associated with the habitations of man since before the dawn of history. It is no new land disease, but has only recently been diagnosed and named for what it is. The removal of vegetation, whether grass, brush, or forest, exposes soils to the dash of rain or the blast of wind, against which they had been protected for thousands of years. Topsoils blow away or wash away, or both. Unprotected sloping lands are usually bared to hard and tight subsoils, which drain off the water as from a tiled roof. The perennial streams, deprived of their reservoirs of supply, dry up except in rainy seasons, when they become torrential floods and sweep boulders and debris down the slopes to deposit them on otherwise fertile lowlands. Then starving wild or domestic herds clean the devastated areas of all palatable vegetation, only to reduce the effectiveness of beneficent rains and to accentuate aridity. This is the cycle which, whether ancient or modern, has transformed vast areas of good lands into extensive arid bad lands, or actual desert. With the loss of vegetation and soils, which had developed interdependently for unknown ages, near-desert to desert conditions have been brought about in the old inhabited portions of the world.

According to archaeologists the Sahara, the Central Asian deserts,

11

arid parts of Palestine, Mesopotamia and the Gobi and North China were once teeming with human life. Traditions of peoples descended from ancient cultures tell of immigration to their present habitation from what are now desert regions of Central Asia. The origin of European peoples was in the East. The Hindus came from the north, the Chinese from the west. Yet this land from which they came is today an immense desert where only very limited regions are still able to nourish a scanty population. Sir Aurel Stein's discoveries of sand-buried ruins in Central Asia revealed numerous towns a square mile or more in size, in a region now depopulated. There were ruins of cities, castles, aqueducts, reservoirs and all the evidences of lost cultures, of vanished populations. Gibbon declared that 500 cities once flourished in what are now the dry depopulated plains of Asia Minor. The recently discovered ruin of Tepe Gawra in northern Mesopotamia is claimed to be the oldest remaining town in the world. The ruins show that in B. C. 3700 this was a well planned city, which must have represented long ages of prior development. The peninsula of Arabia contained an enormous population, called Sealand, which at times annoyed Babylon from B. C. 2500 to 616. Now, a few fierce nomadic Bedouins, remnants of former cultures, fight for existence over every drop of water and every sign of vegetation. The great Sahara desert has recently revealed monuments, ruins of cities, temples, implements and unearthed cut trees. Campalion, the famous Egyptologist, says of it "and so the astonishing fact dawns upon us that this desert was once a region of groves and fountains and the abode of happy millions." The very gradual climatic changes due to the present age of retreating ice do not appear sufficient to account for the excessively rapid desiccation of the vast areas known to have sustained at one time enormous populations.

WE HAVE a written record of encroaching deserts. When Zenobia was overthrown by the Romans under Aurelian, its capital, Palmyra, was the metropolis of a mighty empire. Now the sands of the Sahara almost hide the ruins of that stupendous city of marble and gold. As late as the rise of Mohammed Tripoli, on the north coast of Africa, had a population of six million. It was then clothed with vineyards, orchards and forests. It is now bare of vegetation. The streams are dried up and the population reduced to about forty-five thousand. Archaeologists now claim to have discovered, under shifting sands, the capital of the rich kingdom of the Queen of Sheba. Dr. Breasted, the flying archaeologist, attempted to take twelve thousand feet of film over ancient ruins now being excavated. He encountered fierce and choking dust storms making it necessary to rise to 12,000 feet in order to breathe clean air. These dust blizzards are an exhibition of wind erosion at work on denuded areas. Such dust storms occurred on a stupendous scale for the first time in western United States in 1934 and 1935.

The aerial photographs taken by the Shipley Johnson expedition in Peru portray a similar condition. On the west coast of South America, where the oldest known civilization developed on the western continent, the entire region photographed was shown to be in a treeless, barren, denuded condition. Ancient walls were in some instances entirely drifted over with sand, so that only from the air could they be recognized as huge mute reminders of former civilizations. The native population of today seemed very poor specimens of the flourishing civilization and culture of the past, before man and erosion had completed the destruction of vegetation.

If man has turned fertile lands into barren, semi-arid wastes or actual desert, what are the diabolic processes? Soil wastage or erosion, caused by the destruction of vegetation, is not the geologic erosion which with the leisure of land uplift or subsidence has carved out canyons, rounded off hills and filled in alluvial valleys. In this slow process, the balance between vegetation and soil formation and erosion is undisturbed, and soils are formed as rapidly as they are washed away. Geologic erosion does not proceed faster than soils are formed under a protective cover of vegetation. Thus we may use this geologic norm of erosion, responsive to local conditions, as a basis for the measurement of what may be called "accelerated or man-induced erosion," in which soils are washed away far in excess of possible soil formation.

This accelerated erosion is the direct result of destroying the protective vegetable covering of soils, whether by burning forest or grasslands, by over-grazing, or by clearing and cultivation. Erosion of an accelerated order then begins. The fertile topsoil is rapidly washed away, or blown off. Erosion sorts soils with machine precision. The fine and fertile particles are carried far away, either by wind or by flowing water, leaving the coarse and less fertile materials dumped near by. Water erosion begins with little rivulets, which grow into gullies and tear away the fertile topsoil. Denuded areas may soon be washed away to intractable subsoil. Percolation is reduced to an important degree, and the concentrated rain waters rush off the land as if from a tiled roof, carrying rocks and debris with them. These are strewn on the level alluvial bottom lands, impairing or destroying their productivity. Gullies drain out the ground moisture, excavate the soil, and fasten themselves onto a countryside like tentacles reaching out in all directions. Sheet erosion is less spectacular than wind or gully erosion but is perhaps more dangerous, because it works so stealthily, skimming off the fine-textured soil from plowed fields after each shower. Yearly these little rivulets are plowed over and the farmer forgets his loss until finally his fields change to the color of the infertile subsoil, or "grow rocks." Results from one of many experiments at erosion experiment stations indicate that it would require 12,000 years to wash away 12 inches of surface soil on the Marshall Silt loam in the Mississippi Valley (Missouri), when covered with alfalfa, or more than 100,000 years when covered by native sod; whereas it would require only 29 to 36 years to wash away one foot of soil when cultivated to corn on an eight per cent slope.

The destructive forces of erosion are not confined to fields robbed of fertility or lowlands ruined with rocks and debris. Silt-bearing streams become choked and overflow, with inestimable loss of life and property. Irrigation systems and reservoirs are impounded with troublesome silt which destroys the storage of irrigation water. Furthermore, the denuded hill and mountain slopes store less water than was possible in their former thick, humus, spongy soils, and streams become dry rocky beds in dry seasons, when they are most needed, and roaring destructive torrents in rainy periods.

A few years ago I conducted a number of expeditions into Northwest China, the "cradle of Chinese civilization." Tradition of early times said the Chinese had come from the west. The forebears of the race of Han settled in the alluvial plains of the Yellow River. The largest of these, above the delta, is the lower valley of the Wei River, where Ch'angan was the capital city. The Weipei plain, in Shensi, where the Chinese civilization first came to flower, the seat of the golden age of China, is now filled with the ruins of a former opulence and magnificence. The slopes of the surrounding region are riddled with gargantuan gullies, which promptly drain away most of the usual rainfall. The aspect of the drainage areas is one of an approaching desert impoverishment. Erosion of soils from out of the watersheds has put out of commission an ancient irrigation system and left the former populous plain sparsely peopled and subject to drought famines once or more in a decade. The deterioration of this region might well be assumed to have been due to adverse climatic change, if it were not for the temple forests now growing without artificial aid about Buddhist temples.

My first realization of the man-made destruction of Northwest China was a walled city in the upper Fen River valley, almost empty, with deserted homes and a dried up stream bed outside the city gate. Detailed field studies of the surrounding mountains and lands revealed the whole tragic story of accelerated or man-induced erosion. The first inhabitants had found the mountains heavily forested, the valleys fertile and well watered with perennial streams. They built their city and prospered. As the population increased, they destroyed the forests, primarily to cultivate the rich humus soils, and bench terracing was not used to safeguard the soils. As the topsoil washed off the farmers went higher, clearing the lands until the tops of the mountains were reached.

Soil erosion progressively reduced productivity until cultivation was abandoned. Sheep and goats were turned out on the wasting fields to complete the destruction. Gullies started which swept boulders and debris down onto the fertile lowlands. The streams had gradually dried up. Thus the forests were gone; the streams were gone; the soils were gone; and when the soils go, man either starves or migrates to other regions. A land that supported millions with plenty now meagerly supports a small population frequently ravaged by famines. It is roughly estimated that an average of at least twelve inches of topsoil have, by this suicidal agricultural process, been washed off of hundreds of millions of acres of the water-sheds of North China, by this kind of accelerated erosion, above the normal geologic erosion during recent times.

The decadence of North China has often been attributed to adverse climatic changes. Conclusions from my own studies, reported elsewhere, indicate that man-induced soil erosion and its consequences in increased run-off would account for such decadence without climatic change. I have frequently found, in North China, temple forests like green emeralds in an ugly setting of denuded mountains. These forests had been protected throughout the centuries from the ax and the plow, and the teeth and hoofs of sheep and goats. In their cool and refreshing shade, the trees were reproducing naturally within the present prevailing climate and rainfall. These forests as samples are evidence enough that the present climate would support such cover over similar regions. Thus loss of vegetation in North China is not due to increasing aridity; but increasing desiccation has followed the loss of soils, and resulting lack of conservation of moisture.

Erosion is not confined to temperate zones, if man sets in motion soil-destroying forces. The great Mayan civilization, undoubtedly one of the highest of prehistoric America, was destroyed by erosion. Dr. C. Wythe Cooke of the United States Geological Survey declares that "the Maya civilization choked itself to death with mud washed from its own hillside corn patches. The Maya cities were built near small lakes which are now silted up with sticky clay soil. These lakes were used for transportation. On the nearby hills, the farmers grew their corn. With continued cultivation of the slopes, the soils washed off; transportation on the lakes was made impossible and they were then forced to migrate as is recorded in history."

A man-made desert is not as fantastic as it sounds. At least we may call deserts the regions of aridity and desolation where the recklessness, ignorance or hunger-drive of man have supplemented the forces of wind and water erosion in destroying vegetation and soils, resulting in regional suicide. Many nations are now awakening to the menace of the prodigal wastage of soil erosion. South

Africa has thus lost the productivity of millions of acres. France, Greece, Spain, Australia, Madagascar, Italy and the United States, all show the destroying forces of erosion. In sharp contrast Germany and Japan, particularly, have provided for prevention and control of soil erosion and for the preservation of forest and grass resources.

America has been developing desiccated and unproductive lands more rapidly than probably ever before occurred. About three hundred years ago the colonists entered this continent of vast untouched resources with a burst of energy and began an unexampled period of exploitation. There were reservoirs of population in Europe which supplied millions of vigorous and daring people to clear away the forests and cultivate the soil at an astonishing rate, in their westward march of agricultural occupation. It was a rapid advance over a wide front by farmers and stockmen with their plows and herds, until today all frontiers have been pushed westward to the Pacific. Lands had been free and the supply seemed inexhaustible. Farmers exploited the best of a farm and then abandoned it to a race between erosion and the healing agencies of nature, and moved on west to clear new lands.

During the Presidency of Theodore Roosevelt, forests of the United States were being slaughtered at an appalling rate. Gifford Pinchot showed the President a painting, done in the fifteenth century, of a beautiful, populous and prosperous well-watered valley at the foot of forested mountains in North China, and with it a photograph of the same valley, taken about 1900. The photograph showed the mountains treeless, glaring and sterile; the stream bed empty and dry; boulders and rocks from the mountains covering the fertile valley lands. The depopulated city had fallen in ruins. The President illustrated his message to Congress with these pictures and caused the establishment of the U. S. Forest Service for the protection of forest lands.

In their brief destructive period of occupancy, the American people on the North American continent have, by the same methods of suicidal use of lands, utterly destroyed and abandoned, through loss of vegetation and soil erosion, 51,000,000 acres of good lands; and 200,000,000 acres more are in the clutches of erosion. Now, in a land of so-called inexhaustible land resources, all good lands are largely under cultivation and millions of farmers are eking out a privation existence on farms whose topsoils have washed away. No longer is it possible to move on to new lands to the west. The day has come for the conservation of remaining soil resources.

To the United States, doubtless, goes the speed record in time and extent, for man-made desert conditions. The dust storms of the old world, long occupied by man, have appeared in the new world—and for the same reasons. Great dust clouds obscuring the sun at midday swept out of the western plains eastward to the Atlantic seaboard for the first time in May 1934. Over large areas, in the central and southwest plains, every living thing choked in the dust-filled atmosphere. Pasture vegetation was coated with dust and made inedible for stock. Fields were turned into sand dunes. It has been a tragic experience, but it is the price that a whole nation is paying for the rapid exploitation of its prairie grazing lands about the close of the war. For centuries, nature had anchored these soils with a thick sod of buffalo or native grasses. Then came the war boom and high prices, which stimulated the plowing up of millions of acres of the western grass lands. Grazing lands were attacked with tractor-drawn plows. The rich humus soils first yielded abundant crops. Rains were plentiful. The same crops were planted year after year.

Then came the drought. The soil-binding quality of the humus had been depleted by continuous cropping. The stubble of poor, unharvested crops was pastured by livestock. The ground was pulverized by their hoofs. The usual strong winds began to blow in the spring. They were dry, and there was no vegetation nor roots to anchor the soils, which were blown aloft in the upper wind currents to form gigantic dust clouds. The machine-like sorting process of wind erosion began. Fine and fertile particles were blown to parts unknown and the heavy material was left behind as drifts or hummocks forming sand dunes, some of them twenty feet high. Since May 1934 wind erosion, set in motion by man-made forces, has transformed 5,000,000 acres of formerly good land into waste areas and great stretches of sand dunes. More than 60,000,000 acres more are in the process of wind erosion destruction by the same cause and will follow the desert condition of the 5,000,000 destroyed acres, unless adequate control methods are undertaken.

We boast of a modern civilization and its progress, but we have been following suicidal methods in treatment of soil resources. With high-powered implements we have been rapidly destroying the vegetation and forests, with resulting loss of productive soils and increasing desiccation. Whether ancient or modern, destruction of vegetation on sloping lands, by whatever cause, exposes fertile soils to wind and water erosion so that soils are destroyed greatly in excess of soil formation, until complete destruction of fertility is accomplished. The capacity of humus soils as reservoirs to conserve rain and snow waters is thus reduced, so that springs and streams dry up. With percolation much reduced on denuded slopes, rain waters concentrate to form destructive gullies, which further destroy land utility. An old writer asserts that "the skin of the animal is not more necessary to its wellbeing than is the vegetative cover of the earth essential to the proper condition of the soil."

But it is not necessary for mankind to destroy the good earth upon which he is dependent for sustenance. Some primitive peoples have discovered means of conserving soils. Ideas of conservation on a national scale, however, have been conceived only in recent times. Erosion can be checked and it can be controlled. The description of erosion control methods is beyond the scope of this paper. It seems clear that man and his animals may extend desert conditions, by processes of man-induced desiccation, into regions formerly capable of supporting large populations. Climate does change, but not at the comparatively rapid rate of the decadence of vast areas of habitable regions. Experimental studies within the past two decades in the character and degree of acceleration of erosion, above the normal rates of geologic processes, have given a better understanding of how deserts may be man-made. With this understanding there may be worked out and put into effect measures adequate to the conservation of soil resources and with them moisture, and therewith a restoration of vegetation, suitable crops, and grass and forests. The lands of the earth are occupied; frontiers of new lands have disappeared. The only new frontier that appears is underfoot, in the maintenance of productivity of lands now occupied.

# 2

## SOIL EROSION

### G. V. Jacks

---

*Soil and civilization. The modern phenomenon of soil erosion. Normal and accelerated erosion. Accelerated erosion a vital problem in the New World. The destruction of ancient empires by erosion. Western Europe's immunity. The rapid development of the New World as a cause of contemporary erosion. The course of erosion. Waste of water by erosion can be more serious than loss of soil. Floods caused by erosion. The destructive power of soil-laden water. Siltation. Wind erosion. Erosion as a problem of human ecology. Fundamentals of erosion control and their political and economic impacts.*

---

To gain control over the soil is the greatest achievement of which mankind is capable. The organization of civilized societies is founded upon the measures taken to wrest control of the soil from wild Nature, and not until complete control has passed into human hands can a stable superstructure of what we call civilization be erected on the land. The great advances in science and particularly in transport during the last hundred years have enabled civilized men to penetrate into every corner of the earth, carrying with them the mixed blessings and curses of civilization. They have planted the tree of European civilization over four continents, and in some places it has appeared to thrive. But scarcely anywhere has it taken firm root in the soil, for European men, despite their skill and power over Nature, have learnt only how to cultivate European soils in a European climate. Modern civilization, outside Europe, is more like a plant that will burst its bud and blossom for a short time in a vase than like a tree that will grow indefinitely with its roots in fertile soil. All seems well with the plant while it continues to blossom; but the flower soon fades and the plant on which it blossomed dies.

All seemed well with civilization in the century of expansion that followed the Industrial Revolution. It was a beneficent growth destined to take possession of the world, and to receive a new lease of life in the countries of the New World as it grew old and declined in Europe. But for some years before the Great War, faint and usually unnoticed signs were appearing that the early promises of the new countries were not to be fulfilled. The economic depression of the late nineteen-twenties brought matters to a head. The depression made the peoples of the world take stock of their position. They felt that the structure of civilization was tottering, and those that lived on the ground floor, near to the earth, began to investigate the stability of the foundations in the soil. Those in the upper stories, devoted to industry, politics, sciences and nearer to the light and sun, were fully occupied in propping up their tottering dwelling places and gave no thought to the foundations, without which all their efforts would be vain. Only within the last few years has some of the shakiness in the upper stories been traced to the crumbling foundations.

For, as the result solely of human mismanagement, the soils upon which men have attempted to found new civilizations are disappearing, washed away by water and blown away by wind. To-day, destruction of the earth's thin living cover is proceeding at a rate and on a scale unparalleled in history, and when that thin cover—the soil—is gone, the fertile regions where it formerly lay will be uninhabitable deserts. Already, indeed, probably nearly a million square miles of new desert have been formed, a far larger area is approaching desert conditions and throughout the New World erosion is taking its relentless toll of soil fertility with incredible and ever increasing speed. Science produces new aids to agriculture—new machines that do the work of a score of men, new crop varieties that thrive in climates formerly considered too harsh for agriculture, new fertilizers that double and treble yields—yet taken the world over the average output per unit area of land is falling. There is a limit to the extent to which applied science can temporarily force up soil productivity, but there is no limit except zero to the extent to which erosion can permanently reduce it. A nation cannot survive in a desert, nor enjoy more than a hollow and shortlived prosperity if it exists by consuming its soil. That is what all the new lands of promise have been doing for the last hundred years, though few as yet realize the full consequences of their past actions or that soil erosion is altering the course of world history more radically than any war or revolution. Erosion is humbling mighty nations, re-shaping their domestic and external policies and once and for all it has barred the way to the El Dorado that a few years ago seemed almost within reach.

Erosion in Nature is a beneficent process without which the world would have died long ago. The same process, accelerated by human mismanagement, has become one of the most vicious and destructive forces that have ever been released by man. What is usually known as 'geological erosion' or 'denudation' is a universal phenomenon which through thousands of years has carved the earth into its present shape. Denudation is an early and important process in soil formation, whereby the original rock material is continuously broken down and sorted out by wind and water until it becomes suitable for colonization by plants. Plants, by the binding effects of their roots, by the protection they afford against rain and wind and by the fertility they impart to the soil, bring denudation almost to a standstill. Everybody must have compared the rugged and irregular shape of bare mountain peaks where denudation is still active with the smooth and harmonious curves of slopes that have long been protected by a mantle of vegetation. Nevertheless, some slight denudation is always occurring. As each superficial film of plant-covered soil becomes exhausted it is removed by rain or wind, to be deposited mainly in the rivers and sea, and a corresponding thin layer of new soil forms by slow weathering of the underlying rock. The earth is continuously discarding its old, worn-out skin and renewing its living sheath of soil from the dead rock beneath. In this way an equilibrium is reached between denudation and soil formation so that, unless the equilibrium is disturbed, a mature soil preserves a more or less constant depth and character indefinitely. The depth is sometimes only a few inches, occasionally several feet, but within it lies the whole capacity of the earth to produce life. Below that thin layer comprising the delicate organism known as soil is a planet as lifeless as the moon.

The equilibrium between denudation and soil formation is easily disturbed by the activities of man. Cultivation, deforestation or the destruction of the natural vegetation by grazing or other means, unless carried out according to certain immutable conditions imposed by each region, may so accelerate denudation that soil, which would normally be washed or blown away in a century, disappears within a year or even within a day. But no human ingenuity can accelerate the soil-renewing process from lifeless rock to an extent at all comparable to the acceleration of denudation. This man-accelerated denudation is what is now known as soil erosion. It is the almost inevitable result of reducing below a certain limit the natural fertility of the soil—of man betraying his most sacred trust when he assumes dominion over the land.

Man-induced soil erosion is taking place to-day in almost every country inhabited by civilized man, except north-western Europe. It is a disease to which any civilization founded on the European model seems liable when it attempts to grow outside Europe. Scarcely any climate or environment is immune from erosion, but it is most virulent in the semi-arid continental grasslands—the steppes, prairies and velds of North and South America, Australia, South Africa and Russia which offer the greatest promise as future homes of civilization. It is also the gravest danger threatening the security of the white man and the well-being of the coloured man in the tropical and sub-tropical lands of Africa and India. Until quite recently erosion was regarded as a matter of merely local concern, ruining a few fields and farmsteads here and there, and compelling the occupiers to abandon their homes and move on to new land, but it is now recognized as a contagious disease spreading destruction far and wide irrespective of private, county, state or national boundaries. Like other contagious diseases, erosion is most easily checked in its early stages; when it has advanced to the stage when it threatens the entire social structure, its control is extremely difficult. In the main, unimportant individuals have started erosion and been crushed by it, until the cumulative losses in property and widespread suffering and want have brought governments and nations, with their immense powers for good or evil, into the fray.

In the United States, the problem of erosion has become a dominant factor in national life; in South Africa, according to General J. C. Smuts, 'erosion is the biggest problem confronting the country, bigger than any politics.' In these two countries erosion has already assumed the proportions of a national disaster of the first magnitude, and has sapped their life blood to such a degree that only a tremendous and single-minded effort from a united nation can prevent a further rapid and irreparable decline. Fortunately, there are signs that the effort will be made in time. Elsewhere, the same destructive processes are at work, but owing to less intense exploitation in the past, they have not advanced so far as in the United States and South Africa. Nevertheless, governments, warned by the example of the United States in particular, are everywhere being compelled to take note of erosion, and when government stirs it means that the question involved is no longer the concern of one section, but of the whole of the community.

That the ultimate consequence of unchecked soil erosion, when it sweeps over whole countries as it is doing to-day, must be national extinction is obvious, for whatever other essential raw material a nation may dispense with, it cannot exist without fertile soil. Nor is extinction of a nation by erosion merely a hypothetical occurrence that may occur at some future date; it has occurred several times in the past. Erosion has, indeed, been one of the most potent factors causing the downfall of former civilizations and empires whose ruined cities now lie amid barren wastes that once were the world's most fertile lands. The deserts of North China, Persia, Mesopotamia and North Africa tell the same story of the gradual exhaustion of the soil as the increasing demands made upon it by expanding civilization exceeded its recuperative powers. Soil erosion, then as now, followed soil exhaustion. The early home of Chinese civilization in the north-west loessial region now resembles a huge battle-field scarred by forces far more destructive than any modern engines of war. The sculpturing of that fantastic landscape is the greatest work of Chinese civilization. Over vast areas the once deep and fertile soil has gone completely, and as it was washed away it tore gaping chasms, sometimes hundreds of feet wide and deep, through the underlying loess and deposited the eroded material on the valley plains and in the rivers and sea. The Yellow River and the Yellow Sea are aptly named, for they are coloured with the yellow subsoil that still pours into them from the now barren loessial hinterland. Hundreds of miles from the eroding region, and for hundreds of miles along its course, the bed of the Yellow River is raised higher and higher above the surrounding country by the continual deposition of eroded soil, the headwaters, no longer absorbed by a porous soil, tear down the hillsides in increasing torrents, and the most disastrous floods in the world, which were once regarded as visitations from Heaven, are now normal and expected occurrences. The Yellow River transports an annual load of 2,500 million tons of soil. There are other rapidly eroding regions and great muddy rivers in China, but the gutted North-West and the Yellow River are the outstanding and eternal symbols of the mortality of civilization.

So in Mesopotamia, the River Tigris which once irrigated and enriched the empire of Babylon and Assyria now flows menacingly on a raised bed of eroded soil brought down from the hills when the plainsmen, seeking more water for their irrigated crops and more land to replace their exhausted soils, cut down the hill forests and were rewarded with uncontrollable floods that overwhelmed their fields and swept away their irrigation works, those great feats of engineering that assured man's passing supremacy over the soil.

On the desert fringes of the Persian and Carthaginian empires, soil exhaustion, crop failures and land abandonment allowed the desert sands to encroach relentlessly. The peoples, weakened at the source of their strength, fell a prey to human conquerors, themselves doomed to be overwhelmed by the drifting sands that now cover the Persian fields where Darius ruled, and the North African plains where Hannibal defended the rich lands of Carthage against the Romans.

But as the soils upon which these ancient civilizations were founded were being washed away or covered over with sand and mud, other civilizations possessing new knowledge and powers were growing up along the eastern and northern shores of the Mediterranean. The might of Greece and Rome had its origin in the mastery of the art of continuous cultivation on forest land

—a notable advance on the shifting cultivation (still widely practised by primitive communities) whereby the forest re-occupied the land after a few years and compelled the cultivators to adopt, at best, a nomadic tribal existence. Continuous cultivation of forest soils enabled permanent, organized communities to develop in a more favourable environment and to attain greater heights than the arid environment where civilization could only exist by irrigation. Mankind, however, never completely masters an art. Shifting cultivation, although it kept men as unimportant servants of wild Nature, maintained soil fertility indefinitely, since the forest drove the cultivator out and re-assumed its beneficent control as soon as any sign of soil exhaustion appeared. Continuous cultivation meant continuous depletion of the soil and always more deforestation to secure new land for the rapidly growing community and to replace worn-out soils. The soil, it is true, was recompensed to some extent by manuring and careful cultivation, but Nemesis could not be delayed indefinitely. The decline of the Roman Empire is a story of deforestation, soil exhaustion and erosion. Judea, likewise, was overcome by erosion. From Spain to Palestine there are no forests left on the Mediterranean littoral, the region is pronouncedly arid instead of having the mild humid character of forest-clad land, and most of its former bounteously rich topsoil is lying at the bottom of the sea. No people, however great and powerful in arms, could maintain its virility and dominance under the conditions that must have prevailed 1,500 years ago in the Mediterranean, and no dictator except Nature can restore the conditions that might allow another world power to arise there. That the Mediterranean countries have not suffered complete annihilation like the earlier empires bordering the deserts is due to the comparative rapidity with which new soil forms from the rock beneath. But soil formation has not kept pace with soil erosion.

In Central America, Colombia, Ceylon and other tropical regions the ruins of bygone civilizations, that struggled for a time to cultivate the tropical forests, have been unearthed. Those civilizations were exterminated completely when men could no longer hold their own on the soil. They succumbed without having contributed anything permanent to the advancement of mankind. When the forests were destroyed, no human power could control the destructive force of the torrential rains that swept the unprotected soils from the hills and flooded the plains, and to-day the forest again rules supreme, showing scarcely a trace of the overlordship once precariously held by man.

One after another, the great empires and civilizations of the past have been swept out of existence by soil erosion. Modern civilization in its birthplace in Western Europe, although subjected to many perilous stresses, seems immune from the danger of slow disintegration from soil erosion. It is dangerous to forecast the future from present conditions, but there are indications that Western Europe may already have passed its zenith while its soils remain the most productive and the most stable under cultivation in the world. The system of continuous cultivation of forest soils, evolved around the Mediterranean, has been gradually perfected in the colder and moister regions of the temperate deciduous and coniferous forest zones. Elsewhere, either in its original or in modified forms, it has proven almost always unsuitable as a basis of civilization, and sometimes immediately disastrous, for no sooner has European civilization established itself in a new country than soil erosion, the invariable destroyer of past civilizations, has set in.

The soils of Western Europe have not eroded in spite of being subjected to the most intensive cultivation in history because the system of cultivation evolved in Europe in the course of centuries has, *under European conditions*, enormously increased soil fertility—to such an extent that to-day the agricultural production of even such a small and industrial country as Britain equals in value that of all Canada. It has been argued that the European climate, with its absence of violent storms or excessive droughts, is particularly inconducive to erosion, but it would be more correct to say that Europe owes its immunity from erosion to the adaptation of its agriculture to its climate. The countryside has been desecrated and scarred with ugliness in many places, but the one inviolable condition on which man holds the lease of land from Nature— that soil fertility be preserved—has in the main been respected. The European farmer has regarded land as an inheritance to be handed on to the next generation in at least as good a condition as when he received it. His attitude has not been born of altruism or necessarily of an innate pride in his work, but is attributable to the character of the natural forest soils which were originally unsuited for intensive agriculture and yielded the occupier a profit proportionate to the agricultural improvements effected in them.

The circumstances in which the New World has been opened up and colonized have been entirely different. With few exceptions, profit and wealth have been most easily won by exploiting and exhausting the virgin soils. In particular, grassland soils required merely a superficial cultivation to convert them immediately into almost ideal arable soils, rich in plant food, perfect in tilth, and apparently incapable of further improvement. Or they afforded rich and extensive pastures without having to be touched at all. The occupier of land was, moreover, a member of an embryonic community in which few restrictions were placed, or seemed necessary, on the use to which the land was put, while the insatiable demands of the Old World and the progress of agricultural science and machinery offered immense profits and further opportunities for exploitation to the man who cashed his soil fertility for labour-saving and yield-increasing devices. Over forty million acres of new land in the United States were brought under the plough during the war and immediate post-war period. They were exploited to the utmost to secure the high profits obtainable, and afterwards they were exhausted with no hope of improvement when debts, tariffs and nationalism barred their produce from oversea markets and ruined the occupiers. To-day much of those forty million acres has been eroded beyond repair or has become 'sub-marginal' land to be left for time and Nature to restore to fertility.

Examples could be multiplied indefinitely to show how in every part of the world circumstances beyond human control have compelled men to exploit their newly acquired dominions beyond the limit of safety. In almost every instance where the limit has been passed, soil erosion has started, slowly and inconspicuously at first but gathering momentum with ever increasing speed and becoming more difficult to check as its destructiveness spreads. Were soil erosion an occasional isolated phenomenon confined to a few mismanaged farms, it could easily be dealt with, by changing the management or converting the managers of the farms to other ways. When it threatens a whole country with ultimate extinction, it can only be checked by changing fundamentally the country's management and the way of life of its people. When erosion becomes a national problem (and there are but few regions where it is not one now) it affects all classes and interests adversely, causes a progressive lowering of the general standard of living, and introduces into the community a feeling of insecurity that is inexplicable to those who see only outward signs of prosperity and are blind to the fact that the only sure foundation upon which a superstructure of civilization can be built is a stable soil. The civilizations of the New World, grown with unparalleled rapidity from European seed and nurtured largely on the political, economic and social traditions gradually evolved over a thousand years in Europe, are proving entirely incompatible with their natural environments.

For erosion is the modern symptom of maladjustment between human society and its environment. It is a warning that Nature is in full revolt against the sudden incursion of an exotic civilization into her ordered domains. Men are permitted to dominate Nature on precisely the same condition as trees and plants, namely on condition that they improve the soil and leave it a little better for their posterity than they found it. Agriculture in Europe, whatever its other weaknesses, has been, and perhaps still is, a practice tending on the whole to increase soil fertility. When adopted and adapted elsewhere it has resulted, almost invariably, in a catastrophic decrease in fertility. The illusion that fertility can always be restored by applying some of the huge amounts of artificial fertilizers now available has been shattered by the recognition that fertility is not merely a matter of plant-food supply (for even exhausted soils usually contain ample reserves of plant food), but is also closely connected with soil *stability*. An exhausted soil is an unstable soil; Nature has no further use for it and removes it bodily. The process is the same as denudation, but whereas under normal conditions a fraction of an inch of soil may become exhausted and be removed in a century, under human control the entire depth of soil may become exhausted and be eroded in a few years.

Differences in the climates of the Old and New Worlds have undoubtedly greatly influenced the onset and spread of erosion, but climate is never (or only very seldom) the cause of erosion. To-day the most actively eroding regions are the semi-arid continental grasslands and the tropics, but erosion has only become a serious factor in their existence during the last few decades; for thousands of years before, these regions had preserved a perfect equilibrium between denudation and soil formation. It is a humbling thought that modern men, with their immense knowledge and capabilities, have been defeated by grass, whereas their ancestors succeeded in the more difficult task of subduing the forest. But they took a thousand years or more over the job; their modern descendants have acted as though, with their machines, they could conquer the grasslands in a century.

Leaving aside for the moment the question of how present agricultural systems and methods of land utilization have produced such disastrous consequences, we may enquire why these malpractices, which seem to threaten the whole future of the human race, should have been adopted and have become so prevalent in the newer countries. In the first place, the general principles and methods of land management that had been found eminently suitable for European conditions were the only ones fully understood by the colonizing peoples. Thereafter, the necessary modifications introduced in different countries into land-management practices were dictated not so much by natural environmental factors as by external economic circumstances, particularly those created by the rapidly developing opportunities for international commerce throughout the world. Thus the development of land in new countries has not been a gradual evolutionary process dependent upon local conditions, but part of a sudden and explosive surge of immense and uncoordinated human power into unprepared territory.

While the present world-wide despoliation of the earth and irreparable soil erosion may be ascribed to a general maladjustment of land-utilization practices to the natural environment, the reason for that maladjustment is to be found as much in distant countries as on the sites of erosion. Men have not been ruining their newly gained inheritance for amusement, or from some innate destructive instinct or even from thoughtlessness, but because under the circumstances and with the knowledge they possessed they could do little else. The immediate needs of the rapidly increasing European population in the nineteenth century necessitated an unrestrainable exploitation of new virgin lands without regard to ultimate consequences, which were in any case unforeseeable. Europe took everything that the new countries could send, and the latter willingly bartered their life blood for the amenities of civilization and the opportunities offered for national and personal advancement. Nineteenth-century economy, especially within the British Empire, was based on the mutual exchange of agricultural and industrial produce. That the New World was being robbed of its soil and was being paid in coin that brought no recompense to the land never entered the heads of either partner to a bargain which seemed, and at the time was, natural, sensible, and highly satisfactory to all concerned. The price that has been, and still must be, paid in soil and in the social security, prosperity, health, contentment and aesthetic values that go with it, for the outward show of civilization and wealth, is incalculable.

[*Editor's Note:* Material has been omitted at this point.]

# 3

# *Two Costly Lessons: The Dust Bowl and the Virgin Lands*

## E. P. Eckholm

A CLOUD OF DUST thousands of feet high, which came from drought-ridden states as far west as Montana, 1,500 miles away, filtered the rays of the sun for five hours yesterday," reported the *New York Times* on May 12, 1934, with the paper's characteristic sang-froid. The account continued: "New York was obscured in a half-light similar to the light cast by the sun in a partial eclipse. A count of the dust particles in the air showed that there were 2.7 times the usual number, and much of the excess seemed to have lodged itself in the eyes and throats of weeping and coughing New Yorkers."

Washington, D.C., the nation's capital, was overhung with a thick cloud of dust denser than any seen before. The whole eastern coast of the United States appeared to be covered with a heavy fog —a fog composed of 350 million tons of rich topsoil swept up into the transcontinental jet streams from the Great Plains, half a continent away. Even ships three hundred miles out in the Atlantic Ocean found themselves showered with Great Plains dust.

As the winds subsided the particles settled, covering half the nation with a thin layer of grit. Chicago, closer to the source of dust than New York, found itself deluged by twelve million tons of soil from the storm of May 11 alone, or four pounds for every person living there. In New York, Chicago, and Kansas City, businessmen expecting to get rich from their investments in Great Plains wheat or cattle empires saw their prospects dissipate first-hand, as their topsoil seeped around doors and windows to settle on top of their shiny office desks. The United States was learning an expensive lesson in ecology.

The Great Plains stretch northward from Texas through Montana and the Dakotas into Canada, including parts of ten states. Before the European invasions of the sixteenth through nineteenth centuries, these grass-covered, wind-swept plains were occupied by nomadic Indians and great herds of buffalo. Though Indian-set fires pushed back the forest in some areas and may have altered the grass species in others, land use before the nineteenth century was generally well-adapted to the region's environmental constraints.

The soils of the plains tend to be rich; their full exploitation is limited by water shortages. Throughout much of the region, grains can be grown in the wetter years, but when the rains fail—as they do periodically, usually in clusters of drought years—unirrigated crops can be devastated. And in no other region of the United States is the prevailing wind velocity so high. From both economic and ecological viewpoints, most of the Great Plains is better suited to grazing than to farming. Yet several times over the last century, prolonged periods of good rainfall have encouraged a counterfeit optimism about the capacities of the Plains among newcomers, and even among many long-time residents. Humans have repeatedly extended their farms and expanded their cattle herds beyond safe levels during the wet years, to be brought rudely back to reality when the drought cycle returned.

Perceptions of the environmental potentials and limitations of the Plains have varied with the climatic trends of each decade. They have also been actively influenced by shortsighted commercial interests and even by national moods. The semi-arid grasslands of the west-central United States were not always known as the Great Plains; early in the nineteenth century, the region was known as the Great American Desert. If this label exaggerated the area's aridity, it was probably less inaccurate and certainly less dangerous than the myth of fecundity that followed it. The term "Great Plains" came into general use only after the American Civil War of the 1860s, in part because of the deliberate propaganda of railroad companies and land speculators who wished to sell off their holdings and also to see the Plains settled, creating business for the trains. Newspaper adver-

tisements designed to lure Easterners westward pictured lush corn
fields, rich pastures and prosperous families, and helped to fabricate
an image of the Plains as a garden spot. The claims of profiteers
coincided conveniently with the prevailing national mythology of the
American West as a potential agricultural empire, a place where the
good life would come easily to those enterprising enough to move
there and take advantage of it.[1]

And settle the Plains they did. In the 1880s, the first wave of
farmers moved into the region, their barbed wire fences marking the
closing of the earlier era of cattle empires based on unfettered access
to a vast open pasture. The federal government's offer of 160-acre
homesteads sounded attractive indeed; homesteaders in moister
woodlands and grasslands to the east had prospered with the same
acreage allotment. It was not long before the first drought revealed
the impossibility of supporting a family on a parcel of that size in the
Great Plains. Farmers continually resorted to planting grains on fields
that needed a year-long fallow period to permit the accumulation of
moisture in the soil, and there was inadequate space left for livestock.
Poor rainfall in the 1890s drove many of the homesteaders and other
settlers away, forcing them to sell their land to neighbors or speculators. Others remained to become tenant farmers on the holdings of
absentee landlords. After another drought in 1910 brought ominous
clouds of dust and drove more farmers off the land, two new forces
began to influence developments on the Plains. The tractor, the
combine, and other power machinery permitted a farmer to plant and
harvest a larger area than ever before. For many, the costly new
machines were a mixed blessing; farmers went heavily into debt to
buy them and then were compelled to maximize crop production in
order to obtain enough cash to meet their payments. Second, as
World War I disrupted European agriculture, international wheat
prices soared, enticing farmers to convert more and more pastures
into wheat fields.

In the 1920s, which environmentalist Paul Sears has labeled the
"cloud nine decade" of the Plains, the extension of wheat farming
continued. Smaller farmers desperately trying to pay off their debts
planted as much wheat as they could, even after the price plummeted
in the postwar period. Big-time mechanized wheat operations, as well
as huge cattle ranches, attracted investments from Wall Street, and

in some cases from as far away as Europe. "Suitcase farmers" plowed and planted huge holdings with their tractors in the fall, retired to Florida or California for the winter, then returned for the quick harvest and sale of their crop in the summer; if prices were too low, they did not bother to return, instead leaving their dying crops to the winds.[2]

Though periodic droughts and strong winds had always plagued the Great Plains, in the early thirties their effects on the land became visibly worse. The Plains normally turned brown when the rains failed, but the intertwined roots of the hardy native grasses shielded the earth from the wind. Now, however, on huge areas of former grassland plowed under for grain planting, the gales of the Great Plains lashed against the unprotected ground. Soils desiccated by drought ceased to support plant life; instead, their lighter particles became dust in the air, while the heavier particles rolled about until they formed snow-like drifts of sand that smothered fields and machinery. By the summer of 1933, after two successive years of drought, millions of acres planted to grains were not harvested because the plants were too shriveled to repay the effort. Cattle were dying on the parched rangelands, and thousands of farm families found themselves broke and able to fight starvation only by signing on for government make-work programs. Others gave up, or found themselves displaced by the tractors landlords were purchasing in growing numbers. Many of these refugees made the westward trek, so well-chronicled by John Steinbeck in *The Grapes of Wrath*, to another mythical land of bounty—California.

In the spring of 1934, as hot, dry winds introduced another failure of the grain crops, the situation reached an awesome climax. On April 14, a month before the storm that inundated the eastern half of the country with dust, families of the Plains learned in no uncertain terms that their region was entering a new era. Winds from the north swept through Kansas and eastern Colorado, skimming soil off the roots of crops as they went and building into a massive ebony cloud that avalanched southward. The cloud was sufficiently thick to cause hours of eerie total darkness at midday, and the litter of dead birds and rabbits in its wake revealed the dangers of breathing the palpable air. People in its path could only tie handkerchiefs around their faces to help ease the pain of breathing.

As the dust storms continued intermittently over the next several years, cases of pneumonia, nicknamed "dust pneumonia," were to double in some counties, and other respiratory ailments increased as well. Many found the dark color of the dust storms of the thirties to be especially frightening, and with good reason. Soil scientists determined that it was a disproportionately high share of the topsoil's richly fertile organic matter that gave the clouds their black hue.[3]

President Franklin D. Roosevelt chose drought as the subject of his first "fireside chat" with the nation in the 1936 presidential campaign. His account of the devastation he saw in a tour of the Plains could easily have been drawn from 1973 news accounts of the West African drought: "I talked with families who had lost their wheat crop, lost their corn crop, lost their livestock, lost the water in their well, lost their garden and come through to the end of the summer without one dollar of cash resources, facing a winter without feed or food—facing a planting season without seed to put in the ground."[4]

The rain of dust over half the nation, the vivid tales of human suffering, and the images of a potential new desert being created by Americans in the heart of their own country captured national attention far more effectively than the urgent warnings that had been offered for decades by scientists. Paul Sears has recounted what surely was one of the most convincing political lobbying efforts ever undertaken in Washington. The land argued its own case as dust from a thousand miles away visibly filtered into the congressional hearing room where Hugh H. Bennett of the U.S. Department of Agriculture, then one of the world's leading soil conservationists, pleaded for new funds and programs to protect the nation's topsoil.[5]

The Dust Bowl of the thirties catalyzed a major turn in the ecological history of the United States. A century and a half dominated by the empty legend of endless frontiers and boundless resources had left much of the country's land in deplorable shape. Indeed, many experts agreed with Walter Lowdermilk, a noted colleague of Bennett's, when he observed in 1935 that "America has been developing desiccated and unproductive lands more rapidly than probably ever before occurred." Research in the 1920s had shown at least eighty million hectares in the country to be suffering from accelerated erosion, with twenty million formerly productive hectares already abandoned.[6] Though erosion was by then a critical national problem, it took the spectacular collapse of an entire region to spur the political leadership into action on a meaningful scale.

The unprecedented dust storms were soon followed by the formation, in 1935, of a Soil Conservation Service, with Bennett as its first chief. Federal promotion of soil conservation practices in the Great Plains, and soon in every state, helped to arrest and, in many areas, reverse the accelerating trend of soil degradation which many at the time believed was jeopardizing the nation's survival. In the Great Plains, a federally appointed Great Plains Committee studied the state of the land and recommended solutions. The committee's eloquent description of trends in the Plains, issued in 1936, is a classic statement of the continuing plight of enormous semi-arid regions around the world today:

> Current methods of cultivation were so injuring the land that large areas were decreasingly productive even in good years, while in bad years they tended more and more to lapse into desert. . . .
> The steady progress which we have come to look for in American communities was beginning to reverse itself. Instead of becoming more productive, the Great Plains were becoming less so. Instead of giving their population a better standard of living, they were tending to give them a poorer one. The people were energetic and courageous, and they loved their land. Yet they were increasingly less secure on it.[7]

As rains returned to the Plains in the late 1930s, soil conservation officials urged upon farmers a variety of practices to restore and preserve the land's utility. Millions of hectares of especially vulnerable cropland were returned to pasture. Fallowing, supplemented by the calculated use of crop stubble and plant residues left on the unplanted fields to protect them from the wind, was advocated to conserve moisture and thus ensure a better crop in the years of planting. Strip cropping—the alternation of bands of grain with bands of other more soil-protective crops or grasses—helped cut wind erosion, as did the wholesale planting of trees in windbreaks between fields. Terracing of fields, and plowing along the contour of the land, were encouraged to reduce erosion by water. Herd sizes were limited to prevent the destruction of pastures by overgrazing.

Drought plagued the Plains again in the early 1950s, visiting some areas with even greater severity than two decades before. Improved farming practices helped prevent the large-scale regional debacle of the 1930s, but land damage was widespread enough to provoke another federal inquiry. This time a special new program for the Great Plains was established, under which the federal government provides financial assistance to farmers who wish to implement conservation measures.

Surging world grain prices in the mid-1970s, and consequent all-out production efforts by U.S. farmers in 1974 and 1975, have inculcated new fears for the land and provoked strong warnings from agricultural scientists. Government surveys show that of the nearly four million hectares of former pastures, woodlands, and idle fields converted to crops nationwide in late 1973 and early 1974, over two million hectares had inadequate conservation treatment. The average loss of topsoil to water and wind on the latter lands in 1974 was twenty-seven tons per hectare, more than double the twelve tons considered "tolerable" by government soil conservation officials (a figure still above the rate of natural soil regeneration in many regions). In the southern portion of the Great Plains, the surveys showed, the year's soil loss on twenty thousand hectares of newly planted land ranged from thirty-four to 314 tons per hectare.[8]

In a report to the U.S. Senate in early 1975, a committee of leading experts on the private Council for Agricultural Science and Technology warned that "the farm community may be creating another dust bowl." Despite the substantial progress of the last four decades, they noted, several recent trends could lead to another surge of land devastation should prolonged drought reoccur on the plains. In response to high wheat prices, strip cropping and fallowing have often been neglected in favor of continuous wheat cultivation. Together with low cattle prices, high grain prices have also encouraged a limited shift of rangeland into crops in drier, riskier areas. Many farmers, faced with a growing weed problem on continuously planted wheat fields, have returned to once-abandoned plowing techniques that bury the weeds—but that also leave the soil surface bare of wind-resistant plant residues. The leveling of sandy hills to permit new center-pivot irrigation techniques, the committee observed, often exposes easily blown barren sands. Finally, high land and commodity prices are promoting a reduction in the area planted to trees as windbreaks.[9]

Whether the market conditions encouraging these dangerous trends will persist, and whether the alarms being sounded will spur corrective actions, remain to be seen. Meanwhile, some basic lessons emerge from the history of the Great Plains. Hearteningly, the human ability to reclaim devastated semi-arid regions and, through proper management, to exploit them in a sustainable fashion has been proven. Less encouraging is the continuing widespread erosion in the Plains and the lurking possibility of another Dust Bowl emerging despite the ordeals of the past and the ready availability of conservation technologies. Apparently, neither the hard-bought and soon-forgotten lesson of the Dust Bowl, nor the unregulated forces of the free market, is sufficient to safeguard the soil.

Ecological lessons seem to travel the oceans slowly. The Soviet Union, like the United States, has learned through experience that efforts to stretch natural systems too far will eventually backfire. Nikita Khrushchev inherited some difficult agricultural choices when he assumed leadership of the Soviet Union in 1953. Food production in the previous year had barely been above the pre-revolutionary level of four decades earlier, while growing numbers of consumers were clamoring for a better diet. For decades, the country's agricultural sector had been drained of capital to finance the building of factories, but by now it was apparent that grain production had to be boosted rapidly and massively.

Raising the yields substantially on existing farmlands, Soviet leaders calculated, would be tremendously expensive. Multi-billion-dollar irrigation schemes and costly fertilizer factories would have to be built, channeling resources away from the drive to catch up with their Cold War competitor nations in heavy industry and the production of consumer goods. Instead, they looked to the east. In northern Kazakhstan, western Siberia, and eastern Russia, they saw vast grasslands untouched by the plow, and millions of hectares of established farmlands intentionally left idle each year by farmers. Perhaps plowing these "Virgin Lands" and reducing these wasteful fallow periods would make it possible to multiply the country's grain production at comparatively little cost. . . . The Soviet weather roulette, with its frequent droughts in one region or another, meant there was some inherent risk, but even two good harvests out of three would repay the investment many times over.

A gigantic program to farm the Virgin Lands and ameliorate the Soviet Union's chronic food problem quickly and cheaply was vigorously undertaken. Between 1954 and 1960, hundreds of thousands of enthusiastic settlers brought forty million hectares of new land, an area over three times the size of England, under cultivation. They faced great hardships, often living in tents and using inadequate equipment, but many took to heart Premier Khrushchev's exhortation to work hard, and the initial results were impressive. National grain output climbed by 50 percent over the six-year period, mainly because of these additions to the country's cultivated area.[10]

By 1958, however, some farmers in the fields were openly questioning the government's strategy. The directors of one state farm sent their protest to *Izvestia,* the official government newspaper: "Why are we given sowing plans that require us to cut the ground from under our feet—to use a figure of speech—to fulfill them? Can it be that the officials of the republic organization are not aware that if grain crops are planted in the same soil for the fourth year running, that soil will be exhausted?" Although the planted area had increased fourfold on their farms, the directors complained, they had been instructed to hold fallow 429,000 hectares fewer than they had before the expansion began. "Perennial grasses," they warned, "have been eliminated completely."[11]

Alarmed by mounting wind erosion in the Virgin Lands, A. Barayev, director of the Grain Farming Research Institute at Shortandy, experimented with farming methods quite different from those recommended by Moscow. Keeping a third of the farmland out of crop production each year, he figured, would bring the greatest total grain production over the long run. Moisture could build up in soil as it rested in "clean fallow," encouraging higher yields in the times it was cropped, and reducing the chances of severe wind erosion of desiccated soils. In contrast, the official "crop fallow" system specified that corn, rather than grass or leftover stubble, was to be planted in what were supposed to be rest years for the land between wheat plantings.

Khrushchev advocated deep plowing that would remove the stubble left from previous crops, thus allowing traditional planting equipment to proceed more quickly, and planting to be completed earlier in the season. But Barayev preferred shallow plowing with equipment designed to leave stubble on the ground. Stubble, which he called the soil's "armor," would prevent the vicious winter gales from blowing the snow cover off the ground. Otherwise, he observed, the soil tended to freeze to great depths, and, when snow melted in the spring, it would run off the surface rather than soaking in for later use by crops. In spring and summer, stubble would help protect the soil from the persistent winds, and it was an essential guard of topsoil on fallowed fields. The early planting Khrushchev urged would reduce the chance of widespread crop losses to an early snow at harvest time, such as occurred in the fall of 1959, but this practice ensured major crop and soil damages whenever the unreliable May rains came late—a far greater threat to long-term productivity, in Barayev's eyes. Finally, Barayev saw the alternation of belts of perennial grasses with belts of annual grains as essential to holding down wind erosion, though this strip cropping meant sacrificing grain output in any given year.[12]

With his unorthodox ideas, Barayev was not popular among the national leadership in the early sixties. Those in power found more appealing the methods of A. Nalivaiko, director of the Altai Institute, who supported early planting and crop fallowing rather than clean fallowing. Premier Khrushchev's preferences were clear when he spoke to Virgin Lands agricultural workers in late 1961:

A conversion to the correct system of farming is of great importance both for the Virgin-Land areas and for the country as a whole. At the 22nd Party Congress I spoke of the advisability of having clean fallow wherever it is needed. There is a dispute on this matter between the Altai Research Institute, of which Comrade Nalivaiko is director, and the Grain Farming Research Institute, directed by Comrade Barayev. I must side with Comrade Nalivaiko in this dispute.

You yourselves understand that this is not a matter of personal sympathy. The system recommended by the Altai Institute is more effective and gives better results. I have probably heard Comrade Nalivaiko speak four times. I have listened to his arguments, and I think he is on the right track.

. . . What is our approach to clean fallow? If we are speaking from the point of view of farm development, the answer would be that the less clean fallow there is, the better. If there is no clean fallow and a good harvest is obtained, this is best of all. I call upon you to study and decide for yourselves to what extent this is possible. . . . But, comrades, take risks, because it is not always possible to achieve the desired goal without taking risks.

. . . If you can prove in practice that you can obtain more grain per
hectare of plowland than Comrade Nalivaiko gets, if you win the competition, Comrade Barayev, we will quickly turn to you. . . . I do not think you
will win.

Put Comrade Barayev into conditions of capitalist competition and his
farm, with its present system of plantings, probably would not survive. Could
he ever compete with a large capitalist farm if he keeps 32 percent of his
plowed land in clean fallow?[13]

The new state farms of the Virgin Lands followed the premier's
advice and took risks, but events of 1963 revealed the heavy odds
against this gamble with nature. As a dry spring became a dust-filled
summer, desolation crept over the Virgin Lands. The Virgin Lands
had no John Steinbeck to record indelibly the social consequences of
the emerging dust-bowl conditions, but the direct agricultural consequences were more easily registered. Crops on three million hectares
were lost altogether to the drought.[14] Yields were slashed throughout
the region, and the normally savage winds carried precious topsoil,
now dehydrated and easily torn from the earth, off the farms.

The dramatic crop failures and land devastation of 1963 were a
serious challenge to the national leadership. Questions about the
wisdom of Khrushchev's agricultural strategy grew louder and, with
supporting evidence readily visible in the fields, more pointed. In a
May, 1963, issue of *Izvestia*, Barayev outlined the steps needed to
combat the erosion menace—practices that the premier had derided
two years earlier.

Other analysts were adding up the costs of unsound farming
techniques, costs that had begun accumulating well before the 1963
debacle. In one district of Kazakhstan, for example, up to 180,000
hectares of grain had annually been lost to the wind between 1955
and 1959. In 1960, 450,000 hectares were decimated; in 1961,
700,000; and in 1962, 1.5 million. The total rose still further in the
next year. Not surprisingly, as the losses of planted crops climbed, the
yields on fields still worth harvesting fell to dismal lows. The government had originally projected Kazakhstan's average grain yield to
reach two tons per hectare by 1963, but the yields peaked in 1956
at only 1.1 tons, and plummeted to 0.64 tons in the exceptional
drought year of 1963. F. A. Morgun, a senior Soviet agricultural
official, later calculated that in the 1962–65 period a total of seven-

42

teen million hectares were damaged by wind erosion in the Virgin
Lands, with four million of these lost to production altogether.[15]

In 1963, Premier Khrushchev appointed a commission to study
the year's crop failure, but its findings were never made public. In
December of that year, Barayev, while retaining his research post,
was elected to the high position of deputy of the Supreme Soviet.
Three months later, he was invited to address the national Commu-
nist Party's Central Committee—a tangible indication that support
for his ideas was accumulating among powerful national leaders.
Khrushchev himself admitted, in an early 1964 interview with an
Italian publisher, that the exhaustion and erosion of the Virgin Lands
did indeed require the return of some portions to grazing land or
fallow. But later in the year he was again exhorting the state farms
to engage in all-out production.[16] In October, 1964, Khrushchev was
deposed by his fellow party leaders, his disastrous agricultural policies
part of the mosaic of grievances that led to his political demise.

Khrushchev's successors proved more erosion-conscious. Beginning in 1965, new machinery designed to leave crop stubble on the
ground was employed and the annual fallow area was increased. Tree
planting was stepped up to create windbreaks on the bleak plains of
the Virgin Lands.

The dreams that originally inspired the Virgin Lands campaign
will never be fulfilled, and erosion will always be a serious menace in
this region's inhospitable conditions. Once human practices were
adapted to the environment, however, the program did evolve from
an embarrassing liability to a moderate success. These new farmlands
provide the Soviet Union with an important back-up harvest that can
ameliorate the impact of the periodic crop failures on the more
productive western farmlands, though massive agricultural investments elsewhere in the country have also proved necessary. And, as
the world learned in 1972, and again in 1975, massive grain imports
from abroad are also necessary in years when nature is uncharitable
to the Soviet Union.

**43**

## 3. *Two Costly Lessons: The Dust Bowl and the Virgin Lands*

1. Thomas Frederick Saarinen, *Perception of the Drought Hazard on the Great Plains*, Department of Geography Research Paper No. 106 (University of Chicago, 1966), esp. pp. 14–18.
2. See Paul Sears, "A Empire of Dust," in John Harte and Robert H. Socolow, eds., *Patient Earth* (New York: Holt, Rinehart & Winston, 1971); and *The Future of the Great Plains*, Report of the Great Plains Committee (Washington, D.C.: Government Printing Office, December, 1936), pp. 3–5.
3. Vance Johnson, *Heaven's Tableland: The Dust Bowl Story* (New York: Farrar, Straus, 1947) vividly describes the dust storms of the 1930s. H. H. Finnell, *Depletion of High Plains Wheatlands*, U.S. Department of Agriculture Circular No. 871 (June, 1951), discusses the organic contents of the dust storms.
4. In Frank E. Smith, ed., *Conservation in the United States, A Documentary History: Land and Water, 1900–1970* (New York: Chelsea House, in association with Van Nostrand Reinhold, 1971), p. 438.
5. Paul Sears, *op. cit.*, p. 10.
6. W. C. Lowdermilk, "Man-Made Deserts," *Pacific Affairs*, Vol. 8, No. 4 (1935), p. 417.
7. *The Future of the Great Plains, op. cit.*, p. 1.
8. Kenneth E. Grant, "Erosion in 1973–74: The Record and the Challenge," *Journal of Soil and Water Conservation*, Vol. 30, No. 1 (January–February, 1975).
9. Council for Agricultural Science and Technology, *Conservation of the Land, and the Use of Waste Materials for Man's Benefits*. Prepared for Committee on Agriculture and Forestry, United States Senate, March 25, 1975. (Washington, D.C.: Government Printing Office, 1975), esp. pp. 16, 17. On windbreaks, also see General Accounting Office, *Action Needed to Discourage Removal of Trees That Shelter Cropland in the Great Plains*, RED–75–375 (Washington, D.C.: General Accounting Office, June 20, 1975).
10. The dimensions of the program are outlined in William A. Dando, *Grain or Dust: A Study of the Soviet New Lands Program, 1954–1963*, Ph. D. dissertation, University of Minnesota, Department of Geography, 1969; and Frank A. Durgin, Jr., "The Virgin Lands Programme, 1954–1960," *Soviet Studies*, Vol. 13, No. 3 (1961–62). I am grateful to John Tidd for his assistance with research and analysis of the Soviet Virgin Lands program.
11. W. A. Douglas Jackson, "The Virgin and Idle Lands Program Reappraised," *Annals of the Association of American Geographers*, Vol. 52, No. 1 (March, 1962), p. 76.
12. Basic issues of the debate over proper farming techniques in the Virgin Lands are presented in Werner G. Hahn, *The Politics of Soviet Agriculture, 1960–1970* (Baltimore: Johns Hopkins University Press, 1972), and in an article by A. Barayev, *Izvestia*, May 31, 1963.
13. *Izvestia*, November 22, 1961. Translation by John Tidd.
14. Hahn, *op. cit.*, p. 112.
15. Details of erosion damages in S. S. Kabysh, "The Permanent Crisis in Soviet Agriculture," in Roy D. Laird, ed., *Soviet Agriculture: The Permanent Crisis* (New York: Frederick A. Praeger, 1965), pp. 166–168; and Hahn, *op. cit.*, p. 112.
16. See Hahn, *op. cit.*, for discussion of the political implications of the 1963 Virgin Lands failure. Khrushchev's Italian interview is cited in Naum Jasny, *Khrushchev's Crop Policy* (London: George Outram, 1964), p. 23.

Part II

# MEASUREMENT AND INVENTORY

# Editor's Comments
# on Papers 4, 5, and 6

**4    HUDSON**
*The Design of Field Experiments on Soil Erosion*

**5    DUNNE**
*Studying Patterns of Soil Erosion in Kenya*

**6    RAPP**
*Soil Erosion and Sedimentation in Tanzania and Lesotho*

Knowledge of the global state of agricultural soil erosion can be obtained only through an inventory of its areal extent and measurement of its rate. Reliable data are difficult to collect, however. The techniques for inventory and measurement have been developed independently of each other so that the two sets of data are often at different scales, based on different assumptions, and, therefore, incompatible. Measurements of erosion are sometimes difficult to interpret because they are made as part of erosion experiments rather than as determinations of erosion rates for typical conditions. Also, measurements may be made using a variety of methods so that comparisons between them may not be meaningful.

In Paper 4, Hudson describes what may be regarded as the nearest to a standard method of measuring soil erosion in the field using erosion plots. The method was developed in the United States where the first plots were installed in Utah in 1915 (Moldenhauer and Foster, 1981). During the 1930s, plots were established at a number of Experimental Stations but the plot dimensions varied from 2 to 7 m wide and 10 to 200 m long. Eventually, a standard size was established at 22 m long and 1.8 m wide. The plots set up by Hudson at the Henderson Research Station in Zimbabwe are nonstandard by these dimensions, but they form the first comprehensive attempt to measure and experiment on soil erosion in Africa. Hudson's paper deals with the plots, siting, design, methods of construction, and implementation.

Basically, the technique is to isolate a part of a hillside using metal or wooden borders and to collect and analyze the runoff and sediment coming off the bottom of the plot in a certain time period. Where quantities of runoff and sediment are likely to be large, a method of taking samples is employed. Errors in measurement arise because of the effects of the plot boundaries, difficulties of maintaining a connection between the hillslope and the collecting gutter, silting of the collecting apparatus, inefficient hydraulic design of the divisors used to sample the runoff, failure to cover the collecting tanks and gutters from the direct entry of rainfall, and problems of emptying large amounts of water and sediment from the tanks. Hudson explains the various procedures adopted to deal with these problems, but, although the solutions are generally satisfactory, no guidance is given as to their effects on the accuracy of the measured erosion rates.

In fact, despite the worldwide use of erosion plots and recognition of the likely sources of error, few studies appear to have been made of the magnitude of the errors involved. In practice, it is virtually impossible, not only with erosion plots but with any method of erosion measurement, to quantify the errors because no absolute measure of erosion rate can be obtained for comparison. It is possible that detailed geodetic surveys of changes in the microrelief of the ground surface would provide such a measure, but these have the disadvantage that the effects of erosion cannot be easily separated from other causes of microtopographic change such as freezing and thawing, swelling and shrinking, surface compaction, and cultivation. In addition to the problems of assessing errors, little attention has been given to what level of accuracy of measurement is desirable. Should the erosion rate be measured plus or minus a selected percentage error, or should the acceptable error be varied according to the rate? Perhaps larger percentage errors are permissible at very low rates because accuracy may be less important if erosion is not a problem. Also, a high level of accuracy may be less achievable if the erosion rates are similar in magnitude to the experimental errors associated with the measurement technique being tested.

Although the reliability of erosion plot data may be judged according to whether sensible precautions have been taken against likely errors, until the size of these errors has been established, all plot measurements must be viewed as suspect. Roels and Jonker (1983 and 1985) found with unbounded runoff plots that coefficients of variation for measurements of interrill erosion varied from 20-68% depending upon the sample size. Attempts to estimate soil loss from sample plots may result in deviations of 50-100% from a mean value based on measurements from 60 plots. Since most field data are based on only a few supposedly representative plots, they give, at best, only a general indication of what the absolute value of erosion under any given condition is likely to be. This assessment of data reliability has implications for comparing measured erosion rates against a soil loss tolerance level. Since the errors involved in judging tolerance levels are also likely to be considerable (Mannering, 1981), the comparison is between two dubious rates. Providing this is realized, however, it is a helpful comparison though more in a qualitative sense than quantitative.

Data reliability affects the use of erosion plots as an experimental technique. Experiments are designed to enhance understanding and provide explanation. Hudson investigates the causes and mechanics of runoff and erosion. Even if the effects of only two or three factors are studied, a large number of plots is required to cover a reasonable range of values in each factor. Replication is generally accepted as essential and using three replicate plots for each condition is often considered the norm. Recent studies on the variability of plot data from both laboratory and field experiments show typical values of 13–40% for the coefficient of variation (Luk, 1981). Extreme values of soil loss, expressed by three standard deviations from the mean, therefore range from ±39 to ±120%. As many as 25 to 30 replications may be required to measure soil loss to ±10%. While most experiments are woefully inadequate in terms of replication, it is unlikely that a sufficiently large number of replicate tests could ever be performed in the field. The variability of soil and slope conditions means that plots can rarely be located to give sufficient control if these are the factors that are not being varied.

Given the problems associated with them, it is unlikely that field plots will yield more than a partial explanation of erosion. Since they are also very expensive to install and maintain, it is worthwhile investigating cheaper methods of measurement. In Paper 5, Dunne describes several approaches to the measurement of ground lowering using tree roots, but quality control is difficult to maintain. Erosion pins and simple sediment traps or Gerlach troughs (De Ploey and Gabriels, 1980; Zachar, 1982) have been used. Although the sources of error in these techniques are well documented, no comparison has been made between their results and data from erosion plots.

Extrapolating hillslope measurements of erosion to larger areas is difficult because while rates of runoff and soil loss generally increase with increasing catchment size, they do so at a rate less than unity. Sediment and runoff sinks occur in the landscape, particularly on hillside footslopes and flood plains, so that not all the material washed off hillslopes finds its way into the river system. Erosion measurements over larger areas are made, as shown by Rapp in Paper 6, by determining sediment concentrations in rivers and the rates of sediment accumulation in reservoirs. Linking these with hillslope erosion data requires a knowledge of sediment delivery ratios. As explained by Dunne, however, little information exists on these.

Experimental use of erosion data from hillslopes and from catchments of different sizes to investigate spatial variability reveals that the nature of the factors influencing erosion is scale-dependent. This is best illustrated by studies of drainage density, which can be considered as a surrogate index of the degree of erosion of the land. Regional and global variations are largely dictated by climate whereas local differences relate to relief and, at a still smaller scale, to land use (Morgan, 1973; Gregory and Gardiner, 1975). The importance of land use is emphasized by Dunne who not only shows how it influences the absolute rate of erosion but also how it interacts to affect the rate at which erosion increases with greater runoff and steeper slopes. Although the significance of interactions has been recognized in a number of studies, it has usually emerged from an analysis of the experimental data rather than its being the main objective of the investigation. Quansah (1982), however, designed a series of laboratory experiments specifically to investigate interactions. He found that the detachment of soil particles by raindrop impact was influenced by a soil–rainfall interaction so that detachment increased with the intensity of the rain more rapidly on sandy soils than on clays. The transport capacity of runoff when disturbed by raindrop impact was influenced by a slope–soil interaction; it increased with slope steepness more rapidly on clay soils where the particles are transported as aggregates than on sandy soils where primary particles are carried. These interactions are not accounted for in field experiments using erosion plots.

As an indicator of the severity of soil loss, measurements of erosion are rather limiting in that they are made at selected localities. Also, erosion plots only allow the study of erosion by raindrops, overland flow, and rills. They provide little information on gullies, slides, and debris flows. The value of erosion measurements is enhanced if they are combined with an inventory of the area affected by erosion of different types and of different degrees. The first broad-scale erosion inventory was carried out in the United States by the Soil Conservation Service in the late 1930s using field observations. The difficulties of conducting a nationwide survey in the field meant that the results were never afforded more than reconnaissance status (Bennett, 1939). Nevertheless, the erosion inventory was vital in stimulating public awareness of the unforeseen scale of the problem and in securing federal funds for conservation. Rapp illustrates in Paper 6 how erosion inventory has progressed and can be carried out more quickly. Using aerial photograph interpretation and the techniques of geomorphological mapping, erosion surveys like that of the Matumbula catchment in the Dodoma district of Tanzania can be implemented with fieldwork being used for checking rather than the main source of information. Broader-scale surveys of the extent of eroded land can be made from satellite imagery. Rapp used LANDSAT A, formerly known as ERTS-1, and was therefore limited in the details he could show by its resolution of about 70 m. With the 30-m resolution of LANDSAT D and continuous improvemens in image-processing techniques, it should soon be possible to map quickly large areas at scales as large as 1:50,000 and quantify the erosion according to the presence or absence of gullies and, where present, gully density.

The three papers reviewed in Part II provide the basis for developing a standardized system of inventory and monitoring on a world scale. Surveys of satellite imagery allow the extent of eroded areas to be mapped. More detailed assessments can be made using aerial photography and field checks. These, in turn, form the framework for selecting small catchments, 10 to 50 $km^2$ in size, for the measurement of sediment yield from a range of land use, soil, and relief conditions. Sites can be selected within these catchments for the measurement of erosion on hillslopes. Where erosion plots already exist, they may be used. Otherwise, it will be more cost-effective to use field rainfall simulators of the type described by Dunne to supply a controlled rainstorm to a plot, $5 \times 2$ m, and collect the runoff and sediment arising from it. By analyzing rainfall records and selecting a number of storms of different frequencies and magnitudes, the results can be integrated to supply an estimate of mean annual soil loss. The portable rainfall simulator combined with the small size of the plot will give sufficient flexibility to study the erosion on different soils and slopes with a variety of land uses and will allow sufficient replications to be made to determine the likely variability in erosion rate from sites of similar conditions. If the measurements of erosion are carried out in areas representative of those identified on the aerial photographs and satellite images, the data can be extrapolated, at least in theory, to give regional or even global estimates of soil loss. Sequential monitoring can be achieved by repeating the whole exercise at, for example, ten yearly intervals.

Although such a scheme is probably workable at the present time, it will prove more effective if priority is given to the following areas of research: methods of mapping erosion classes from satellite imagery; techniques for routing estimates of soil loss from rainfall simulation plots down hillsides and into rivers; development of field equipment, analogous to the rainfall simulator, for measuring wind erosion; methods for routing windblown sediment across the landscape; and the development of a statistically sound experimental procedure for linking satellite-image interpretation to the location of 10 $m^2$

erosion plots. Rapp (1960) and Iveronova (1969) have shown that regional evaluations of erosion can be successfully based on localized field measurements. Since these studies were carried out, satellite imagery has become available and the techniques of erosion measurement have improved. The technology now exists for providing much needed data on the severity of the world's soil erosion problem.

## REFERENCES

Bennett, H. H., 1939, *Soil Conservation,* McGraw-Hill, New York.

De Ploey, J., and D. Gabriels, 1980, Measuring soil loss and experimental studies, in *Soil Erosion,* M. J. Kirkby, and Morgan, R. P. C., eds., Wiley, Chichester, pp. 63–108.

Gregory, K. J., and V. Gardiner, 1975, Drainage density and climate, *Zeitschr. Geomorphologie* **19:**287–298.

Iveronova, M. I., 1969, Opyt kolichestvennogo analiza protsessov sovremmenoi denudatsii, *Izvestiya AN SSR, Ser. Geograf.,* 1969-2:13–24, transl., An attempt at the quantitative analysis of contemporary denudation process, Russian Translating Programme, National Lending Library, Boston Spa, RTS 7436, 1972.

Luk, S. H., 1981, Variability of rainwash erosion within small sample areas, in *Twelfth Binghamton Geomorphology Symposium Proc.,* Urbana, pp. 243–268.

Mannering, J. V., 1981, The use of soil loss tolerances as a strategy for soil conservation, in *Soil Conservation: Problems and Prospects,* Morgan, R. P. C., ed., Wiley, Chichester, pp. 337–349.

Moldenhauer, W. C., and G. R. Foster, 1981, Empirical studies of soil conservation techniques and design procedures, in Morgan, R. P. C., ed., *Soil Conservation: Problems and Prospects,* Wiley, Chichester, pp. 13–29.

Morgan, R. P. C., 1973, The influence of scale in climatic geomorphology: a case study of drainage density in West Malaysia, *Geogr. Annaler* **55-A:**107–115.

Quansah, C., 1982, *Laboratory Experimentation for the Statistical Derivation of Equations for Soil Erosion Modelling and Soil Conservation Design,* Ph.D.Thesis, National College of Agricultural Engineering, Cranfield Institute of Technology.

Rapp, A., 1960, Recent developments of mountain slopes in Kärkevagge and surroundings, northern Scandinavia, *Geogr. Annaler* **42:**65–200.

Roels, J. M., and P. J. Jonker, 1983, Probability Sampling Techniques for Estimating Soil Erosion, *Jour. Soil Science Soc. of Am.* **47:**1224–1228.

Roels, J. M., and P. J. Jonker, 1985, Representativity and accuracy of measurements of soil loss from runoff plots, *Am. Soc. Agric. Engineers Trans.* **28:**in press.

Zachar, D., 1982, *Soil Erosion,* Elsevier, Amsterdam, 547p.

# 4

Reprinted from *Jour. Agric. Eng. Research* **2**:56–65 (1957)

# The Design of Field Experiments on Soil Erosion

N. W. Hudson*

## Introduction

Work started in May 1953 at Henderson Research Station, Mazoe, Southern Rhodesia, on a long term programme of research with the object of measuring and investigating the causes and mechanics of run-off and erosion, and studying control methods. This account describes the design, construction methods and equipment used. A full description of the experiments, treatments and results to date will be published shortly in the Rhodesian Agricultural Journal. The initial three year development programme has been completed and consisted of installing, in three successive dry seasons, a series of experiments on each of the three main soil types of the district. Only the major or field scale experiments of each series are described here.

## Part I

### Series I Experiments

The Series I experiments are sited on six acres of fairly uniform soil on the eastern slopes of the hill which forms the centre of the station. It is particularly suitable as the slope changes from $7\frac{1}{2}$ per cent at the top to 3 per cent near the base of the hill, allowing plots on different slopes to be placed sufficiently close to ensure reasonable uniformity of soil. The soil is a yellowish brown silty clay of the Tatagura series, derived from schistose sedimentary rocks. The chief features of this soil are relatively high silt content, and shallow depth. The underlying rock here consists of decomposed sedimentary rocks of the Iron Mask series, and while the depth of soil decreases from about 5 ft on the lower slopes to about 3 ft at the top of the site the main characteristic of free drainage is found over the whole site. From the

*footnote:*
\* Department of Conservation and Extension, Southern Rhodesia.

agricultural point of view it is a difficult soil. It has an unstable structure, is readily pulverised, and when wet is readily compacted tending to form a hard surface crust. Apart from these poor physical characteristics the fertility is only slightly less than the red soils of the Mazoe and Shamva series. The past history of the site is not exactly known. It had been partly cleared of trees and contour ridged in the past, but had been lying fallow for at least seven years before the experiments started, and judging by the tree regrowth probably for considerably longer. In order to fit in plots on the required slopes it was necessary to plough out some of the old contour banks, but apart from this and levelling two anthills, little disturbance of the soil was required. A complete range of soil samples is taken annually, both composite samples by auger for mechanical and chemical analysis, and clods for structure determinations.

### Experiment I

The experiment consists of the measurement of run-off and soil loss under various cropping practices on three slopes of $6\frac{1}{2}$ per cent, (1 in 15), $4\frac{1}{2}$ per cent, (1 in 22) and 3 per cent, (1 in 33). There are 16 plots, six on each of the steeper slopes and four on the flatter slope. Each plot is 1/20 acre, this being the minimum size which can be comfortably worked with tractor mounted implements. It was considered essential to carry out all farming operations in the way in which they would actually be done in the field, and this practice is followed as far as possible using a Ferguson tractor and mounted implements. The aim is to use hand work only in operations which would normally be done by hand. To allow room for tractor turning and to avoid border effects the plots were set out with a space of 35 ft between them. Half of this intermediate area next to a plot is cropped and treated in the same manner as the plot itself, and in fact forms part of the plot except for collection of run-off and measurement of yields.

Since the experiment is designed to measure erosion between contour ridges the length of each plot down the slope is that distance at which contour ridges would normally be spaced. This is calculated from the formula:

$$\text{Vertical Interval (V.I.)} = \frac{\% \text{ slope} + \text{factor}}{2}$$

The factor varies with soil type, past treatment, and cropping practice. In this case the factor taken was 4.

Thus for $6\frac{1}{2}$ per cent slope

$$\text{V.I.} = \frac{6\frac{1}{2} + 4}{2} = 5 \cdot 25 \text{ ft}$$

H.I. (Horizontal Interval or spacing)

$$= \frac{5 \cdot 25}{6 \cdot 5} \times 100 = 80 \cdot 5 \text{ ft}$$

The plots are actually 80 ft 0 in. long × 27 ft 3 in. wide.

Using the same formula the plots on $4\frac{1}{2}$ per cent and 3 per cent slope are 90 ft × 24 ft 3 in. and 100 ft × 21 ft 9 in. respectively.

The length of the plots is in each case slightly less than the theoretical H.I. because in practice a proportion of the interval between contours is taken up by the contour channel, and the uphill face of the contour bank itself. This difference between calculated H.I. and actual plot length was increased on the flatter slopes where broad based contours are more usual.

## Plot Boundaries

Various devices such as shallow drains, earth mounds, sheet metal strips and wooden planks have been used on run-off experiments elsewhere to isolate the actual plot from its border areas. The disadvantages are (a) of drains, that plot run-off may be diverted into them, (b) of earth mounds, that heaping up the earth leaves a channel which concentrates run-off and tends to scour, or alternatively to avoid forming the channel, extra soil may be carted onto the plots, but this is a laborious process when several cultivations are required during the season (c) that sheet metal is expensive and (d) that wood is attacked by termites. The material adopted in these experiments was flat asbestos-cement planks 6 in. deep and $\frac{1}{2}$ in. thick.

The widths of the actual plot from which run-off is collected are those calculated above, and the sides and top of the plots are demarcated by the asbestos planks set on edge with 3 in. projecting above ground level. Steel $\frac{1}{2}$ in. round pegs driven vertically hold the planks upright, and the ends of the planks are butted tight together and a sheet metal spring clip 3 in. wide is driven home to make the joint watertight. Inside the plot the soil is levelled carefully and on the outside a small fillet of earth a few inches high is thrown up to make sure no seepage occurs. At the lower end of the plot boundary the board is keyed into the brick collecting trough and a small heap of earth packed round to prevent leaks (*Fig. 2*).

At the top boundary of the plots the earth is well heaped up against the outer face of the planks and a shallow drain leads away on either side. This drain is only to carry water which runs off the road immediately above each plot, the main protection being a large storm water drain above the road.

When ploughing, cultivating and other operations are carried out, the boundary planks are taken up in front of the tractor and replaced behind it (*Fig. 3*).

*Fig. 1. Series I. Experimental sites*

One disadvantage of these strips is that being only 6 in. deep they cannot be put in on deep ploughed land without first breaking the clods down. However, the usual practice on these soils is to plough in autumn, by which time there is little likelihood of storms heavy enough to cause run-off. In spring the first harrowing is usually carried out before erosive rains occur, except when the land is left rough ploughed for late crops, and in this case some hand work is necessary to put in the planks.

It is essential to carry out periodic checks on all the plot boundaries during the rains to make sure there are no leaks or seepage.

### Collection of Run-off and eroded soil

Maximum rates and quantity of rainfall to be expected were given by the Department of Meteorological Services as follows:

(1) A maximum intensity of 10 in/hr sustained for 20 minutes and

(2) A maximum daily fall of 8 in.

In calculating the maximum probable run-off rate it was assumed that such a storm could occur when the ground was completely saturated and would give 100 per cent run-off, and the collecting troughs etc. were accordingly designed to take a peak flow of 0·5 cusec, i.e. 100 per cent run-off at 10 in/hr from 1/20 acre. In calculating the capacity of the tanks a maximum figure was assumed of 80 per cent run-off

of the maximum daily fall of 8 in., i.e. a maximum daily run-off of 6·4 in. which gives a total of 7250 gallons from 1/20 acre.

The run-off is collected in a brick channel 9 in. wide × 6 in. deep on a grade of 1/100 (capacity 0·75 cusec) leading to the measuring tanks through a 6 in. spun concrete pipe, also on 1/100 grade. Silting has occurred in the furrows and a grade of 1/50 was used in the later experiments. The collecting furrows were at first uncovered but in a gentle rain the error due to the 100 per cent run-off from the furrow is considerable compared with the run-off from the plots. Light sheet-metal covers have been added to prevent this error.

### Sill of Collecting Trough

A serious problem arises in the installation of any fixed collecting channel below erosion plots, in that the upper lip of the channel must be resistant, but after erosion has taken place the ground surface is lowered over the whole plot, yet the lip remains the same height and acts as an artificial control, upsetting the formation of the natural eroded profile. The solution of this problem has been to make this lip of a weak lime plaster and when the erosion on a particular plot warrants it, this lip or sill is scraped down to the correct height.

In the absence of data on the alteration of the profile by erosion it has been assumed that the shape of the layer of eroded soil lies halfway between a uniform layer over the whole plot,

*Fig. 2. Collecting channels and plot boundary plants*

*Fig. 3. Temporary removal of plot boundaries allows tractor cultivation*

and a wedge tapering from zero at the top of the slope to a maximum at the bottom (*Fig. 4*).

On this basis and assuming the weight of soil to be 100 lb/cu. ft, a soil loss of 15 tons/acre gives a soil loss from the plot of 1500 lb or 15 cu. ft. This gives x as 0·0825 in. The amount the sill is to be reduced is $\frac{3x}{2}$ or 0·125 in. $=\frac{1}{8}$ in.

Since $\frac{1}{8}$ in. is the minimum amount which it is practicable to remove accurately the practice adopted is to lower the sills $\frac{1}{8}$ in. for each 15 tons/acre of soil loss.

The soil level immediately next to the sill is most important. If it is raised above the sill e.g. by ploughing, excessive erosion will take place as the water drops into the collecting trough. If it is below the sill, ponding will occur and unduly low run-off and erosion will be recorded. After the initial construction of the collecting trough the surface of the soil was carefully restored by hand and a strip about a foot wide next to the sill is now left undisturbed. The change from this narrow strip to the ploughed land is tapered off by handwork. When the sill level is lowered this narrow strip is also lowered.

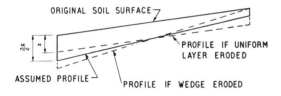

*Fig. 4. Assumed form of erosion profile*

## Hydrograph Recorders and Flumes

As the run-off comes into the first collecting tank it is passed through an " H.S." flume where the rate of flow is recorded. This device was designed specifically for this purpose by the American Conservation Service. It is a type of Cipoletti notch, rectangular at the bottom with sloping sides. The bottom of the flume is level straight through so there is no deposition of silt. A stilling well, at a specified distance upstream from the notch, operates a float con-

trolled recorder. Special charts are used, prepared from rating tables which convert the depth of flow to rate in cusecs. The recorded flow in cusecs is converted to run-off in inches per hour from the plot by multiplying by 20. (For all practical purposes 100 per cent run-off of 1 in. per hour over one acre gives one cusec).

The hydrograph obtained from the recorder is used to give rates of run-off only, not quantities as this would require individual calibration of the flumes and recorders, and also the type of recorder chosen on account of the simplicity and cost has circular charts which cannot be integrated without a polar planimeter.

The flow recorders used have been specially manufactured with provision for converting to operate over alternative ranges of depths of flow, so that in several years time, when sufficient hydrographs have been obtained, the recorders can be used on other experiments to record run-off from large plots or small catchments.

## Collecting Tanks and Divisors

The collecting tanks have been installed in pits each containing the tanks for two plots. This is to simplify construction, recording and drainage and is preferable to separate pits below each plot. Where site conditions allow it is desirable to concentrate the tanks more into larger groups as has been done in later experiments. This was not possible on this site as the space between the plots on different slopes was barely adequate to get the pits in at all. It was necessary to have the pits close up under the plots, and to keep the pipes leading from plot to collecting tank of minimum length so that the levels allowed the surplus run-off to be disposed of in surface drains. The pits are not symmetrical, which means that the gathering times of the two plots running to a pit are not identical, but the difference is not significant over such short distances.

The tank system for each plot consists of a first tank (A) into which all the run-off is passed, then a fraction of the overflow from (A) is passed to a second tank (B), and a fraction of the overflow from the second tank is passed to a third tank (C). The tank sizes are based on capacity to store eroded soil for the first tank and volume of water for the third tank. Bearing in mind the need for standardisation and the fact that the tanks form a large proportion of the total equipment cost, the second tank is the same size as the third.

*Fig. 5. Series I. Tank installation*

Assuming a probable maximum annual soil loss (after several years) of 60 tons/acre or 60 cu. ft from 1/20 acre plots on the $6\frac{1}{2}$ per cent slope, and assuming that the tank can if necessary be emptied after a storm causing half this loss, then the capacity required is 30 cu. ft. To allow space for water above the settled soil 50 cu. ft is needed and a tank 4 ft diameter and 4 ft high was chosen.

Assuming a maximum annual soil loss of 30 tons/acre on the $4\frac{1}{2}$ per cent and 3 per cent slopes, a 4 ft diameter tank, 2 ft high is required. Assuming that if necessary all the third tanks can be drained daily, the capacity of the third tank must be 1/49 of the maximum daily run-off of 7250 gallons or 150 gallons. The second tanks also have a capacity of 150 gallons.

The overflow from the first tank passes through a Geib divisor. This is a device designed for this purpose by the United States Conservation Service and consists of a rectangular sheet-metal box with a series of identical vertical slots which accurately divide the flow. The model used has seven slots each 6 in. × 1 in. (capacity 0·77 cusecs) and the flow from the central slot is passed through a spout into the second tank. The overflow from this tank is passed through a second divisor of seven slots 4 in. × $\frac{1}{2}$ in. (capacity 0·21 cusecs).

To ensure accurate division of the flow through the divisors it is essential (a) that floating material be trapped by screens and (b) that the divisor be exactly level. Adjustments to the level are made by means of the device shown in *Fig. 6*. Two $\frac{1}{2}$ in. bolts set in concrete fit loosely into tubes fixed to the underside of the divisor box. An accurate spirit level is placed inside the divisor along the bottom edge of the slots, and the divisor levelled by adjustment of nuts on the $\frac{1}{2}$ in. bolts. The slight give in the box allows this adjustment to be done without disturbing the joint of the divisor box to the tank, which is sealed with bituminous putty. Periodic checks of this adjustment are necessary.

*Fig. 6. " Geib " divisor, with levelling device and trough for collecting surplus 6/7*

### Screens and Shape of Tanks

Circular corrugated galvanized iron tanks were used throughout this experiment, but make the fitting of screens awkward in the first or (A) tanks. Floating debris, leaves, roots, frogs, snakes and rats are often found in the run-off, and choke the divisors if not retained on screens. The first season's results must be treated as indications only as during most of the season various types of screen were being tried out, and none was completely successful. The result when the screens fail or overflow is for some of the slots to partially or completely choke up, and so a false proportion is passed. Two screens are now used, the first $\frac{1}{2}$ in. mesh, and the second 64 mesh. The screens fit into channels at the sides of the tanks and slide out for cleaning and emptying the tanks.

*Fig. 7. Partial obstruction of divisor due to inadequate screen*

## Drainage

The design of the tank layout in the pits is fairly satisfactory but has been improved in later experiments. The best scheme where site conditions allow is for all the surplus run-off to be led off in surface drains. On this site, however, the three slopes were so close together that sufficient fall could not be found to lead the water out of the bottom of the pits by open drains. Two separate drainage schemes are therefore used. The surplus 6/7 is collected in raised concrete troughs just below the divisors and led off in open drains. The direct rain, and water drained from the tanks when they are emptied, is drained from the pits by a system of 4 in. glazed earthenware pipe drains, at a minimum depth of 2 ft 6 in. to avoid interference with ploughing.

The floor of the pits was not made continuous, each tank resting on an independent concrete slab 5 ft in diameter, and the levels of all six slabs in each pit vary—the object being to ensure that a minimum fall was used to get the water from the collecting channel to the third tank, since any unnecessary depth at the third tank would make the levels of both the surface and the sub-surface drains lower, and increase the already considerable difficulty of getting water away without large excavations.

The whole problem of providing adequate head through the flumes and tanks and for drainage from the tanks has been complicated by the fact that the three slopes used are so close together. The alternative on this site would have been to install a network of 12 in. pipes to dispose of all run-off from the plots. This was not done because of the great expense, but in the light of experience would have been worth while. This system has been used on Series II.

The pits are in all cases only a few feet away from the plots, and to prevent drainage interfering with the plots, the sides are built up in brickwork to above ground level.

## Techniques of Draining and Emptying

In similar American installations the usual technique for recording the water and soil in the tanks is to measure the volumes of muddy water and sludge, and to take representative samples of each for laboratory determination of the proportions of soil and water. This method is not practicable without a properly equipped and staffed laboratory and another method was devised. This consists of flocculating the suspended material and allowing it to settle, then draining off the supernatant clear water and removing the whole of the sludge.

The amount of solid material remaining in the water after flocculating is negligible, and the volume of water displaced by the soil is allowed for. First, the clear water is drained and recorded, then if the silt level is below the outlet more clear water is baled out and the volume recorded. When no more water can be removed without stirring up the soil, it is deliberately stirred and the whole baled out as a thin mud into 1, 4, 10 or 45 gallon drums as required. The tank is now empty and ready for the next storm. The sludge in the drums is left for a day or two when more clear water can be decanted off at intervals, always being recorded. When the material is a thick mud and no more water can be decanted, it is taken to the store for drying and weighing. At this stage allowance is made for the water still with the soil by assuming the water is 50 per cent of the total volume. This volume is so small in comparison with the total volume of run-off that errors in this assumption are negligible.

A laboratory has now been staffed and equipped and this system will be replaced by direct analysis of small samples. It is hoped to combine this with chemical analysis to determine loss of plant nutrients and organic matter.

Originally an adjustable outlet was fitted into the tanks so that the clear supernatant liquid could be drained down to sludge level. For the introduction of the sub-sampling technique with laboratory analysis, this has been replaced by conical floors allowing rapid draining both of the clear water and the sludge. A tube of suitable length is inserted in the outlet while the relatively clear run-off is drained, and removed for drainage of the settled sludge. The tanks are calibrated for full capacities, and when partly full drainage is through calibrated $1\frac{1}{2}$ in. water meters.

## Frequency of Draining and Emptying

To measure annual soil and water losses was the original intention, but a most important piece of information is how erosion varies in individual storms. It is not possible to drain and empty all the tanks in a few hours using the flocculation method and so emptying after each storm is done only on selected plots. The introduction of the laboratory facilities will permit recording of all plots on a storm or daily basis.

# Part II

## Series II Experiments

The Series II experiments were installed during the winter of 1954 on red clay soil of the Mazoe series, typical of the main maize producing area of the country. It is a very deep uniform soil with good structure and free drainage, derived from sedimentary rocks, chiefly banded iron-stone. The main experiment (No. 6) is very similar to Experiment 1 of Series I in that it consists of measuring run-off and soil loss under various cropping practices on a particular soil type. All the plots are on a 3 per cent slope, except one on a $4\frac{1}{2}$ per cent slope.

## Plot Size and Layout

The plots are 1/20 acre and as in Experiment 1 the length is based on the normal spacing between contour drains. As this soil is supposed

to be less erodible than that of Series I a higher factor of 4·5 is used in the formula.

Thus V.I. $= \dfrac{3+4\cdot5}{2} = 3\cdot75$ ft

H.I. $= \dfrac{3\cdot75}{3} \times 100 = 125$ ft

Making an allowance for the width of the contour bank, the plots are actually 116 ft 6 in. $\times$ 18 ft 8 in.

Asbestos-cement planks are used for the plot boundaries as for Experiment 1.

*Fig. 8. Series II. Experimental site*

## Collection of Run-off and Eroded Soil

In designing the collecting tanks the same probable maximum rates and amounts of run-off were used as for Series I, i.e. 0·5 cusec maximum rate of flow and $6\frac{1}{2}$ in. maximum daily quantity of run-off.

The collecting channels are similar to those in Experiment 1 but with a steeper gradient of 1/50. In addition they are given a fall from front to back to concentrate low flows and reduce silting. Metal covers are provided. No adjustments of sill levels have been necessary after two seasons. Type H.S. flumes and recorders are installed at the intake of each tank.

## Collecting Tanks and Divisors

The general arrangement of collecting tanks is similar to that of Experiment 1, that is the plots are arranged in pairs and a single pit contains the tanks for both plots, but the tanks and divisors used in Experiment 1 are expensive and require precision sheet metal work for which few local firms are equipped, so a more simple system was designed. This consists of two brick collecting tanks, and the overflow from the first or (A) tank is passed through a divisor consisting of 10 V notches in a metal plate set into one wall of the tank. The aliquot passing through one notch is collected and passed into the second or (B) tank, the remaining 9/10 flowing to waste. The V notches are designed to take the maximum flow of 0·5 cusecs, and the brick tanks to store the maximum daily quantity of run-off. Each tank has a capacity of 700 gallons and the combined capacity is 7 700 gallons. The V notch divisor is made in two parts, the main plate and the V notches which are spot welded to it. The dimensions of the individual notches are not important, but they must be identical so they are cut out of small plates as exactly as possible, then bolted up together in a block of 10 and milled and filed to exactly the same shape. The 10 small notched plates are then spot welded to the main plate so that the tops are in a dead straight line, and the top edge is used to level up the divisor when installing it in the side of the tank.

The advantages of this system as opposed to that of Experiment 1 are (a) it is cheaper to build, (b) it is simpler to operate having only two tanks to empty instead of three. The disadvantages are (a) it is difficult to devise a satisfactory method of installing the metal divisor plate so that it can be levelled very

accurately, and then fixed rigidly (b) there is always the danger of cracks and leaks developing in the brickwork of the tanks. To minimise the danger of cracking, a reinforced concrete slab was cast in the floor of the pit, and independent slabs cast on top of this to form the floor of each pair of tanks. An open walkway is left all round the tanks so that leaks can be observed.

Instead of individual covers for the tanks, a roof is built over the whole pit, allowing observations to be made at leisure during heavy storms. Two screens of $\frac{1}{2}$ in. mesh and 64 mesh to trap floating debris fit in channels in the brickwork and slide out for cleaning.

### Drainage

Only one drainage scheme is used, as opposed to the dual system of Experiment 1 and this consists of 8 in. and 12 in. concrete pipes which dispose of the rain falling in the pits, and the 9/10 waste from the divisor, and also the contents of the tanks when they are emptied. Storm water from the areas surrounding the plots is also carried by this drainage system.

The technique of flocculating, draining and emptying has been the same as for Series I, but will also be replaced by laboratory determinations on samples.

## Part III

### Series III Experiments

During the winter of 1955 two more series of experiments were laid down using a new design which takes advantage of the information gained from the construction and operation of the previous experiments. Both these groups of experiments are on coarse sandy soil derived from granite, Series III on a shallow poorly drained soil and Series IV on a deep well drained soil.

The main experiments of these two series follow the general plan of the previous groups except that the cropping treatments concentrate chiefly on tobacco as this is the most important crop grown on this soil type. The plots (nine on Series III, and four on Series IV) are again 1/20 acre, and using the previous formula and

*Fig. 9. Series II. Tank installation during construction*

allowing for contour banks, on Series III with a slope of 5·5 per cent have a length of 80 ft, and on Series IV with a slope of 8 per cent have a length of 70 ft.

The plot boundaries and collecting troughs are as in the earlier experiments, but all the run-off is carried through closed pipes to a central building which houses all the collecting tanks and a store room. This greatly simplifies recording and allows simultaneous observations of run-off from all the plots to be made during heavy storms. The pipes are 6 in. spun concrete, on a footing to avoid cracks due to settlement, with a minimum grade of 1/50. At each change of direction circular chambers provide both access for rodding if necessary and a trap in which the gravel and coarse sand settle. This removal of coarse particles simplifies the taking of sub-samples when the remaining finer fractions are drained from the tanks.

Circular corrugated iron tanks are used as considerable difficulty was experienced in eliminating leakage from the brick tanks of Series II. Two tanks are used for each plot, each 7 ft 3 in. diameter and 2 ft 0 in. deep with conical concrete bottoms, the second or (B) tank storing 1/15 of the overflow from the (A) tank. The tanks have no metal bottom and are set into a circular slot cast into the concrete base. This slot is filled with hot poured bitumen and covered by the topping coat on the concrete. The concrete base slab is cast round 1½ in. steel pipes leading from the bottom of the cone which forms the bottom of each tank. The four pipes from two adjacent pairs of tanks all lead through gate valves to a circular pit, drained by a 4 in. underground pipe. It is possible for one operator in this pit to drain the four tanks simultaneously and take sub-samples at suitable intervals while the tanks are draining.

The improved divisor consists of a flat stainless steel plate let into the side of the (A) tank and having drilled in it eight horizontal rows each of fifteen ½ in. diameter holes. The water passing through the central vertical row of holes is carried by a light sheet-metal flume to the second tank. This type of divisor is very simple and inexpensive to make to the high standard of accuracy required. To attain this the holes must not vary in diameter, and each horizontal row must be exactly parallel. A template

*Fig. 10. Series III. Experimental site*

*Fig. 11. Series III. Collecting tanks*

*Fig. 12. Tank and divisions used on Series III and IV experiments*

is used to drill all the plates. The divisor is easily levelled accurately by inserting shims under the tank sides before sealing with bitumen. One large mesh and one fine mesh screen are fitted in each (A) tank to prevent floating debris from choking the holes, and are removable for cleaning. To prevent ripples or surge from the inflow affecting the divisor, a baffle extends below full supply level on the frame which carries the fine mesh screen. Provision has been made for the later addition of flumes and flow recorders, also inside the central building.

This layout and equipment used for Series III and IV appears after a full seasons operation to be most satisfactory and a great improvement on the earlier groups of experiments.

## Summary

The measurement of soil erosion and run-off under field conditions requires special equipment for the collection, division and storage of large volumes of run-off and eroded soil. Plots must be large if practical field scale cultural practices are followed, but in the high rainfall area of Southern Rhodesia very severe tropical storms cause very high rates of both run-off and erosion. Annual soil losses as high as 225 tons per acre have been recorded.

In order to reduce the collected volumes to manageable proportions a known and accurate aliquot of the total run-off must be taken, but the division of a flow of water which may carry widely varying amounts of suspended material and bed load required special apparatus. The removal, sampling and recording of this heterogeneous mixture of soil, water and floating debris presents further problems.

The author has designed and installed four groups of experiments which now form the most comprehensive erosion research unit in Africa, and this account describes the different methods and equipment used in successive experiments, showing how the design was improved by information gained in the earlier experiments.

# 5

Reprinted from *FAO Soils Bull.* **33**:109-122 (1977)

## STUDYING PATTERNS OF SOIL EROSION IN KENYA

by

Thomas Dunne

Department of Geological Sciences and Quaternary Research Center
University of Washington, Seattle, USA

### INTRODUCTION

Soil erosion is intense in many areas of developing countries because of erosive climatic conditions, rugged terrain, and heavy land use, yet little quantitative information is available about the patterns of erosion. This ignorance limits an assessment of the real magnitude of erosion problems, or the ranking of priority regions for the most urgent soil conservation programmes. When sites are chosen for reservoir impoundment or other water resource development, a rapid assessment of sediment transport is usually made but very few such studies are continued for a long enough period to sample the vagaries of weather and flow which characterize the hydrologic regimes of most developing countries, The accumulation of information on soil erosion and sediment yields is usually too meagre and too late for adequate design (Dunne and Ongweny, 1976).

The situation requires the training and support of a small number of field scientists in developing nations who will be concerned with field assessment of erosion and sediment transport. They should be able to take advantage of hydrologic records which have sometimes been accumulated for years and have lain unanalysed in the files of water-resource agencies. Such scientists should also be capable of setting up networks of stations for the collection of data on erosion and sedimentation in hitherto ungauged areas, and of carrying out field experiments on soil erosion. The cost of such work would not be high, relative to its value, but the success of the work requires a commitment to field work and continuity of purpose which is frequently lacking.

In this paper, I review some work on patterns of soil erosion in Kenya and refer to relatively inexpensive methods of studying erosion processes and patterns. It will also indicate some of the gaps in our knowledge of soil erosion, particularly in developing countries where physical and land use conditions differ from the more intensively studied, commercial agricultural region of developed countries.

It is appropriate to stress from the outset, however, that the accumulation of field measurements is not enough. In developed countries, large amounts of money have been spent on monitoring programmes and the data have. lain unanalysed for long periods of time. Developing countries cannot afford this waste, and the field scientist must be encouraged to analyse data as they accumulate. This part of the scientific training is as important as instruction in field methods.

### Methods of Quantifying Erosion Rates

There are two basic approaches to the study of erosion patterns. The first involves sampling the rate of sediment transport past some point on a river channel at the outlet of a drainage basin. This method is relatively cheap, and it is easy to monitor soil loss rates from large, representative areas by installing gauging stations on a few rivers. Because the measurement of sediment loss is made at a single point, however, it is not possible to interpret much about the spatial pattern of erosion within the catchment. Nevertheless, sediment monitoring is the most widely used method of assessing soil erosion rates and many water-resource agencies collect suspended sediment records routinely. It is useful, therefore, to consider methods of extracting the maximum possible information from such records.

The second method of quantifying soil erosion involves direct measurement of soil removal by individual processes at a number of sampling sites within the drainage basin. By strategic location of plots, erosion pins, surveyed cross-sections of gullies and river channels, it is possible to define the spatial pattern of soil loss, and to study the local controls of erosion. If measurement sites are distributed so as to sample a range of hillslope gradient, soil types, land use, and conservation practices, for example, the effects of these variables on soil erosion can be isolated and quantified. This kind of information is necessary in the design of land-use and conservation strategies for developing countries. Yet very few measurements of hillslope erosion processes are presently being made in these lands. There is a need to encourage scientists in these countries to use the techniques that are now available. The most useful field methods are described in another paper (Dunne, 1976 a), which includes a bibliography of original sources.

Both of the approaches referred to above include systematic monitoring. The concept of environmental monitoring is gaining acceptance and support ( U.N. Conference on the Human Environment, Stockholm, 1972) and we can reasonably look forward to an increase of erosion measurements in developing countries in the near future. In order to interpret the results from monitoring networks, however, it is usually necessary to carry out some controlled experiments of erosion under different condi-conditions of hillslope gradient, land use, conservation practice or other variables of interest. The most common type of controlled experiment involves measuring soil loss from small hillside plots under natural or artificial rainfall (Battawar and Rao, 1969; Dunne, 1976b; Fournier, 1967; Goel et al, 1968; Hudson, 1971; Vasudevaiah et al, 1965). The plots can be subjected to various treatments, such as removal of vegetation, trampling, or the growing of various crops. They are useful for previewing the soil erosion consequences of a range of management options.

Each of these approaches is presently being used to study the pattern of soil erosion in Kenya.

SEDIMENT YIELDS OF KENYAN RIVERS

During the period 1948-68, suspended sediment concentrations were measured by the depth-integrating method at a large number of river gauging stations throughout south and central Kenya (the only regions of the country which support perennial streams). At 63 stations, the data were adequate for constructing sediment rating curves. Daily discharge records from the same stations were then used in conjunction with the sediment rating curves to calculate suspended sediment yields for drainage basins covering a wide range of climate, topography, and land use. A map of mean annual suspended sediment yields was constructed from the data (Dunne, ms in preparation). Sediment yields range from 8 to 19 520 t/km$^2$/year. The results of this national survey can be used directly for estimating potential rates of sedimentation of proposed reservoir sites. They can also be used for an analysis of the major controls of basin sediment yields.

A great deal of attention has been directed toward quantifying general relationships between basin sediment yield and climate (Langbein and Schumm, 1958; Fournier, 1960; Douglas, 1967; Wilson, 1973). The climatic parameter generally used is mean annual rainfall, either obtained from direct measurements or calculated from mean annual runoff and air temperature. Each of the publications listed above proposes a different reationship between sediment yield and climate. Wilson, who analysed the most comprehensive set of data, concluded that differences in climatic regime and land use make it impossible to define a single rule relating sediment yield to rainfall or runoff.

The Kenyan data confirm the suggestion of Wilson, and of Douglas (1967) that land use is the dominant variable which confounds the establishment of general relations of sediment yield and climate. In Kenya, as in many other countries, land use depends partly upon climate but there are important differences of land use in each climatic zone.

In Figure 1, mean annual sediment yield per unit area of catchment is plotted against mean annual runoff. The dominant land use in each catchment is indicated by a symbol. In the absence of a detailed quantitative analysis of land use, the classification was confined to four classes: completely forested; forest covering more than 50 percent of the basin; agriculture covering more than 50 percent of the basin and the remainder under forest, and grazing covering more than 50 percent of the basin. A fifth class, lightly grazed scrub forest, contained only two basins. Even with such a coarse classification of land use, however, a pattern is evident.

The lines in Figure 1 are approximate envelopes for each set of land use symbols, and very few points fall outside the appropriate region of the graph. The envelopes do not separate the symbols completely because of differences in the ruggedness of topography, the degree to which the major land use dominates a basin, the duration of records, and the quality of the original data.

There are dramatic differences of sediment yield between land use types. For a fixed value of runoff in the figure, differences in sediment yields between land use types can vary over two orders of magnitude or more. The graph shows, however, that land use is not the only important variable. Agricultural catchments with heavy runoff may have sediment yields which are far greater than the driest grazing lands.

For each land use type, there is a general increase of sediment yields with annual runoff. The higher runoff yields are associated with heavier rainfalls and therefore with greater kinetic energy for hillslope erosion and stream transport of eroded sediment. Regression analysis for basins in each land use category yielded the following equations, all of which are significant at the 0.05 level:

| | | |
|---|---|---|
| Forest | Sed. yield $= 2.67$ Runoff$^{0.38}$ | $r = 0.98$, $n = 4$ |
| Forest $>$ Agriculture | Sed. yield $= 0.042$ Runoff$^{1.18}$ | $r = 0.75$, $n = 10$ |
| Agriculture $>$ Forest | Sed. yield $= 0.038$ Runoff$^{1.41}$ | $r = 0.73$, $n = 39$ |
| Grazing dominant | Sed. yield $= 0.002$ Runoff$^{2.74}$ | $r = 0.87$, $n = 7$ |

The regressions are plotted in Figure 2.

Although only four forested basins were available for this analysis, the results are almost exactly the same over the range of the data as those from a similar analysis of sediment yields from 27 catchments in eastern Australia made by Douglas (1967). The Australian catchments were "selected to avoid as much human disturbance as possible". His results are shown in Figure 2.

For the other land use types, sediment yields are higher than under the complete forest cover. The exponents in the regression equations above also show that sediment yield increases with runoff less rapidly in regions with a forest cover than in cultivated lands, which in turn are less sensitive than rangelands.

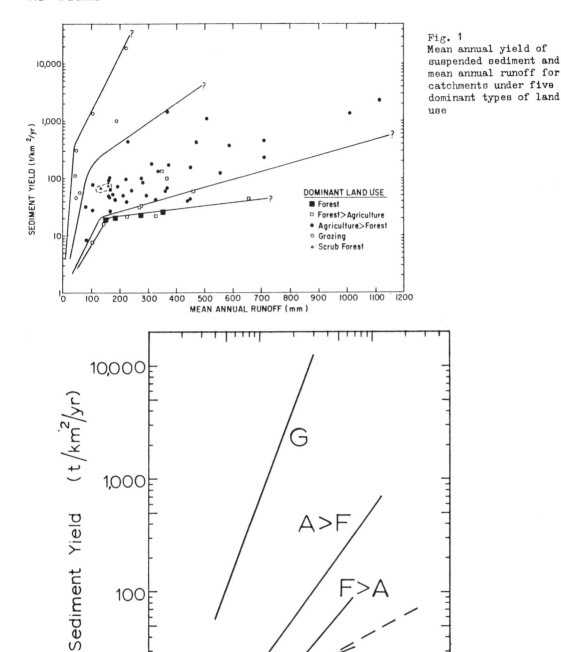

Fig. 1
Mean annual yield of
suspended sediment and
mean annual runoff for
catchments under five
dominant types of land
use

Fig. 2  Comparison of the regression lines computed for the relationship between sediment
yield and mean annual runoff for each land use type.  The dashed line was computed
by Douglas (1967) for forest catchments in eastern Australia

65

Field observations and visual examination of the data suggested that topographic steepness is a significant factor affecting sediment yields. Measurement of the frequency distribution of hillslope angles in each drainage basin was not feasible with the resources available for this study, and a surrogate measure of basin steepness had to be used. Schumm (1955) showed that the relief·ratio of a catchment (its maximum relief divided by the length of the main stream) was positively correlated with sediment loss in Colorado.

The relief ratio was used with mean annual runoff in a stepwise multiple regression of the Kenyan sediment yields. The results were limited because most land use categories contained few points. Runoff proved to be the dominant variable in each case, but only on agricultural lands did relief ratio add significantly to the explanation of the variance in sediment yield. In the other land use classes, however, there was a positive relationship between relief ratio and sediment yield when relief ratio entered the multiple regression as a second variable, and it is likely that the effect of topography would have been demonstrated with a larger sample. The limited data also suggest that in a logarithmic multiple regression equation the exponent of relief ratio increases in the same order as that for runoff. In other words, the effect of basin steepness on sediment yield increases as the vegetation cover becomes sparser. Correction of sediment yields for the effect of catchment area by the method of Brune (1948) did not alter the general form of the results, except by increasing the sediment yields.

No bedload data are available for Kenya and so the yields referred to above underestimate the true soil loss. Field observations suggest that bedload transport is small in the volcanic uplands, where most of the eroded sediment is fine grained. The larger rivers draining the lowlands of Eastern Kenya receive considerable amounts of coarse sand from erosion of soils on schists and gneisses. Some of this material moves as bedload, but its contribution to the basin sediment yield will not be known until a programme of bedload transport measurements is undertaken.

HILLSLOPE MEASUREMENTS OF EROSION

In sparsely populated dry regions, where stream flow is rare, there is little likelihood that developing nations can bear the cost of maintaining stream gauging stations for the purpose of assessing sediment yields. Under these conditions, soil erosion can be monitored directly on hillslopes. This can be done by installing plots or networks of erosion pins. Leopold et al, (1966) demonstrated how various techniques for measuring hillslope erosion processes could be used to obtain a sediment budget for a small rangeland catchment. A major problem with all field methods which involve installing even simple equipment, however, is its susceptibility to theft or disturbance.

Soil erosion rates can also be evaluated by measuring recent lowering of the surface against some dateable reference. Judson (1968) obtained rates of soil removal from the depth of exposure of Roman archaeological sites. Fence posts often show marks indicating the position of the soil surface at the time of installation. The difference between this height and the present soil surface divided by the age of the fence-line gives the soil erosion rate.

The most widespread indicators of surface lowering in some areas where erosion is intense are exposed tree roots or mounds of residual soil protected under the canopy of trees or bushes while the surrounding soil is lowered (see Fig. 3). If the tree or bush can be aged by counting growth rings (as many tropical species can, in spite of the popular misconception that tropical woody plants do not produce annual or seasonal growth rings), the height of the mound divided by the age of the plant indicates the average rate of surface lowering. In some areas the dating problem is simplified dramatically if there is evidence that soil erosion was accelerated after a period of intensive vegetation clearing. The height of the root exposure or mound can be measured simply and quickly as shown in Figure 4.

Fig. 3   An erosion mound protected by a tree canopy while the surrounding land surface
is lowered by erosion.   The height of this particular mound is 60 cm

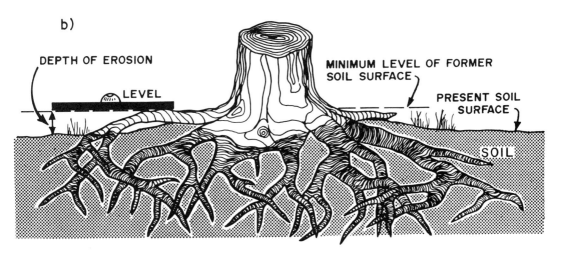

Fig. 4   Measurements of erosion rates from tree root exposures.   On suitable tree
species the height of the former ground surface is located by examining the
tree for signs such as the position of the basal flare or the boundary between
trunk bark and root bark.   This should be done only after examining trees in
relatively uneroded sites.   A carpenter's level is then placed at the estimated
level of the former land surface and its height above the present soil surface
is measured with a ruler

**67**

Problems of interpretation arise with this method, and a great deal of care should be taken to check for potential problems in each region before the method is used there. A dateable tree species must be found and the tree-ring chronology established, or the onset of accelerated erosion must.be dated from aerial photography or other local information. Growth rings can be counted on each tree for which root exposures are measured, but this can be very time-consuming. An alternative method involves cutting down or coring only a sample of trees and constructing a graph of trunk diameter versus age for each species and region (see Figure 5). Each tree used for measuring the erosion rate can then be aged from its stem diameter.

Other sources of uncertainty arise with this method. Some trees produce their own mound by developing a wide basal flare or even by developing buttress roots above the ground surface as they grow. This problem can be avoided by choosing a species which does not have these characteristics. Careful examination of trees in sites which are not undergoing intense erosion (such as plateaux or heavily vegetated areas) should suggest the most useful tree species to use as an erosion indicator in each region. We also compare plants with a range of ages to observe how the plant, its root, and the mound or root exposure develop as the tree or bush grows.

Species, or at least individual trees, which regenerate from old stumps or root stocks should be avoided because the mound is more likely to be related to the age of the older plant than to the new stem. Termites often build mounds around trees and these must also be avoided. Recognition of this problem is not always easy, especially if the mound is no longer colonized and has been eroded. Small termite mounds can usually be recognized by their looser texture and higher organic content than surrounding eroded soils. They also lack pedogenic structures. Mounds produced by wind deposition also have a different structure and texture from the surrounding eroded area, and can be recognized through careful examination. Other sources of uncertainty are described by Eardley (1967) and by Lamarche (1968), who pioneered the method on Bristlecone pines in Utah and in the White Mountains of California.

We incorporate measurements of tree-root exposures into a general hillslope survey of topography, vegetation cover and soils, as described by Leopold and Dunne (1971). At intervals of 100 meters along the hillslope profile we measure the height of the root exposure or erosion mound under the five or ten nearest trees or bushes of the species being used in that area. The procedure illustrated in Figure 4 is carried out on opposite sides of the tree along the contour. The plant is also aged. The average erosion rate for the 5-10 plants is then computed for each site.

The data can be used for mapping the variation of erosion depth along a hillside (Dunne, 1976a, Figure 10) and therefore for computing the total amount of soil lost from a sample of hillslopes in each region. They can also be used for studying the effect of gradient on erosion, as shown for a single rock/soil complex in Figure 6. Measurements of this kind were used to quantify differences in rates of soil loss on three rock/soil complexes in Kajiado District, a heavily grazed rangeland in southern Kenya. I have quantified differences of soil erosion rates on hillslopes with differing gradients, soil types, and intensity of vegetation removal in the Maralal area of northern Kenya. The results are illustrated in Figure 7.

These field measurements show that the rate of soil erosion on even gentle gradients in Kenyan rangelands is extremely high by comparison with the rates compiled by Young (1969) for a variety of regions throughout the world. Over the last 10-20 years, soil has been lost at rates in the range of 0.1 to 0.5 cm/yr on the Athi-Kapiti plain and 0.4 to 1.2 cm/yr in Northern Kenya. These values are equivalent to yields of 1000 - 18 000 $t/km^2/yr$ depending on the bulk density of the soil. It is difficult to compare these values directly with basin sediment yields, because a portion of the soil mobilized from hillsides comes to rest in swales,

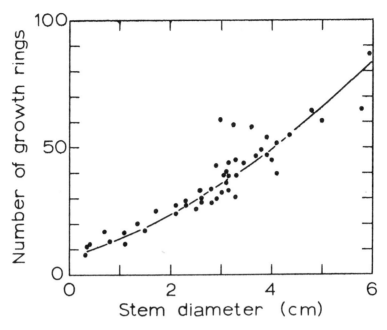

**Fig. 5**   Relationship between number of seasonal growth rings and trunk diameter 0.25 m above the ground surface for <u>Acacia drepanolobium</u> trees on the Athi-Kapiti plains of Kenya. The rings have not yet been counted under a microscope and so their numbers are still tentative. Biologists measuring plant growth in the region tell us that there are two strong growth periods in each year, even during times of low rainfall

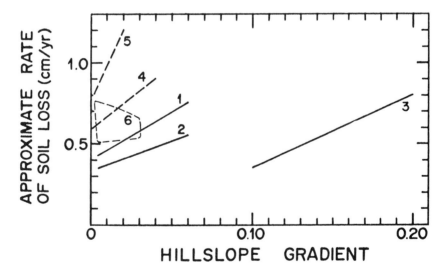

**Fig. 7**   Average annual rate of soil loss as a function of gradient for hill-sides on two rock types near Baringo and Maralal, N. Kenya. The granitic and gneissic Basement rocks (solid lines) weather to sandier soils than those which develop on the lavas (dashed lines). The former are more resistant than the latter to erosion

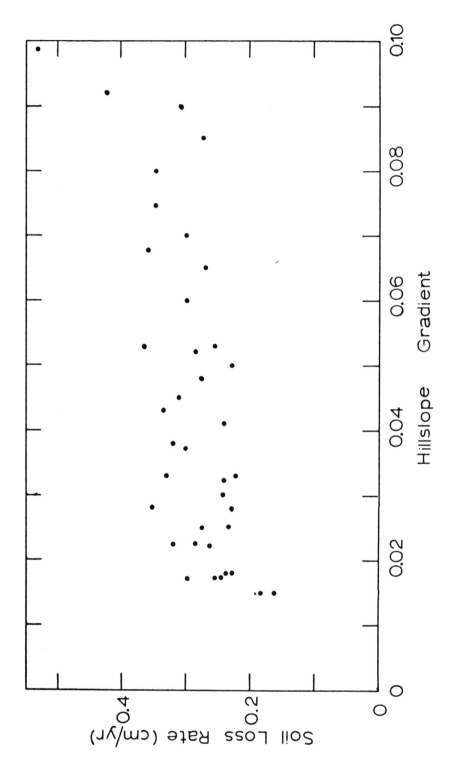

Fig. 6 Average annual rates of soil loss as a function of gradient for 38 sites on
7 grazed hillsides on the Athi-Kapiti plains of Kenya. The rates are
tentative until the tree rings from which they were calculated can be checked
under a high-powered microscope

floodplains, and other storage sites. The measured sediment yield per unit area for catchments usually declines, therefore, as the size of the basin increases. I know of no published information which indicates the rate at which this decline occurs for arid regions, but in moister agricultural regions of the United States, Brune (1948), Maner(1958), and Roehl (1962) have shown that sediment yield per unit area is proportional to the catchment size raised to a power of approximately − 0.15 to −0.20. Using a value of −0.20 the Kenyan rangeland sediment yields would be approximately 150-2 700 t/km$^2$/yr at 100 km$^2$ and of 90-1 620 t/km$^2$/year at 2 000 km$^2$. These values are in the same range as most of the basin values for the drier grazing lands in Figure 1, and confirm the evidence that soil erosion there is extreme.

This kind of simple measurement could profitably be made more widely in the rapidly eroding regions of developing countries. Two people with only a hand level, tape, carpenter's level, and rule can make an erosion survey of one to two kilometers per day, and in doing so collect a great deal of information on erosion rates and their controlling factors, In addition to costing little, the method has some other advantages over installing plots to monitor soil loss. The tree-root measurements yield data immediately, rather than the investigator having to wait three or more years to obtain usable data. Secondly the resulting calculations of erosion rate average out inter-annual fluctuations which may distort the picture over a short measurement period on a plot. Thirdly, there are no installations to be disturbed or stolen. On the other hand, monitoring of soil loss from plots or by erosion pins can yield more detailed information, such as the contribution of rainstorms of various sizes. Plots are particularly useful where the rate of erosion is less than the high values shown above, or where trees and bushes are rare. In other words, use of the two methods can be complementary.

In addition to collecting information on soil loss from hillsides we need to know   more about the fate of the eroded sediment. There is very little information on this topic even in regions where soil erosion has been studied intensively and almost none for developing countries. Sediment is temporarily stored at many locations as it moves down a drainage basin after its initial release from a hillside. Such locations include footslopes, unchannelled swales, channels and floodplains, lakes and reservoirs. The amount of sediment accumulating at each of these sites is important from both an economic point of view (rates of filling of reservoirs and stock ponds) and an ecological point of view for those interested in the nutrient supply and depth of water holding sediment delivered to swales and floodplains. Our ignorance of the fate of eroded sediment is important to a full understanding of the effects of soil erosion, and could be remedied by a programme of simple, repeated topographic surveys at sites where the sediment accumulates.

CONTROLLED PLOT EXPERIMENTS

To provide quantitative information on the controls of soil erosion on Kenyan rangelands, we have begun a set of controlled experiments using a portable sprinkler system which generates artificial rainstorms over a 5m by 2m plot (see Figure 8). With this system a storm of, say, one hour's duration and intensity of 7cm/hr can be applied to plots on a range of hillslope gradients on wet or dry antecedent conditions, with the grass cover in various states. With repeated irrigations of a plot to simulate a whole wet season, we can grow and cut grass to various cover densities.

But runoff and soil loss rates are monitored during the storm, and a sample of the results from one experiment are given in Figure 9. The results can be used to compare plots on the basis of infiltration capacity, total runoff, or total soil loss. Figure 10, for example, compares soil loss from three soil vegetation complexes in their typical conditions at the end of a dry season. These and similar results will be described in a set of forthcoming papers.

71

Fig. 8  The sprinkler system used for generating artificial rainstorms on hillside plots in Kajiado District, Kenya

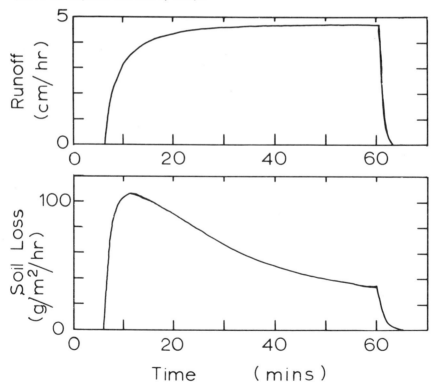

Fig. 9  Hydrographs of runoff and soil loss from a 5m x 2m plot during an artificial rainstorm.  Storm duration was one hour and the intensity was 6.9 cm/hr

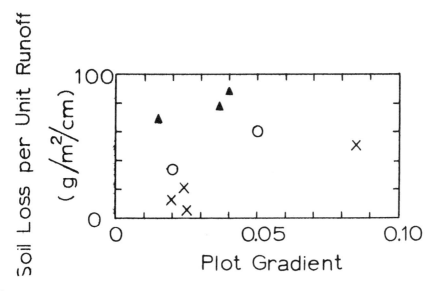

Fig. 10    Total soil loss in a one-hour rainstorm on dry conditions from plots in
their usual condition at the end of a dry season. The crosses represent
vertisolic clay soils on volcanic rocks with a ground cover averaging 65 –
85%; the circles represent sandy clay loams on schists with a ground cover
of about 10%, and the triangles indicate sandy clay loams with covers of
0 – 7% developed on volcanic rocks

## SUMMARY

The purpose of the present paper is to indicate the range of approaches
available for studying soil erosion in developing countries. Most of the techniques
are simple and can be carried out by a small team of field scientists in each country.
The cost of such a programme would be small relative to its value, which was reviewed
at the beginning of this paper. There are many gaps in our knowledge of the magnitude,
distribution, and controls of soil erosion in the tropics. We do not know a great
deal about the degree to which various conservation techniques presently reduce soil
loss. In view of the present concern about "desertification" and the many pessimistic
reviews of the status of eroded lands in some developing countries, it would be
worthwhile to collect some quantitative information to form an objective basis for
decision making about soil conservation.

## ACKNOWLEDGMENTS

The work reported here was begun with the funding of the McGill-Rockefeller
Programme (Prof. T. L. Hills, Director) and the cooperation of the Ministry of
Water Development, Nairobi. It was continued with support of the American Philosophical
Society and the Explorers' Club. The sprinkling experiments were carried out while
the writer was employed as a consultant to the Kenya Wildlife Management Programme
(FAO) and the Ministry of Tourism and Wildlife, Government of Kenya.

REFERENCES

Battawar, H.B. and Rao, V.P.  Effectiveness of crop-cover for reducing runoff and soil loss.
1969     J. Soil and Water Conservation in India  17(3 and 4): 39-50.

Brune, G.M.  Rates of sediment production in the midwestern United States.  U.S. Soil
1948     Conservation Service Technical Paper  65.

Douglas, I.  Man, vegetation and the sediment yields of rivers.  Nature   215: 925-928.
1967

Dunne, T.  Evaluation of erosion conditions and trends.  In:  Watershed management:  field
1976a     guidelines.  S.H. Kunkle (ed.) FAO, Rome (in press).

Dunne, T.  Intensity and controls of soil erosion in Kajiado District, Kenya.  Consultancy
1976b     Report to the Kenya Wildlife Management Project, Nairobi.  FAO, Rome.

Dunne, T. and Ongweny, G.S.O.  A new estimate of the rate of sedimentation in reservoirs
1976     on the Upper Tana R., Kenya.  The Kenyan Geographical Journal, December (in press).

Eardley, A.J.  Rates of denudation as measured by bristlecone pines, Cedar Breaks, Utah.
1967     Utah Geological and Mineralogical Survey Special Studies 21, 13 p.

Fournier, F.  Climat et érosion:  la relation entre l'érosion du sol par l'eau et les
1960     précipitations atmosphériques.  Presses Universitaire, Paris, 201 p.

Fournier, F.  Research on soil erosion and conservation on the continent of Africa.
1967     African Soils, 12(1): 5-51.

Goel, K.N., Khanna, M.L. and Gupta, R.N.  Effect of degree and length of slope and soil
1968     type on plant nutrient losses by water erosion in the alluvial tracts of Uttar
         Pradesh.  J. Soil and Water Conservation in India  16(1 and 2): 1-6.

Hudson, N.H.  Soil Conservation.  Cornell University Press, Ithaca, N.Y., 320 p.
1971

Judson, S.  Erosion rates near Rome, Italy.  Science, 160: 1444-1446.
1968

Langbein, W.B. and Schumm, S.A.  Yield of sediment in relation to mean annual precipitation.
1958     American Geophysical Union, Transac. 39: 1076-1084.

Leopold, L.B. and Dunne, T.  Field method for hillslope description.  Tech. Bull. 7,
1971     British Geomorphological Research Group, 24p.

Maner, S.B. Factors affecting sediment delivery ratios in the Red Hills physiographic area.
1958     American Geophysical Union, Transac. 39(4): 669-675.

Roehl, J.W.  Sediment source areas, delivery ratios and influencing morphological factors.
1962     International Association of Scientific Hydrology Publication 59, p. 202-213.

Schumm, S.A.  The relation of drainage basin relief to sediment loss.  International
1955     Association of Scientific Hydrology Publication 36, p. 216-219.

Vasudevaiah, R.D., Singh Teotia, S.P. and Guha, D.P.   Runoff-soil loss determination
1965        studies at Deochanda Experiment Station:   II Effect of annual cultivated grain
            crops and perennial grasses on 5 percent slope;   J. Soil and Water Conservation
            in India   13(3 and 4): 30-36.

Wilson, L.   Variation in mean annual sediment yield as a function of mean annual
1973        precipitation.   American Journal of Science   273: 335-349.

Young, A.   Present rate of land erosion.   Nature   224: 851-852.
1969.

75

# 6

# SOIL EROSION AND SEDIMENTATION IN TANZANIA AND LESOTHO

## Anders Rapp

*In many African countries, the pressures of overcultivation, overgrazing and wood collection have led to serious erosion damage, increased sedimentation and silting up of reservoirs. The author stresses the need for catchment studies and recommends that reclamation and development plans always be adapted to local ecological and social conditions.*

The repeated shortages of food and water in Africa are caused by many contributing factors which can be identified as political, socio-economic and ecological. One of the necessary lines of action to change the negative trend in many African countries is an improved and ecologically sound use of the natural resources of water, vegetation, and productive soil. The increasing population and the increasing pressure of overcultivation and overgrazing tend to accelerate soil erosion and other forms of land degradation. The situation today is that we do not know how fast Africa's forests, grazing lands and soil are being destroyed. Nor do we know to what extent this manifold degradation is irreversible or where and how to concentrate reclamation efforts.

A first step toward rational and ecologically sound land use, in Africa or elsewhere, is to inventory the types and extent of present degradation of vegetation and soil and to monitor the changes. This is true with regard to both traditional and mechanized or "modernized" forms of land use. A second step is interdisciplinary ecological research and training for better land use.

In order to study types of soil erosion, monitor present rates of erosion and determine the consequences of erosion, it is necessary to carry out continuous studies of water and sediment flow in natural watersheds. This is called the "catchment approach" or "watershed approach". Such studies must be performed in a variety of environments. So far, only a few quantitative catchment studies of this kind have been performed in developing countries. Examples of such studies are those undertaken by Bormann,

Likens and Eaton (1) in New England, Douglas (2) in Malaysia, and Rapp *et al* (3, 4) in Tanzania. Some aspects of the latter studies will be presented below, together with examples from Lesotho in southern Africa. Both Tanzania and Lesotho are facing very serious erosion and water shortage problems.

## SOIL EROSION AND ITS EFFECTS

The most common forms of soil erosion are caused by running water or by wind, or by a combination of these two. Both water erosion and wind erosion are particularly effective in climates where long-lasting dry seasons weaken the vegetation cover and the soil is exposed to later attack by intensive rainstorms or dust storms. Vegetation cover is the best general protection against the washing away of soil particles by runoff or the blowing away of soil as dust. A dense grass cover can in this respect be as efficient as a forest cover in protecting the soil (5). Fire and overgrazing contribute very much to the erosion hazard in African semi-arid areas by exposing the soil to the assault of water and wind as well as to the heat of direct sunlight, reducing the biological activity in the near-surface zone of the soil.

Tropical and subtropical *semi-arid areas* are thus, in general, zones of great potential erosion hazard, due to the combination of sparse vegetation at the end of the dry season and periodically very intense rainstorms or dust storms. Other environments with potentially great erosion hazard are *mountains*. Many mountain areas in Africa have been subjected to a high degree of exploitation by

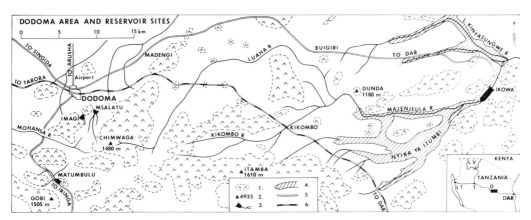

Figure 1. Location map of reservoir sites, Dodoma area, Tanzania. Key: 1. Bedrock hills of granite or migmatic granite. 2. Summit with altitude in meters. 3. Reservoir investigated in the Duser project. 4. Floodplain of "mbuga" type. 5. Road. 6. Railway. On insert map: L V = Lake Victoria, L T = Lake Tanganyika, D = Dodoma, Dar = Dar es Salaam, black rectangle marks area covered by main map.

man, with deforestation, burning, cultivation of steep slopes, soil erosion by slope wash, landsliding and other forms of degradation.

Erosion causes losses of productive topsoil, organic matter, nutrients and water. Up to 50 percent of the annual rainfall can be lost from eroded slopes due to decreased infiltration and high surface runoff. The downstream effects of this are increasing floodpeaks with higher sediment loads from denuded source areas of rivers. The increasing sediment loads result in faster filling of reservoirs and thus shorten the useful life of these reservoirs.

## FORMS OF EROSION

The water erosion on a naked, soil-covered slope increases with intensity of rainfall, slope gradient, length of slope, surface runoff, and erodibility of the soil. *Splash erosion,* caused by the direct impact of water drops on the soil, affects the entire slope, from the crest to the valley bottom. The splash makes the fine soil particles start moving. It also creates a wet crust on the soil; small splashed particles of soil enter and seal the surface pores, thus decreasing the infiltration rate of the water into the soil and increasing the surface runoff.

*Sheet erosion* is the washing away of a thin surface layer of soil. *Rill erosion* causes erosion furrows of several cm in depth. *Gully erosion* creates furrows of more than one meter in depth. All these forms of erosion can be created by temporarily high surface runoff. *River erosion,* either by bank cutting or bed scouring, does not come under the heading of soil erosion (interfluvial erosion), but is nevertheless of very great importance and is closely linked with soil erosion.

On hillslopes of some steepness, *landslide erosion* may occur, generally as a sudden movement of soil and weathered rock, creating a *slide scar.* Some landslide masses that begin with a sliding movement change into viscous flows as they move down the slope. Such flowing masses of loose material composed of soil or rock debris are called *debris flows* or *mudflows* (6).

The common forms of splash, sheet and rill erosion can be studied in erosion plots of a few meters in dimension or even in a laboratory. The larger forms of erosion such as gullies, landslides and debris flows have to be studied mainly in natural catchments, due to their less frequent occurrence and large size. Water erosion and slides have downstream effects on sediment transport and deposition. Thus a catchment basin is a useful unit for studies of the erosion/sedimentation system.

## EROSION AND SEDIMENTATION IN FIVE CATCHMENTS IN SEMI-ARID TANZANIA

Man-made and natural lakes act as sediment traps and store part of the material eroded and transported by running water from the catchment. In semi-arid areas, streams flow only occasionally and the flow of water and sediment is difficult to measure. However, the sediment trapped and deposited on the bottom of a reservoir can be measured and analyzed and can thus reveal the history of sedimentation since the time of the dam construction. It can also tell the history of erosion in the catchment, although only in a generalized way, as much of the material eroded from the slopes of the catchment is temporarily deposited upstream of the reservoir, eg some parts are deposited at the base of the eroded slope, other portions on the floodplains and still others on the riverbeds. But some material reaches the reservoir and is deposited in a delta or as fine-grained bottom sediment on the lake floor beyond the delta. Some of the fine-grained suspended material is carried out of the reservoir by the flow.

Sediment surveys of silting reservoirs can be used to forecast how fast the basins will be filled up and made useless.

The sediment-survey approach was used in studies of five small reservoirs and catchments, four in the Dodoma area (4) of central Tanzania and another near Arusha in northern Tanzania (7). The Dodoma area is a semi-arid plain with scattered hills of granites and gneisses, so-called inselbergs (Figure 1). The soils are red and sandy on the

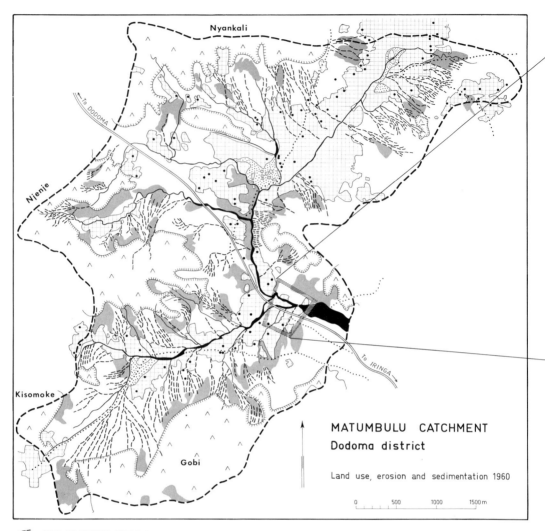

MATUMBULU CATCHMENT
Dodoma district

Land use, erosion and sedimentation 1960

| | |
|---|---|
| 0 | 500 | 1000 | 1500 m |

///  AREA OF INTENSE GULLYING

▓  AREA OF INTENSE SHEET WASH

▒  SAND FAN

∿  STREAM CHANNEL

◆  RESERVOIR

▦  CULTIVATED LAND OR
   RECENTLY ABANDONED FIELDS

∧∧  INSELBERG WITH FOREST OR
   DENSE BUSH

═  ROAD

·····  MAJOR CATTLE TRACK

•ₚ  SETTLEMENTS

‒‒‒  DRAINAGE DIVIDE

Figure 2. Map of land use, erosion and sedimentation, Matumbulu catchment, Tanzania. Based on air photographs from 1960 and field checking during 1969—71. Note the zones of erosion and deposition: gullied upper pediments with intense sheet wash, cultivated lower pediments, stream channels with three sand fans and reservoir with heavy sedimentation. Map by C Christiansson. Area of Figure 3 indicated by frame.

| Place | Catchment area km² | Relief ratio m/km | Annual sediment yield m³/km² | Soil denudation rate* mm/year | Period | Expected total life of reservoir |
|---|---|---|---|---|---|---|
| 1. Ikowa | 640 | 730/50 | 195 | 0.20 | 1957—69 | 30 years |
| 2. Matum-bulu | 18.1 | 257/4.4 | 729 | 0.73 | 1962—71 | 30 years |
| 3. Msalatu | 8.7 | 183/4.1 | 406 | 0.41 | 1944—71 | 110 years |
| 4. Imagi | 1.5 | 122/1.6 | 601 | 0.60 | 1930—71 | 190 years |
| 5. Kisongo | 9.3 | 225/5.7 | 481 | 0.48 | 1960—71 | 25 years |
| 6. Morogoro | 19 | 1598/6.8 | 260 | 0.26 | 1966—70 | no reservoir |
| 7. Mgeta-Mzinga | 20 | 1325/6.2 | 13 500 | 14 | 1970-02-23 | no reservoir |

* Data on denudation rates are based on reservoir sediment surveys [Nos 1—5], sampling of suspended load in streams [No. 6] (3) and total volume of landslide scars resulting from one rainstorm of two hours duration [No. 7]. No. 7 is the only case based on volumes of erosion scars (8). Hence it is not directly comparable to the other cases which are based on sediment deposits or sediment load in streams. Relief ratio is maximum relief of catchment in meters, divided by length of catchment in kilometers. Economic life of silting reservoirs is shorter than expected total life until 100 percent filling (from ref 4, p 311).

Table 1. Soil denudation rates in seven catchment basins in Tanzania. Nos 1—5 are catchments with reservoirs in semi-arid areas. Nos 6—7 are catchments in the Uluguru mountains with high precipitation and no reservoirs.

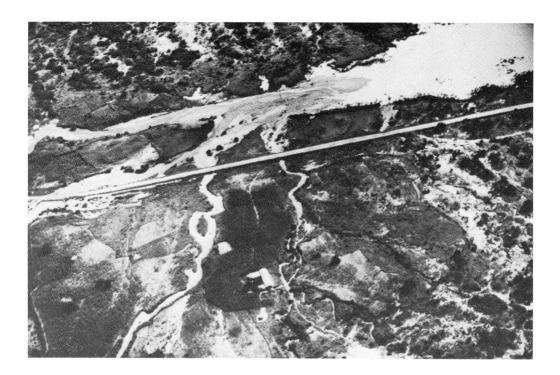

gently sloping plains (= pediment slopes), and black, cracking clays on the floodplains. The annual rainfall is less than 600 mm. The dry season is about 7—8 months long, lasting from May until December.

Kisongo catchment near Arusha is underlain by young volcanic rocks. The soils are dark, silty-sandy and derived from weathered volcanic lavas or ashes. The annual rainfall is about 800 mm and the dry season normally lasts 5—6 months. Some data on the five reservoirs and catchments are given in Table 1. All catchments are under intense land use and are much overgrazed by cattle. Cultivation and collecting of firewood also contribute to erosion.

The sedimentation rate in each of the reservoirs was determined by repeated surveying of cross-profiles. Inventories of the erosion features in the catchments were made by means of air photo interpretation and field checking (Figure 2).

Rainsplash and sheet wash are probably the most important types of erosion in the area that was studied. Gullies appear in distinct zones on the upper pediment slopes near the foot of the inselbergs in the Dodoma catchments. In the Kisongo area the gullies are cut along cattle tracks near the main drainage lines of the catchment. No new areas of gullying could be traced by comparison with old air photographs in the Dodoma catchments but extension of some single gullies had occurred since the early 1950s.

The reservoirs have very high rates of sedimentation. Three of them, Ikowa, Matumbulu and Kisongo, have very short expected total lives of 25—30 years and still shorter

**Figure 3.** Oblique aerial photograph of sand delta in Matumbulu reservoir. Photo by A Rapp, March 1970. Farm and maize fields in foreground. Areas of intense sheet wash and gully erosion at the right side and beyond delta. Sandy channels with aggrading beds tend to choke road culverts, eg at road bridge to the left. Cf Figures 2 and 4.

economic lives. The expected total life of a reservoir is the period from the construction of the dam to the time when the whole storage volume is filled with sediment. The economic life of a reservoir is the period during which the economic advantages of using the reservoir outweigh the investment costs; the reservoir's economic life usually ends when sediment fills 50—75 percent of the total storage volume. The annual sediment yields corresponding to the sedimentation in the surveyed reservoirs vary from 200 to 730 m³/km² per year (mean values for longest period of available data). In addition, large volumes of sediment have been deposited upstream of the reservoirs as thin, sandy sheets on lower pediments, as sand fans along stream channels and as silty-clayey layers on *mbuga* floodplains, the latter occurring particularly in the Ikowa catchment (Figure 1).

The reservoir sedimentation corresponds to a soil denudation rate of 0.2—0.73 mm per year. Sediment yield in m³/km² has been transformed into values of soil denudation rate under the assumption that the dry bulk density of sediments as well as soils is 1.5 g/cm³. A bulk density close to 1.0 is characteristic for topsoils of the area, so the topsoil denudation figures may be more relevant if 50 percent is added to the values given in Table 1. Most of the reservoir floors have been surveyed when they have been dry, with compacted but cracked clayey sediment (4).

Table 1 and Figure 5 summarize data from the five catchments at Dodoma and Kisongo.

The figures for specific sediment yields decrease with increasing drainage area, due to sedimentation in the catchment (Figure 5). Therefore, sediment yields from small catchment basins (a few km² in area) reflect most closely the erosion in the catchment.

## WATERSHED MANAGEMENT TO COMBAT RESERVOIR SEDIMENTATION

Controlling erosion and increasing the life of the reservoirs are mainly questions of better management of grass and other vegetation in the catchments and protection of harvested fields against splash and sheet erosion. Reduction of stock numbers, of overgrazing and of excessive burning of grass and mulch is necessary in order to combat erosion and lengthen the useful life of the reservoirs. The type of sediment which fills a reservoir is of great importance for

its continued use after it has silted up. A reservoir which is filled with sandy or coarser sediments can be used for groundwater in the future. This potential should be further investigated.

Reservoir surveys to document the rate and type of sedimentation and to establish the remaining life of reservoirs should be undertaken as standard practice for all existing and planned reservoirs in semi-arid areas. Reservoir maps and profiles should be made and sedimentation pegs established to make later comparisons possible.

In addition to the reservoir surveys, mapping and monitoring of erosion in the catchment should be performed to define areas and rates of sediment production.

### DEFORESTED MOUNTAIN SLOPES

The Uluguru Mountains in Tanzania were selected for the study of an area representing deforested mountain slopes with intensive cultivation. In one of the study catchments, a three-year sampling and analysis of suspended sediment load in the Morogoro River was carried out (3). The catchment area is 19.1 km². The elevation ranges from 550 m at the stream gauge, to 1450 m at the rainfall recorder, where a montane rainforest reserve begins, to 2100 m altitude on narrow mountain ridges in a rainforest zone (Figure 6). The annual precipitation is about 900 mm at 530 m altitude and 2400 mm at 1450 m altitude. The soils are sandy loams, weathered from the local basement bedrock of gneisses and granulites.

Another study catchment in the Ulugurus is the Mgeta Valley, which is similar to the Morogoro catchment in geology, landforms, climate, vegetation, soils, population density and land use in general. On February 23, 1970, this valley was hit by an intense rainstorm during which 100 mm of rain fell in two hours. It triggered more than 1000 landslides and mudflows and caused serious damage to human life and property, including crops and land (8).

These and other catchment studies in the Uluguru Mountains showed the importance of three kinds of erosion hazard (Figure 6).

1. Sheet and rill erosion occurring every year on cultivated slopes, although highly variable in severity, depending on the intensity of the individual rainstorms.

   Terracing, grass barriers, cover crops, mulch cover on bare soil and reduced burning of vegetation are simple but effective methods of protecting against these kinds of erosion.

2. Numerous small debris slides and mudflows triggered by extremely intensive rainstorms with a periodicity of several years or decades (8).

   Planting of shelter belts of forest in critical zones, to stabilize the soil and regolith with tree roots, is recommended for protection against this kind of hazard.

3. Single, large landslides, many meters deep. These are rare in time and space as compared to category 2, but can have far-reaching destructive effects through stream sedimentation (the Palu example) (9).

As such deep landslides can probably not be avoided, even by tree planting, there must be an awareness of the potential destructive force of large landslides, and this must be taken into account in water supply management.

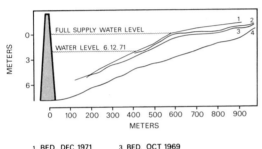

Figure 4. Long profiles of Matumbulu reservoir, 1960—71. A sandy delta is rapidly filling the reservoir. The gradient of the accumulation is slightly less than the original stream channel above full supply level, slightly more below that level. Prediction of 100 percent filling by 30 years after construction. Potential use of filled reservoir as groundwater storage should be investigated.

1 BED, DEC 1971      3 BED, OCT 1969
2 BED, OCT 1970      4 ORIGINAL BED, MARCH 1960

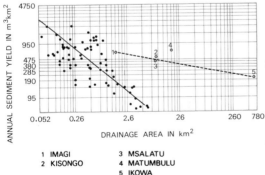

1 IMAGI          3 MSALATU
2 KISONGO        4 MATUMBULU
                 5 IKOWA

Figure 5. Relation of mean annual sediment yield to drainage area for five catchment basins in Tanzania (open circles) and 73 catchments in eastern Wyoming (closed circles). The latter after Schumm & Hadley (16). Decrease of specific sediment yield with increasing catchment area is evident in both groups but is less marked in the Tanzanian cases.

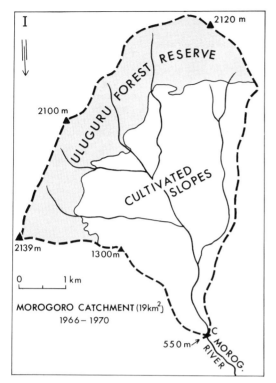

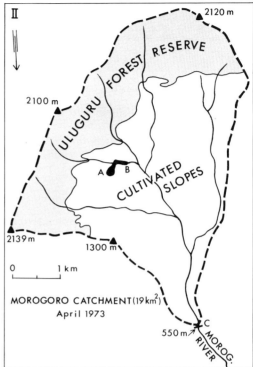

Figure 6. Three maps showing three major types of erosion on deforested and cultivated slopes in tropical mountains, as represented by the Uluguru Mountains in Tanzania.

Map I. Annual "average" slopewash 1966—1970 from cultivated slopes caused a soil loss of 7500 tons/year, measured as suspended load in the Morogoro River (3).

Map II. In April 1973, a large landslide within same watershed area as shown on Map I, started at A, left 26 000 tons mudflow at B, plus unknown sediment load beyond C. It blocked the water intake of the town of Morogoro, silted up the water pipes and damaged the water supply for weeks (9).

Map III. A large number of small landslides triggered by a heavy rainstorm of 100 mm on February 23, 1970. Slides indicated by T-symbols. A few landslides that occurred in the forest reserve, and the more than 800 that occurred on deforested slopes below the forest reserve boundary (8), moved 400 000 tons of soil and regolith into streams.

Annual soil losses by slope wash from cultivated slopes are high. Protection measures: grass barriers, terracing, cover crops, mulching (Map I).

At intervals of one or a few decades, maximum losses are caused by extreme rainstorms and numerous small landslides. Protection measures: planting of forest belts in critical zones (below ridge crests, above roads, near stream sides) (Map III).

At intervals of several decades, single, large landslides of 100 000 tons occur. Protection mainly by awareness of landslide danger in watershed.

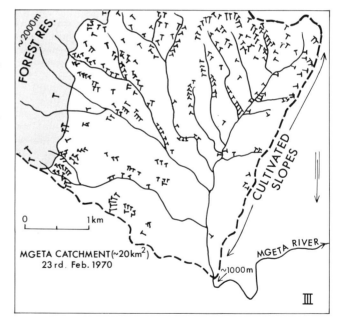

81

## LESOTHO

Another African country with very pressing problems of soil erosion and water losses in both lowland and highland areas is Lesotho, the small mountain state in southern Africa (10). The central and eastern parts of Lesotho consist of high mountain plateaus, *eg* the Central Range and the Drakensberg Mountains, culminating with peaks almost 3500 m above sea level. The western lowland of Lesotho is a narrow fringe of plains and valleys at about 1500—1800 m altitude.

Rainfall in Lesotho varies greatly; from 500 mm annually in the southwestern lowland to over 1600 mm in the eastern mountains. This represents the highest precipitation in southern Africa and is the major source of water for the vast semi-arid and arid regions surrounding Lesotho. The Orange River (or Senqu, as it is called inside Lesotho) is the main water carrier. Part of its upper catchment, with mountains and foreland, is shown in the ERTS satellite image (Figure 7).

The flow of water and sediment in the Orange River

Figure 7. Map of western Lesotho and adjacent lands in South Africa.

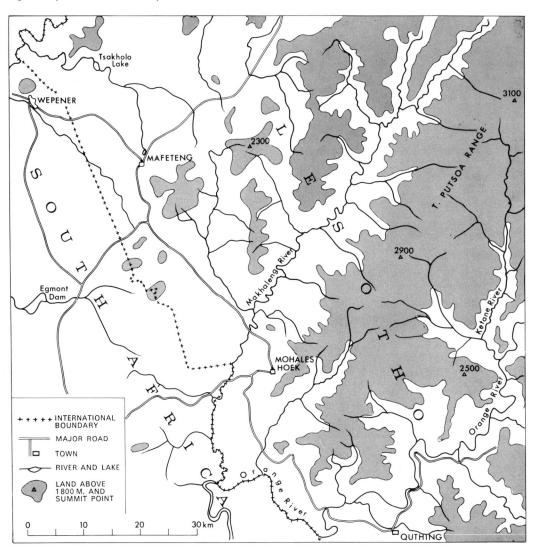

and the many smaller rivers of Lesotho is greatly affected by the land use, both in the main valleys and in the treeless high mountain zones.

## LOWLANDS AND VALLEYS

An ERTS satellite image over western Lesotho and adjacent parts of South Africa is shown here together with a location map (Figures 7 and 8). A black-and-white print of bands 4, 5 and 7 shows the springtime vegetation in grey-to-black tones; the predominantly dark tones are fields of winter wheat and other crops in South Africa, and high mountain pastures in Lesotho. Between the two is a distinct zone of pale tone—the overgrazed, overcultivated and eroded lowlands and valley bottoms of Lesotho (Figures 8 and 9). ERTS satellite pictures, combined with air photographs and ground checks in selected watersheds as reference, show the countrywide extension of the degradation process.

A research project using a three-level approach to catchment studies (ground truth, aerial photographs, satellite

Figure 8. ERTS satellite image of boundary zone in South Africa (left) and Lesotho (right). Same area as shown in Figure 7.
The three main tones are, from left to right:
1. predominantly grey pattern is wheat fields and other vegetation in South Africa.
2. Pale zone is overgrazed, eroded and vegetation-poor farmlands in Lesotho, separated from South Africa by a boundary fence.
3. Dark zones to the right are mountains and plateaus in Lesotho with green mountain pastures.
Date of ERTS image is November 21, 1972. Reproduction by W Arnberg, Stockholm. This and other ERTS pictures clearly show the countrywide extent of soil erosion in Lesotho as contrasted to the high-economy conservation farming in South Africa. Note the pale zone of soil degradation in the Orange River Valley to the right.

pictures) has been started in Lesotho by a joint team from the University of Roma (Lesotho) and the University of Uppsala (Sweden) (11). In this project a number of small catchments have been selected for repeated surveys of erosion, runoff and sedimentation conditions.

Gullying is very spectacular in many lowland areas in Lesotho (Figure 9). This has been emphasized in many reports, eg already in the 1930s by Pim (12): "It has been estimated that 10 percent of the arable land is threatened by existing dongas (gullies), and in the lowlands . . . it is the best land which is the most seriously affected . . .". The current soil erosion study in Lesotho will also focus on the relative importance of gully erosion as compared to sheet erosion in the lowlands and valleys of Lesotho.

In a 1972 report to the United Nations Development Program, the mean annual runoff of the two main rivers of Lesotho are listed as 90 m³/second for the Orange River and 24 m³/second for the Caledon River (13). The Caledon drains the western lowland of Lesotho and adjacent mountains. Part of its water comes from South Africa, but that portion is not included in the figure quoted above. The sediment load of the Caledon is regarded as being very high, but is not known.

> For some of the Caledon tributaries, sediment may be a decisive factor in determining whether or not a particular site can be developed economically for water storage . . . The significance of sediment problems, particularly in the Caledon basin, means that sediment sampling and analyses should be planned as an activity distinct from, and of equal importance to, stream gauging. (13).

Figure 9. Deep gullies in farmlands of St Michael's catchment, near Roma, Lesotho. Reservoirs located at far left in background will be used for sedimentation survey. Mountain plateau with cliff of cave sandstone. Photo by A Rapp, October 1973.

## DEGRADATION OF MOUNTAIN PEATLANDS

The highlands of Lesotho appear as dark mountain pastures on the ERTS image (Figure 7), in contrast to the pale zone of bare and eroded soil in the western lowlands and valleys. However, the mountain areas are also threatened by overgrazing and erosion, even above 2500 m altitude, where the highest peat bogs are located.

Jacot-Guillarmod (14) and van Zinderen Bakker and Werger (15) have pointed specifically to the mountain bogs of Lesotho as important regulators of river flow. "The bogs are of great significance for the regular flow of clear water and for grazing. They are in the process of being eroded through animal trampling, and measures for protection are being proposed." (15). The situation is described as follows:

> Strongly increased human activity in the highest regions of these mountains has recently caused considerable soil erosion and it is evident that this will have a profound impact on the regional ecology. Therefore a detailed knowledge of the area seems of utmost importance ... The bogs vary in size from about 1 ha to several km$^2$ ... Many of them show signs of severe soil erosion. Deep channels have formed into the soft peaty soil up to the rock bottom. Some of these valuable bogs are practically completely washed away. These bogs should be better utilized to the benefit of the Basuto herdsmen, as otherwise their most valuable grazing areas will be destroyed in the foreseeable future ... Besides their value for grazing, the "sponges" are also of great importance for the regular flow of the rivers. They act as filters which produce crystal-clear water. It is therefore necessary that these bogs, including their surrounding slopes, are fenced in such a way that animals like horses are kept outside. Sheep and mohair goats ... which form a very important source of income for the Basuto and which do not cause soil erosion, should be allowed to graze. This matter is of great urgency, as the destruction which has been initiated can otherwise no longer be stopped. (15).

to draw conclusions for better land use from these observations. Reliable reference data—maps, photographs, descriptions—must be obtained for land and water development projects, so that it is possible to compare the situation before, during and after implementation of a plan. By means of repeated comparisons or monitoring, initial mistakes can be recognized and corrected.

**References and Notes**

1. F H Bormann, G E Likens, J S Eaton, *Bioscience* **19**, 600 (1969).
2. I Douglas, *Journal of Tropical Geography* **26**, 1 (1968).
3. A Rapp, V Axelsson, L Berry, D H‑Murray-Rust, *Geografiska Annaler, A,* **54**, 125 (1972).
4. A Rapp, D H Murray-Rust, C Christiansson, L Berry, *ibid* **54**, 255 (1972).
5. N Hudson, *Soil Conservation* (B T Batsford Ltd, London, 1971).
6. D J Varnes, in *Landslides and Engineering Practice,* E B Eckel, Ed (Highway Research Board, USA, Special Report 29, NAS—NRC Publication 544, 1958).
7. D H Murray-Rust, *Geografiska Annaler, A,* **54**, 325 (1972).
8. P H Temple, A Rapp, *ibid* **54**, 157 (1972).
9. L Lundgren, A Rapp, *ibid* **56**, 251 (1974).
10. M G Bawden, K D Carroll, *Land Research Study* No. 3 (Directorate of Overseas Surveys, London 1968).
11. Q Chakela, *Studies of Soil Erosion and Reservoir Sedimentation in Lesotho,* (Department of Physical Geography, Uppsala, Sweden, Report No. 34, 1974).
12. A W Pim, *Financial and Economic Position of Basutoland,* (Report of Commission by Secretary of State for Dominion Affairs, London, 1935).
13. Binnie & Partners, *Lesotho Study of Water Resources Development* Vol 0, Summary (Binnie & Partners, Hunting Technical Services, UNDP and International Bank, 1972).
14. A Jacot-Guillarmod, *Hydrobiologia* **34**, 3 (1969).
15. E M Van Zinderen Bakker, M J A Werger, *Vegetatio* **29**, 37 (1974).
16. S A Schumm, R F Hadley, *Progress in the Application of Landform Analysis in Studies of Semi-arid Erosion* (Geological Survey Circular 437, Washington, 1961).
17. These studies have been financially supported by grants from the Bank of Sweden Tercentenary Fund, the Scandinavian Institute for African Studies, the Swedish International Development Authority (SIDA) and the Secretariat for International Ecology, Sweden (SIES), which support is gratefully acknowledged.

## CONCLUSIONS

Soil erosion by running water is a growing problem in many countries in Africa, due to increasing pressure on marginal lands for grazing, cropping or wood collection. Erosion has very marked effects on the flow of water and sediment in streams. Thus erosion/sedimentation systems have to be studied together in natural catchment basins, under different environmental conditions and different types of land use, to clarify the mechanisms and rates of active processes. Such studies or "catchment diagnoses" provide a sound basis for reclamation plans, as these have to be adapted to the local ecological and social conditions to be successful.

Every land and water development scheme should be combined with critical evaluation of its impact on the environment and on man. It is necessary to observe critically the reactions of the environment to exploitation, and

Anders Rapp, Ph D, is Deputy Director of the Secretariat for International Ecology, Sweden (SIES) and Associate Professor at the Department of Physical Geography, University of Uppsala. During the 1950s and early 1960s he conducted geomorphological research on erosion in arctic mountain areas. Since 1966 he has worked and travelled extensively in Africa, eg during the period 1968—1972 as leader of a research project on soil erosion and sedimentation in Tanzania. His address: SIES, Sveavägen 166, S-113 46 Stockholm, Sweden.

Part III

# ASSESSMENT AND EVALUATION

# Editor's Comments
# on Papers 7 Through 13

**7    GRANT**
*Erosion in 1973–74: The Record and the Challenge*

**8    JIANG, QI, and TAN**
*Soil Erosion and Conservation in the Wuding River Valley, China*

**9    DAS**
Excerpts from *Soil Conservation Practices and Erosion Control in India*

**10   RICHTER**
*On the Soil Erosion Problem in the Temperate Humid Area of Central Europe*

**11   MORGAN**
*Soil Erosion and Conservation in Britain*

**12   TEMPLE**
*Measurements of Runoff and Soil Erosion at an Erosion Plot Scale with Particular Reference to Tanzania*

**13   MAENE and SULAIMAN**
*Status of Soil Conservation Research in Peninsular Malaysia and Its Future Development*

Studies of soil erosion made at a range of scales from local to regional by sampling concentrations and yields of sediment in rivers and by measuring runoff and soil loss from erosion plots generally confirm the global pattern described in Part I. Greatest erosion occurs in the semi-arid and semi-humid areas of the world and least erosion in the humid temperate areas. The papers reviewed in this section show that within this broad pattern conditions particularly vulnerable to erosion are found on loess-derived soils and, to a lesser extent, on sandy and sandy loam soils, and also on mountain footslopes. Within any given area the local variation in erosion rates with land use may be as great as the global variation in rates for a single land use. Broadly, however, erosion rates classified as natural, cultivated, and bare soil follow the same pattern of regional variation several orders of magnitude apart. One exception to this is the humid tropics where erosion under the rain forest is much slower than under the savanna grasslands of the semi-humid tropics (Roose, 1971). When the land is cleared for agriculture, the rain forest areas respond with very rapid increases in erosion so that rates on cultivated and bare soils are much greater than the equivalent accelerated rates in the savanna areas. Continued pressure to clear the tropical rain forest is anticipated by Maene and Sulaiman (Paper 13). The result could well be erosion on an unprecedented scale. Another exception is found in the humid temperate areas where man has disturbed landforms that are relict features of a previous climate. In Paper 10, Richter attributes the extension of gullies in cultivated areas in parts of Germany to the concentration of runoff in the heads of concave valleys or dells formed under periglacial conditions and since stabilized until cleared for farming.

The rates of erosion quoted by Jiang, Qi, and Tan (Paper 8) for the loessial soils of the Huang He catchment, China, are among the highest in the world. The mean annual rate for the 30,217 $km^2$ drainage area upstream of Chuankou is 90 t $ha^{-1}$. Much higher rates are recorded in small gullied watersheds: 185 t $ha^{-1}$ in the Shejia gully and 196 t $ha^{-1}$ in the Tuanshan gully. Mean annual rates as high as 708 t $ha^{-1}$ are recorded on gully beds, 362 t $ha^{-1}$ on farm roads, and 265 t $ha^{-1}$ on steep slopes affected by landslides. Rates due to splash, rill, and shallow gully erosion on farmland range from 158 to 199 t $ha^{-1}$. The high magnitude of these rates is emphasized when they are compared with rates in the United States considered by Grant to be severe (Paper 7). The highest rates for the 1973–74 season occurred for water erosion in the Corn Belt and for wind erosion in the Southern High Plains and were equivalent to annual rates of 247 t $ha^{-1}$ and 346 t $ha^{-1}$ respectively. Generally, however, rates for the season on badly eroded land were equivalent to 70 to 100 t $ha^{-1} y^{-1}$.

Somewhat lower erosion rates are considered typical of areas affected by agricultural soil erosion according to other authors whose papers are reviewed here. Morgan (Paper 11) has measured mean annual rates of erosion by overland flow of 10 t $ha^{-1}$ from bare sandy soils in the United Kingdom, and Richter quotes rates up to 2 t $ha^{-1}$ for erosion from farmland in West Germany (Paper 10). Both authors consider these rates high enough to constitute a problem, especially when they are compared with likely local soil renewal rates and the extent of the eroded areas. Richter indicates that 50% of the area around Trier in Germany is affected by soil erosion. Evans (1982) estimates that 21% of the arable land in England and Wales has a risk of soil erosion. When rill erosion is included, mean annual rates in the United Kingdom rise to 44 t ha$^{-1}$ on bare sandy soils and 20 t ha$^{-1}$ on sandy loam soils under cereals (Morgan, 1985). These rates are similar to those observed in Belgium by Bollinne (1978).

Although the magnitude of erosion differs between, say, China and Europe, soil erosion is no less of a problem in Europe in terms of its effects on soil depth and long-term productivity or in terms of visual evidence in the landscape as rills and gullies. What differs is the frequency at which severe erosive events occur and the level of conservation required to protect the land. Soil erosion is a natural hazard associated with wind and water as geomorphological agents but it is closely linked to the prevailing economic and social conditions, as illustrated by the agreement of all the authors that agricultural soil erosion results from unwise use of the land. Grant refers to poor crop selection associated with the conversion of 22 million hectares of grassland, woodland, and idle land in the United States to cropland in the 1973–74 season—much of the conversion taking place without adequate soil conservation. This change in landuse was a direct response by farmers to United States Government policies. Richter briefly discusses the trends in erosion through the history of central Europe. Erosion increased during periods of misuse of the land but decreased when the area of land under the plow contracted. When the land is misused, the effects of high-magnitude geomorphological events are exacerbated. Thus, soil erosion should be viewed as an integral part of both the natural and cultural landscape.

Generally, soil conservation measures are designed to minimize erosion in bad years and they therefore need to be varied regionally in degree and kind according to the magnitude of the erosive events in those years. The concept of matching soil erosion control systems to the scale of the problem and the economic and social circumstances is stressed by Richter but, more significantly, is used in India as the basis for planning and implementing soil conservation schemes. Das (Paper 9) describes how India is divided into ten Soil Conservation Regions based on physiography, soils, rainfall, and rilling hazard and how the level of conservation treatment recommended for each region relates to the severity of erosion and the likely economic return. These levels of conservation cannot remain static, however, because the agricultural context, the technology of erosion control, and the perception of erosion alter through time. Grant shows how in the United States the severity and extent of erosion is dictated by changing land use in response to agricultural policy, how approaches to conservation have moved in favor of reduced tillage as research has made this technique more feasible, and how conservation is no longer represented in terms solely of loss of soil and therefore loss of a valuable natural resource but now also takes account of sediment as a pollutant.

A common theme to the papers is the role of ground cover in erosion control. Temple (Paper 10), in interpreting data from erosion plots in Africa, and Maene and Sulaiman, in discussing erosion in Malaysia, refer to Hudson's mosquito gauze experiment (Hudson and Jackson, 1959). This experiment illustrated the importance of raindrop impact in initiating soil erosion and showed how erosion could be reduced if this impact was lessened by breaking the force of the raindrops on a fine wire gauze. Over a period of six years, an uncovered plot lost, on average, a hundred times more soil each year than a gauze-covered plot. Temple stresses the importance of procumbent cover either as low-growing plants or as mulch. Such cover reduces raindrop impact on the soil surface and increases surface detention of water which, in turn, may further reduce the effect of raindrop impact as the drop energy is dissipated by the standing water. Temple demonstrates how different plant covers provide different degrees of protection, but also shows that a plant cover or mulch of any sort always reduces erosion compared with bare soil.

Maene and Sulaiman question the protection afforded by tree crops based on measurements of the drop-size distribution of the rain reaching the ground surface under oil palm. Increases in the median volume drop size compared with rain in open ground combined with a fall height of 7 m mean that rainfall energy under the oil palm can be greater than that without a plant cover. Since the soil is shaded and generally without a litter layer, high erosion rates can occur. Recent work has noted similar effects with crops shorter than trees. Rates of detachment of soil particles from the soil mass by raindrop impact under Brussels sprouts can be greater at 50% canopy cover than with no cover at all (Noble and Morgan, 1983). The kinetic energies of a millimeter of rain reaching the ground surface following interception by canopies of Brussels sprouts, sugar beet and potatoes can all be increased compared with rain in open ground (Finney, 1984). De Ploey, Savat, and Moeyersons (1976) show how a plant cover can also interact with surface runoff to increase instead of decrease erosion. Clearly much more research is required to determine under what conditions plants promote erosion. Our present understanding of the role of plant cover in soil conservation is being questioned and appears to be far from satisfactory.

In 1967, 64% of the cropland in the United States needed conservation treatment. The increase in the area under crops in the 1973–74 season resulted in an additional 60 million tons of soil loss. The average rate of erosion on these newly cropped lands was equivalent to 30 t ha$^{-1}$ y$^{-1}$. This increase in erosion occurred even though the season was preceded by an extensive publicity campaign to promote soil conservation, using the slogan "Produce More, Protect More." If American farmers with their history of soil conservation activity cannot be persuaded to protect their land, what hope is there for the rest of the world where the conservation message is less strong and less widely accepted?

Surprisingly, the situation described by some of the other authors is less pessimistic than that indicated by Grant. Jiang, Qi, and Tan show how application of shelterbelts to the wind-drift sands, planting of trees and bushes on the slopes of ravines and gullies, use of check dams, and the building of bench terraces has reduced the annual sediment yield in the Wuding Valley from 272 to 104 million tons. This has been achieved by treating only 20.5% of the area affected by soil erosion. Grain output has also trebled but this increase is from a very low base. For example, similar conservation on the loessial soils of the Dachai commune, Shanxi, one of the "model communes" of the Maoist period, produced increases in the average grain yield from 750 to 7500 kg ha$^{-1}$ (Wen and Liang, 1977). Das believes that the uptake of soil conservation measures depends on their ability to show economic returns on the investment. Some of the work in India is extremely promising in this respect. Reclamation of ravines in Uttar Pradesh has increased crop yields by 16% created employment opportunities and repaid the investment within two to ten years (Tejwani, Gupta, and Mathur, 1975). Economic evaluations of small check dams used in gully reclamation show that they are quite attractive in terms of cost-benefit ratios (Das and Singh, 1981).

The large-scale regional solutions adopted in China and India contrast with the recommendations made by Temple in favor of field-scale methods. Use of low-growing crops combined with tillage practices to increase surface microtopographic roughness are all that is required to reduce erosion under many conditions. Morgan also emphasizes the role that tillage might play in producing an erosion-resistant surface. More recent studies (Dixon and Simanton, 1980; Stuttard, 1984) indicate that erosion could be reduced by restricting tillage to periods of certain soil moisture conditions and using specialized equipment to roll and press the surface to a specified bulk density. Further research is needed to determine the levels of moisture and bulk density required, levels that are also compatible with the conditions needed for seed germination.

All the papers illustrate that there is virtually no area in the world used for agriculture where soil erosion does not occur. In no country is there room for complacency. Modern farming practices cannot be sustained without risk of land degradation, even in western Europe. The semi-arid and semi-humid areas have one of the worst problems because soil conservation and land degradation are inextricably linked with the need for water conservation and the onset of desertification. The humid tropics are also extremely vulnerable because these areas show the greatest change between natural and accelerated rates of erosion on clearance of the rain forest. Soil conservation can work. With good conservation practice, land can be improved within three to four years; with poor practice, severe erosion can set in within one year. The problem at present is that far too small an area of the world is being adequately protected. Further work is required, aimed at increasing the uptake of soil conservation by designing strategies appropriate to the scale of the problem and the cultural environment. Large-scale solutions involving terraces, waterways, and land reclamation are successful when introduced as part of a program of watershed management. They are less successful for in-field control at the farmer level. Thus the optimistic picture painted in India and China is not repeated in the United States where conservation is left more to the initiative of the individual farmer. Farmers are prepared, however, to change cropping and tillage practices, as shown by the recent interest in the United States in reduced tillage systems. The challenge is to extend the area under conservation treatment by developing erosion control methods that the farmer can appreciate as being both ecologically and economically feasible. National campaigns and extension services will only be effective if they are promoting appropriate techniques.

## REFERENCES

Bollinne, A., 1978, Study of the importance of splash and wash on cultivated loamy soils of Hesbaye (Belgium), *Earth Surf. Processes* **3:**71–84.

Das, D. C., and S. Singh, 1981, Small storage works for erosion control and catchment improvement: mini case studies, in *Soil Conservation: Problems and Prospects,* R. P. C. Morgan, ed., Wiley, Chichester, pp. 425–447.

De Ploey, J., J. Savat, and J. Moeyersons, 1976, The differential impact of some soil factors on flow, runoff creep and rainwash, *Earth Surf. Processes* **1:**151–161.

Dixon, R. M., and J. R. Simanton, 1980, Land imprinting for better watershed management, in *Symposium on Watershed Management,* American Society of Civil Engineers, New York, N.Y. pp. 809–826.

Evans, R., 1982, Accelerated water erosion of soils in England and Wales, paper presented to *Workshop on Soil Erosion and Conservation: Assessment of the Problems and State-of-the-Art in EEC Countries,* Firenze, Italy, 19–21 October.

Finney, H. J., 1984, The effect of crop covers on rainfall characteristics and splash detachment, *Jour. Agric. Eng. Res.* **29:**337–343.

Hudson, N. W., and D. C. Jackson, 1959, Results acheived in the Measurement of erosion and runoff in Southern Rhodesia, in *Third Inter-African Soils Conference, Dalaba, Proceedings,* Commission forTechnical Cooperation in Africa South of the Sahara, pp. 575–583.

Morgan, R. P. C., 1985, Soil erosion measurement and Soil conservation research in cultivated areas of the U.K., *Geog. Jour.* **151:**11–20.

Noble, C. A., and R. P. C. Morgan, 1983, Rainfall interception and splash detachment with a Brussels sprouts plant: a laboratory simulation, *Earth Surf. Processes and Landforms* **8:** 569–577.

Roose, E. J. 1971, Influence des modifications du milieu naturel sur l'érosion le bilan hydrique et chimique suite à la mise en culture sous climat tropical, Rapport Multigraphié, ORSTOM, Centre d'Adiopodoumé, Ivory Coast, 22p.

Stuttard, M. J., 1984, Effect of tillage on clod stability to rainfall: laboratory simulation, *Jour. Agric. Eng. Res.* **30:**141–147.

Tejwani, K. G., S. K. Gupta, and H. N. Mathur, 1975, *Soil and Water Conservation Research (1958-71),* ICAR, New Delhi.

Wen, Y., and H. Liang, 1977, *Tachai: The Red Banner,* Foreign Languages Press, Beijing, 199 p.

## 7

Copyright © 1975 by the Soil Conservation Society of America
Reprinted from *Jour. Soil Water Conserv.* **30**:29–32 (1975)

# Erosion in 1973-74: The record and the challenge

KENNETH E. GRANT

Expanded crop acreage and weather extremes last year reemphasized
the ever-present need to protect the soil from wind and water

THE 1973-74 growing season in many ways was not a good one for the nation's soil and water resources. Excessive soil erosion from both wind and water accompanied the efforts of many farmers to increase crop production.

Contributing to the year's poor record was some of the worst weather in years. But severe erosion could also be traced to the unwise selection of certain soils for cultivated crops as well as the existence of too few conservation measures on the land.

The vagaries of the weather ranged from a fall and winter drought in the Southern High Plains to spring floods in the Northern Great Plains. In parts of the Corn Belt, torrential rains deluged fields in the spring—at the worst

*Kenneth E. Grant is administrator of the Soil Conservation Service, U. S. Department of Agriculture, Washington, D. C. 20250.*

possible time—only to be succeeded by prolonged drought.

In all parts of the country there were farmers who ignored the lessons of history—not to mention the data contained in soil surveys—and who planted crops where it was practically impossible to protect soil from washing or blowing. Some farmers lacked the time, money, or incentive to protect their fields with appropriate soil and water conservation practices. It was a year in which too many farmers gambled with their basic resources and lost. The debt will be a long time being repaid.

One fact that emerges clearly from the year's experience is that the need is more urgent than ever to increase permanent conservation practices on America's farms and ranches before even more damage is done to our agricultural resource base.

Another fact is that last year's difficulties already have prompted thousands of farmers and ranchers to step up conservation work. Reports from state conservationists of the Soil Conservation Service reveal a new surge of interest in stubble mulching, minimum tillage, and no-till farming. They also report widespread construction of such conservation measures as terraces and field windbreaks. Evidently many farmers were sufficiently disturbed by the 1973-74 erosion warning signs to respond to the threat with greater soil conservation efforts.

### High Erosion No Surprise

The year's erosion record did not come as a total surprise to conservationists. On February 14, 1974, the Soil Conservation Service delivered an ominous forecast of the erosion outlook for cropland newly converted

from pasture, woodland, and set-aside acreage. Based on U. S. Department of Agriculture surveys in the field, SCS predicted that some 9.5 million acres were going to convert to cropland in the spring of 1974 and that some 4 million acres of this total were going to be subject to erosion losses of more than 4 tons per acre.[1]

The same forecast pinpointed several areas where severe erosion on newly plowed acres seemed likely. Among these regions were the drought-stricken High Plains of Oklahoma, Texas, and New Mexico and the Corn Belt, where thousands of additional acres of land—some of it rolling land—were expected to be planted to crops. Other probable trouble spots, SCS predicted, could be the Palouse in the northwestern United States and the Piedmont and Coastal Plains of the Carolinas. Unfortunately for the nation's resources, these forecasts proved all too correct.

One purpose behind publication of the erosion outlook last February was to stimulate farmers, USDA employees, and local conservation district officials to intensify efforts to get new cropland acres under conservation plans. In the short time remaining before planting, SCS wanted to see as many measures applied as possible to stop excessive soil erosion. The agency also wanted farmers and ranchers to carefully consider the characteristics of their soils before they took acres out of grass and put them into corn, wheat, or soybeans.

**"Produce More, Protect More"**

A major campaign aimed at accomplishing this end had been set in motion by USDA several months prior to the February 14 warning. Shortly after the Department's announcement that farmers would be free to plant as many acres of wheat, corn, and feed grains as they wished during the 1973-74 season, SCS began work on a "Produce More, Protect More" campaign to be pursued at national, state, and local levels.

From the outset, planning and execution was carried on in close cooperation with the Extension Service and the National Association of Conservation Districts. The primary aim of the campaign was to encourage farmers

[1]Up to 5 tons per acre per year for most soils is considered an "allowable loss limit" by SCS.

Reduced tillage combats soil erosion and conserves energy.

to apply sound conservation practices on land newly planted to crops.

The drive was launched on September 10, 1973, with an address by Assistant Secretary of Agriculture Robert Long before the Iowa Association of Soil Conservation District Commissioners.

In that speech, Mr. Long called for higher farm production, but he cautioned that "it is equally important" for farmers to do careful conservation planning.

"This is no time to have a great plow-up followed by a great wash-away," he declared. "If our soils become pollutants instead of a resource, the long investment that we and you have made will go down the drain and a good many futures with it."

Subsequently, in a USDA press release issued October 5, 1973, Secretary of Agriculture Earl Butz called on farmers "to use sound conservation practices on cropland that is being brought back into production in 1974."

Secretary Butz said that while there is enough potential cropland available that can be erosion-proofed, "there are also millions of acres of farm and ranch land with soils so prone to blow-

ing or to water erosion that they should never be used for crops. Such land should remain in grass or under other vegetative cover. We do not want to risk starting another Dust Bowl."

The Secretary's message to farmers was re-emphasized in an editorial published in the November-December 1973 issue of this *Journal* in which he stated: "Soil that cannot adequately be protected from soil erosion by wind or water should be nailed down with grass or trees and kept that way."

Information materials in support of the "Produce More, Protect More" campaign poured into the field during the next few months. They included tens of thousands of color posters exhorting producers to "get your land under conservation plan," public service radio and TV announcements, newspaper and magazine articles, and a picture story for editors, used by many of the country's most influential farm publications. Suggestions to farmers and ranchers also were included in packets to producers from USDA's Extension Service.

The impact of these national campaign efforts was multiplied many times over by publicity and face-to-face meetings with farmers by district-level employees of SCS, county agents, and district officials. Many of these efforts seemed to bear fruit.

To cite but one example, the SCS area conservationist serving northeast Nebraska reported that USDA field employees in his area had discussed the merits of reduced tillage with 938 farmers during February and March. Of those contacted, 399 said they planned to reduce the number of tillage operations this year on a total of 76,000 acres.

During those two winter months, area newspapers published 307 stories on reduced tillage; 22 newspapers carried full-page advertisements on the subject; 157 radio and TV spots and programs were broadcast; and 15 Agricultural Stabilization and Conservation Service newsletters carried campaign material. One natural resource district in the area—the Lower Elkhorn district—alone spent more than $1,600 for radio and TV time to promote minimum tillage.

But laudable as these local efforts were—and there were hundreds comparable to the Lower Elkhorn campaign—they evidently weren't good enough, particularly on acres of "new" cropland.

### A Survey of Soil Losses

In July, thousands of SCS and other USDA field employees carried out a second survey to determine the conservation treatment and soil losses on acres converted to crops in the 1973-74 season.

Their findings showed that 3.6 million acres were switched from grassland to crops, 0.4 million acres from woodland, and 4.9 million acres from idle cropland and land withheld from crop production under federal programs. The total acreage converted to crops, according to the survey, was 8.9 million.

Of this total, about 5.1 million converted acres—more than half—had inadequate conservation treatment and water management. Of these, 4 million acres had inadequate erosion control.

Total soil loss on the 8.9 million acres amounted to about 60 million tons for the 1973-74 season, over and above the estimated soil loss that would have occurred had the land not been converted. Some 13 million tons of soil were eroded by wind; 47 million tons were eroded by water.

Most of this loss occurred on land that had received inadequate conservation treatment; these lands experienced an average soil loss of over 12 tons per acre per year.

Areas with new cropland in trouble were widely scattered over the country. In the Corn Belt, SCS estimated that 390,000 acres of converted land experienced soil losses ranging from 15 to 100 tons per acre per year. These losses were over and above those that would have occurred had the land not been converted to crops.

The western Great Plains of the Dakotas, Montana, Wyoming, and eastern Colorado suffered severe wind and water erosion on about 325,000 converted acres, where soil losses ranged from 5 to 40 tons per acre per year.

The eastern Great Plains, including much of Nebraska, Kansas, and South Dakota, experienced severe erosion ranging from 5 to 55 tons per acre per year on about 260,000 converted acres. Areas around the Great Lakes had soil losses of from 5 to 55 tons per acre per year on about 195,000 converted acres.

The southern Coastal Plains of Florida, Georgia, Louisiana, Alabama, and Mississippi had severe water erosion on about 210,000 acres of new cropland, with losses ranging from 5

to 70 tons per acre per year.

Other trouble spots for converted acres included the Palouse of the Northwest, with losses ranging from 5 to 75 tons per acre per year on 35,000 acres; the eastern Piedmont and Coastal Plains from Pennsylvania south to the Georgia line, with losses of from 5 to 30 tons per acre per year on 70,000 acres; the Mississippi Valley silty uplands of western Kentucky, Tennessee, and Mississippi, with water erosion of from 5 to 60 tons per acre per year on 56,000 acres; and the central farming area of Kentucky and Tennessee, where water erosion on 35,000 acres resulted in losses ranging from 5 to 40 tons per acre per year.

Compared to total erosion in the Southern High Plains, the acreage of new cropland damaged was slight, but that was only because very little acreage was converted to crops in 1973-74. On that acreage, soil losses were very heavy. Fifty thousand acres of new cropland experienced losses from wind erosion of 15 to 140 tons per acre per year.

Of particular concern on acres newly converted to crops from grassland is the fact that erosion is normally lighter the first year after cultivation than in succeeding years. The fibrous root structure of native grasses helps bind soil particles for a year or two or more. As this fiber slowly breaks down with time and tillage, we can expect the erosion hazard on these acres to increase unless measures are taken to protect them.

### Overall Conservation Needs

Soil erosion in excess of allowable loss limits during the past year was not limited, of course, to newly converted cropland, but no nationwide survey was conducted last year with respect to losses on older cropland. The last exhaustive survey of the condition of America's farmlands was the 1967 Conservation Needs Inventory (CNI). That survey, it may be recalled, reported that "judged against current standards, 64 percent of cropland needs additional conservation treatment."

The CNI report further stated: "Although we have abundant soil resources for foreseeable future needs, three-fifths of locally controlled land is not being cared for in a way that protects the soil resource for sustained production."

*Soil Conservation Service photo by Bryan*

**This system of parallel terraces reduces soil erosion and conserves rainfall. Construction of the terraces was cost-shared under the Great Plains Conservation Program.**

It is almost axiomatic that when substantial changes in agriculture take place—as they have in the United States in recent years—there is a time lag between the production changes and the conservation response to those changes. That was certainly apparent in the findings of the CNI, which demonstrated once again that putting conservation measures on the land is not a one-shot affair. It is a continuing task and one in which the techniques are constantly undergoing change as the methods of farm production change. It is no more possible to complete the job of soil and water conservation than it is to complete the education of the young or the conquest of illness.

The inadequacy of conservation on the land—both old and new cropland—showed up all too clearly in two areas during 1973-74 that were hit hard by adverse weather. One was the Southern High Plains, involving parts of Oklahoma, Texas, and New Mexico. The other was Iowa.

The Southern Plains were hit hard by drought, with a total of 2.7 million acres damaged by wind. Texas alone reported more than 2 million acres damaged by erosion. This was more than half the damage reported for all 10 Great Plains states during the 1973-74 blowing season.

Even more land in the Southern High Plains was reported "in condition to blow"—nearly 6.2 million acres—but, fortunately, many of the prolonged 40 to 60 mile-per-hour windstorms that strike the Plains so often in February and March failed to materialize in 1974.

According to SCS field reports, the major contributing factors to wind erosion in Oklahoma, Texas, and New Mexico this past year included, besides lack of moisture, lack of residues, land clean tilled for seedbed preparation, unsatisfactory tillage operations, poor soil structure, and excessive grazing. The reports demonstrate what conservationists learned so often during the last 40 years: Man can do practically nothing to control the weather, but he can do a great deal to mitigate its impact on the land—or compound it.

Water erosion also took a heavy toll of old and new cropland in several parts of the United States, but no state suffered more severe damage during the past year than Iowa. According to the personal observations of State Conservationist Wilson Moon, "Erosion in Iowa was the worst in about a quarter of a century."

More than one-third of the state was declared a disaster area following torrential thunderstorms, floods, and tornadoes during May and June. SCS inspections revealed that about 4.5 million acres of Iowa's 20.7 million acres of corn and soybean land suffered severe erosion of more than 10 tons per acre. Soil losses of 40 to 50 tons per acre were not uncommon, and some farms experienced the staggering loss of 200 tons of soil per acre.

Besides the loss of topsoil, seed was washed out, fertilizer lost, crops covered with mud, and ditches and waterways plugged or eliminated.

### Eliminating Severe Soil Losses

Why did it happen? Can we blame all our troubles on the weather? Don Muhm, farm editor of the *Des Moines Register* and an experienced observer of Iowa agriculture, is one man who doesn't think so.

"There is no way of measuring the damage to this year's crops or to future crops," he wrote in the May 26, 1974, *Sunday Register*, "because landowners failed to construct enough terraces or utilize sufficient strip-cropping or other means to protect the soil resource."

We agree. It is true that really severe soil erosion generally occurs during extreme weather conditions and in a very short period of time. But erosion control measures, if they are going to provide producers with adequate protection, simply cannot be predicated on average rainfall or typical windstorm velocities and durations. Soil conservation measures must be planned and installed to protect resources, not against the adverse weather of an average sort of year, but against the drought or wind or storms of an unusually bad year. And without question, someplace in the United States every year, that "worst weather in years" is going to come.

The important thing, therefore, is to be prepared for the exceptional. Even in Iowa, where one storm last spring delivered 3 inches of rain in 40 minutes, the farms with modern soil conservation practices on the land survived the deluge. Well-built terraces did the job for which they were intended, even during the hardest downpours within many farmers' memories. And conservation tillage and no-till practices, which have gained favor rapidly in Iowa in recent years, saved countless tons of soil from washing downstream. Farmers reported that a good cover of cornstalks and other crop residues on the surface of the soil kept erosion to a minimum, even on gently sloping land. Already the past year's experience has led to a big increase in terrace building and minimum tillage in Iowa as well as many other farming areas.

So while the task ahead for those engaged in soil and water conservation is a challenging one, it is far from impossible. Events of the past year should not be cause for pessimism, but rather should renew the conservationist's determination to move faster with his important work. Commenting on the year's erosion record in my annual message to SCS employees last fall, I said:

"These setbacks underscore the fact that farmers and ranchers simply cannot afford to let up on conservation, and we need to redouble efforts to assist them. We need to help them use soils information to select the right kinds of acres for crops and we need to show them how to protect those acres from erosion. We want to help farmers make their acres as productive as possible, but we want to make sure that the cropping systems leave the land resilient enough to bounce back after flood, drought, or what have you."

I am confident that we have the wherewithal to bring conservation on the American land up to standard by reducing soil erosion to allowable loss limits. Soil conservationists have the knowledge and the expertise. Our program's assets include more than 2 million voluntary district cooperators, more than 15,000 district officials serving nearly 3,000 conservation districts, a growing number of state and county conservationists, and more than 13,000 SCS employees as well as other USDA people in the field who are dedicated to the principles of soil and water conservation. I invite all who share our concern for improved conservation of our basic resources to join us now in an all-out effort to protect and manage America's soil and water. The need was never greater than it is today, nor the stakes so high. □

# 8

# Soil Erosion and Conservation in the Wuding River Valley, China

Jiang Deqi, Qi Leidi *and* Tan Jiesheng

## 1. Introduction

The key problem of the Huang He has been the severe soil erosion of the loess plateau on the middle reaches. The river has been raised in level through aggradation and has become a menace on the lower reaches. According to statistical data, the long term average sediment load at Sanmenxia has been $1.6 \times 10^9$ tons per year of which $300-400 \times 10^6$ tons are coarse sediment ($d > 0.05$ mm). Some 400 million tons of sediment have been deposited every year in the channel all the way down the lower course of the river to Lijin, of which coarse sediment comprises one half (Qian et al, 1980). The Wuding River is one of the large tributaries on the middle reaches of the Huang He. It contributes nearly 1/6 of the sediment load and 1/4 of the coarse materials but has a drainage area of only 1/23 of the total catchment of 688,000 km$^2$ above Sanmenxia. It is obvious that work on soil conservation here is of much significance to the development of agricultural production as well as the alleviation of hazards caused by sedimentation in the Huang He.

## 2. Soil Erosion and Water Loss

The Wuding River runs through northern Shaanxi and southern Nei Mongol, with total length of 491 km and a catchment of 30,260 km$^2$ at 800–1800 m above mean sea level. The average temperature is $5^o - 9^oC$ and the mean annual precipitation is between 350 and 500 mm, with strong winds and heavy storms. The land surface is covered mainly by deep strata of new and old loess and recent wind-drift sand, loose in structure and susceptible to erosion. Owing to the indiscriminate reclamation of land, overgrazing and excessive cutting of trees over the

99

years, the dense natural vegetative cover has been badly damaged and, as
a result, soil erosion and water loss have been very serious. The area
concerned covers 22,500 km$^2$ comprising 74% of the watershed. When
soil conservation was first carried out, the annual sediment load was
272 x 10$^6$ tons (mean value of 1952-59, according to the statistics of the
Huang He Conservancy Commission), the average amount of soil erosion
being 9000 t/km$^2$/yr, and even more than twice as much in areas of
severe erosion. The agricultural production was low and unstable. After
control, the annual sediment load has been reduced to 104 x 10$^6$ tons
(mean value for eight years) and the harvest of grains has been increased.

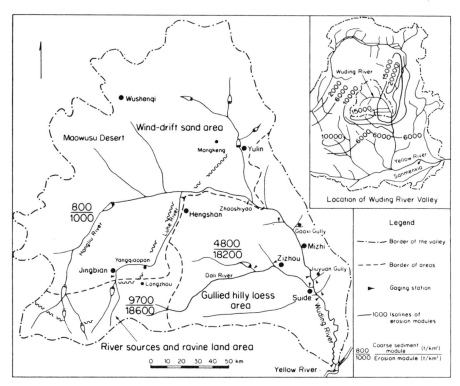

**Figure 1. Typical areas of soil erosion in the Wuding River Valley.**

The whole basin can be divided into three typical areas with respect
to loss of soil and water (Figs 1 and 2). (1) Wind-drift sand area including
mainly the moving, stable and semi-stable sand dunes and floodlands
(Fig 3). Wind erosion is very serious, while water erosion is slight.
Population density is 6–30 per km$^2$ and the people live chiefly by stock-
raising. (2) The area of river sources (upper watersheds) and 'ravine land',

Figure 2. Satellite photography of the Wuding River Valley.

Figure 3. Landform of the wind-drift sand area.

in its early stage of severe dissection. The hilly loess landform here is characterized by flat ridges and large mounds as well as level land at the bottom of broad ravines. Erosion by water, wind and gravitational action is very severe, the principal form being that of gully erosion, which has caused deep grooves and dissected most of the 'ravine lands'. There are 30—50 people per km$^2$, who live on farming and animal husbandry.

**Figure 4. Landform of the gullied hilly loess area.**

(3) The gullied hilly loess area, with hills in the form of ridges and mounds (Fig 4). Here, the gullies are deep and slopes are steep. The density of gullies is 5—7 km/km$^2$. There are 70—150 inhabitants per km$^2$. The index of cultivation is rather high. Both sheet erosion and gully erosion are very severe in this area, which is the main source of sediment of the Wuding River.

The main features of soil erosion of the Wuding River Valley are as follows.

(1) Large sediment load with coarse particles. The overall sediment discharge, soil erosion amount and coarse sediment discharge of different areas of the Wuding River basin are shown in Table 1. In the overall sediment discharge, fine particles (d < 0.025 mm) comprise 29% of the total load, medium particles (d = 0.025—0.05 mm) 34.8%, and coarse sediment (d > 0.05 mm) 36.1%. The median diameter (d$_{50}$) is 0.042 mm (Fig 5).

Soil layers yielding relatively coarse sediment are mainly new loess (sandy), old loess, recent deposits of wind-drift sand and accumulation on 'ravine land'; while the lower strata of river terrace accumulation as well as red sandstone also contribute in smaller amount to the coarse sediment load (Hydrological Team, No 2, 1976).

| Region | Area | | Annual sediment load | | | Annual Coarse Sediment $d > 0.05$ mm | | |
|---|---|---|---|---|---|---|---|---|
| | $km^2$ | % | $10^6 t$ | % | Erosion rate† $(t/km^2)$ | $10^6 t$ | % | Rate $t/km^2$ |
| Wind-drift sand area | 16271 | 53.8 | 16.3 | 6 | 1000 | 13 | 13 | 800 |
| River sources and 'ravine land' area | 3680 | 12.2 | 68.5 | 25 | 18600 | 36 | 37 | 9700 |
| Gullied hilly loess area | 10266 | 34.0 | 187.2 | 69 | 18200 | 49 | 50 | 4800 |
| Drainage area above Chuankou | 30217 | 100.0 | 272.0 | 100 | 9000 | 98 | 100 | 3250 |
| % | | | 100.0 | | | 36 | | |

† Including water erosion, gravitational erosion and a portion of wind erosion which contributes sediment to the water-courses.

Table 1. Annual sediment discharge, soil erosion and coarse sediment discharge of various areas in Wuding River Basin.

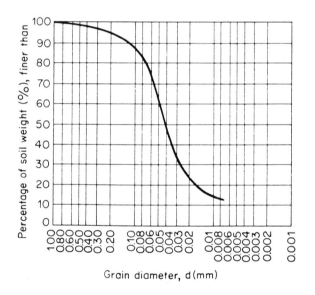

Figure 5. Particle-size distribution of suspended silt at Chuankou gauging station of Wuding River.

(2) Soil erosion and water loss is markedly affected by the characteristics of the rainstorms, concentrating in 3 months of the year and varying
greatly from year to year.

In this region about 65% of annual rainfall concentrates in July, August,
and September, mostly in storms of high intensity, which may be 1−3 mm/
min in 5−10 min. At Shiwan, there occurred on July 23, 1971, a heavy
rain of 212.6 mm within 6 hr 25 min, which was 43% of the rainfall of
the whole year. In small watersheds, the runoff and sediment yield
during one relatively heavy storm may be 35−86% of that for the whole
year. Hence, the sediment yield is also concentrated in the three months
of the flood season during which time 87% of the annual yield is delivered
(Fig 6).

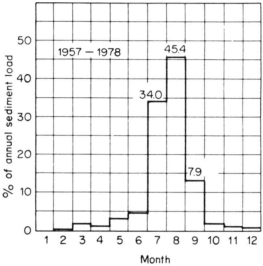

Figure 6. **Monthly distribution of average annual sediment load of**
**Wuding River Valley.**

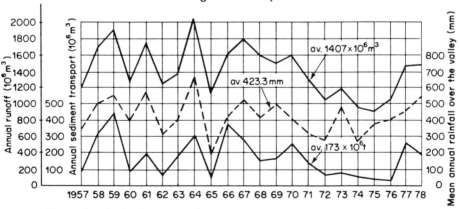

Figure 7. **The hydrograph of annual runoff, sediment transport and**
**mean annual rainfall over the catchment of Wuding River.**

The annual runoff, sediment discharge and average annual rainfall over the valley vary widely from year to year (Fig 7). The annual sediment discharge usually varies with the total runoff in a year and there is a relatively close correlation between them (Fig 8).

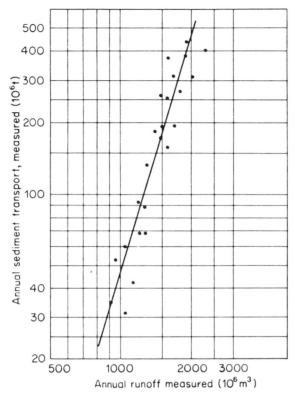

**Figure 8. Plot of annual runoff against annual sediment transport of Wuding River.**

(3) Based on analyses made on the sources of sediment from different small watersheds in the gullied hilly loess area, it has been found that, as regards landform, the total amount of soil erosion is greater on the lands within gullies, which are mainly on gully slopes, than on the lands between gullies; whereas according to land use, the total amount of erosion is greatest on the farmland (Table 2; Jiang et al, 1966; Gong and Jiang, 1977). Hence, the slope and farmland should be considered as points of emphasis of soil conservation.

(4) Sediment transport with hyper-concentration. Large quantities of sediment are carried down by the storm runoff from hillslopes to gullies of different sizes in the small watersheds. The muddy flow is aggravated

| | Type of land | Main types of soil erosion | % of the watershed | | Erosion rate (t/km$^2$) |
|---|---|---|---|---|---|
| | | | Area | Soil Erosion | |
| **Landform** | Land between gullies | splash, rill, shallow gully | 44—74 | 30—62 | 11600—26300 |
| | Land within gullies | | 26—56 | 38—70 | 14200—34500 |
| | Gully slopes | rill, gully, land-fall, landslide | 25—52 | 32—62 | 16600—26100 |
| | Gully bed | gully | 1—4 | 6—8 | 38500—70800 |
| **Land use** | Farmland | splash, rill, shallow gully | 57—67 | 44—59 | 15800—19900 |
| | Grazing and wasteland | gully, sheet | 8—25 | 9—23 | 13400—16100 |
| | Steep slopes | landfall, landslide | 13—21 | 20—25 | 21800—26500 |
| | Roads, farm-yard, gully bed | gully | 4—7 | 7—13 | 28400—36200 |

Small watersheds with areas of 0.18 — 70.7 km$^2$

**Figure 2. Soil erosion on different types of land in small watersheds in gullied hilly loess area.**

by additional contributions on the way, with annual maximum silt content generally of 700—1000 kg/m$^3$ (or 48.7—61.5% by weight), thus resulting in high sediment concentration in the water courses. According to observed data of Chuankou gauging station, the average annual sediment content was 123 kg/m$^3$, while the maximum monthly average value was 383 kg/m$^3$, and the maximum sediment content (mean of the cross section) observed in a year reached a record of 1290 kg/m$^3$.

The main features of such hyperconcentrated sediment transport are as follows: (1) the sediment content remains high even with small discharges; (2) when the discharge reaches a certain level, the sediment content tends to remain   more or less constant; (3) within a certain range, the higher the sediment content of the flow, the larger is the median diameter of the sediment; and (4) when the sediment content exceeds a certain limit, the turbulence of flow becomes weak, there is no differentiation of sediment, and the sediment-carrying capacity increases.

(5) The amount of sediment transported by the river is about as much as the amount of soil erosion in the small gully watershed. Owing to the development of ravines and gullies in recent times, the lower part of the slope in the gullied hilly loess area has become steeper than the top. The gradients of the gullies and river beds are steep and the scours have cut deep into the base rock. As a result, there is practically no deposition of sediment carried by the runoff all the way down the slopes, through gullies, into the ravines and tributaries and finally into the main river. Hence, the amounts of sediment transported by the tributaries and the main river are about the same as the amount of soil erosion from small watersheds (Table 3, Gong and Xiong, 1980).

| River or gully | Station | Years of observation | Drainage area $(km^2)$ | Mean annual rates $(t/km^2)$ | |
| --- | --- | --- | --- | --- | --- |
| | | | | Soil erosion | Sediment transport |
| Tuanshan gully (Dali River) | Tuanshangou | 1959–1969 | 0.18 | 19600 | |
| Shejia gully (Dali River) | Sujiahe | 1959–1969 | 4.26 | 18500 | |
| Xiaoli River (Dali River) | Shejiagou | 1959–1969 | 807 | | 15700 |
| Upper Dali River | Qingyangcha | 1959–1969 | 662 | | 16100 |
| Mouth of River Dali | Suide | 1959–1969 | 3893 | | 16300 |
| A reach of Wuding River | Chuankou-Zhaoshiyao-Yulin | 1960–1969 | 9954 | | 15500 |

**Table 3. A comparison of the mean annual soil erosion and sediment transport measurements in watersheds of various sizes in Wuding River Basin.**

It is obvious that the quantity of sediment entering the Huang He can be reduced if proper work of soil conservation on the basin can be done.

## 3. Implementation of Soil and Water Conservation

Since the founding of the People's Republic of China, the governments at different levels have attached importance to soil and water conservation work. Planning for the control over the entire basin was worked out on the basis of scientific researches and summarizing the experiences of the masses. A large scale programme of soil and water conservation of a mass character has been implemented continually in a comprehensive way, with concentrated efforts on critical areas and focal points. The principles observed are: (1) measures should be appropriate to local conditions; (2) measures taken on the slopes and in the gullies should be coordinated; (3) engineering and vegetative means should be resorted to at the same time; and (4) work should be done both to draw benefits and to eliminate hazards. Also, medium and large reservoirs have been constructed on the tributaries and the main river.

(1) In the wind-drift sand area, the main task is to prevent the wind from deflating and shifting the sand. Shelter-belts, network of wind-breaks and forest to stabilize sand, have been planted in coordination. The main kinds of plantation are willow, poplar, narrow-leaved oleaster, Salix sp, Artemisia sp, Hedysarum sp and Astragalus sp. In addition, measures have been taken to divert water for sluicing the sand to create arable land, and to divert flood-waters to warp and irrigate the land, in order to ameliorate the soil and transform the deserts. The Mangkeng Production Brigade in Yulin county planted wind-breaks 22 km in length on 850 ha of sand dunes and floodland, and some two-thirds of the former drift sand have been stabilized by 420 ha of plantation (Fig 9). The total grain output has trebled. The Yangqiaopan Production Brigade in Qingbian county has created, since 1942, the technique of diverting water to sluice the drift sand to form more than 700 ha of levelled land (Fig 10).

**Figure 9. Network of windbreaks at Mangkeng.**

(2) Erosion in the river sources (upper watersheds) and 'ravine land' area is mainly controlled by stabilizing the gullies and protecting the ravine lands. On slopes, it has been advocated to plant trees, bushes and herbs, to build conservation engineering works and to divert flood water

**Figure 10. Sluicing the sand to create land at Yangqiaopan.**

for warping and improving the soils. In the gullies, soil-saving dams (large and small check-dams) have been constructed to raise the local base levels of erosion to check further scouring of the gullies and to trap soil and silt to form lands for cultivation. Moreover, reservoirs of large and medium size have been built on the main river and tributaries such as Hongliu River and Luhe River to detain water and sediment for irrigation. These have been proved effective in gaining time for control work to be carried out over large areas. The Longzhou Production Brigade of Jingbian county has been the model in adopting the above-mentioned measures of control and has controlled 68% of the area of soil erosion with the result that no runoff has come out of the gully for many years.

(3) The gullied hilly loess area is being improved by con current measures taken on the slopes and in the gullies, with stress laid on the former. Every small watershed is considered as a unit in which the control works are arranged on three types of land, namely, the land between the gullies, the gully slopes and the gully beds (Fig 11). The topography is changed on a small scale by forming bench-terraced lands on relatively gentle slopes. In line with the increase of grain output, rotation of crops and grass has been exercised on a portion of the land. The steep farm lands are used for restoring forests and meadows to increase the vegetative cover and ameliorate the soils as well as to provide more feed and manure. The gully slopes are planted with trees, bushes and herbs to stabilize them and at the same time to meet the needs of

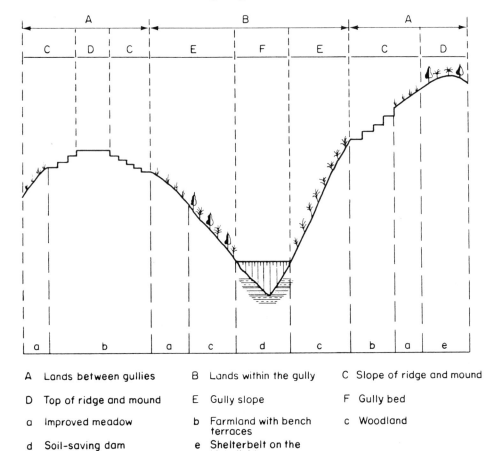

| A | Lands between gullies | B | Lands within the gully | C | Slope of ridge and mound |
|---|---|---|---|---|---|
| D | Top of ridge and mound | E | Gully slope | F | Gully bed |
| a | Improved meadow | b | Farmland with bench terraces | c | Woodland |
| d | Soil-saving dam | e | Shelterbelt on the high divide | | |

Figure 11. Cross-section of soil and water conservation measures in
the small watersheds of gullied hilly loess area.

grazing, fuel and timber. The main kinds of tree and herb are: Caragana
sp, locust, poplar, Chinese pine, jujube, pear, apple, alfalfa, sweet clover
and Astragalus sp. It is advisable to construct a series of check-dams
along the gully beds to detain flood runoff and sediment, to silt up land
and to prevent further gully erosion (Fig 12). In recent years, the pace
of remoulding the gullies has been quickened by extensively using the
method of sluicing-siltation in the construction of small and large check-
dams and by other soil and water conservation measures applied on a
large scale.

Since 1958, the Gaoxigou Production Brigade in Mizhi County has
terraced the fields (Fig 13), silted up land, provided irrigation for dry
farmland and restored plantation of trees and herbs on two-thirds of the

**110**

Figure 12. Soil-saving dam in Hengshan county

Figure 13. Terraced fields at Gaoxigou

sloping farmland. Total grain output has increased 3.4 times and the gross value of products of agriculture, forestry and stock-raising has increased 5.8 times. For a number of years, very little sediment from this area has reached the Huang He. The Jiuyuan Gully in Suide County has a drainage area of 70.7 km$^2$, of which 35.7% has been put under preliminary control by conservation measures on the slopes. There, 317 soil-saving dams were built of different sizes. As a result, 24% of the runoff and 55% of the sediment have been kept from entering the river during 18 years of control. Grain output also trebled (Gong and Jiang, 1977).

According to statistics up to the end of 1978, in the entire catchment 61,800 ha of bench-terraced fields and over 10,000 soil-saving dams of all sizes (silting up 11,500 ha of fertile land) were built; 44,500 ha of land were brought under irrigation; 288,200 ha of land were afforested;

and 55,100 ha of improved grassland were preserved. Altogether, 4618 km$^2$ of land surface are now under control, being 20.5% of the total area of soil erosion. Also, 200 reservoirs, 26 of which are capable of holding more than $10 \times 10^6$ m$^3$ each, have been built, with a total capacity of $1.5 \times 10^9$ m$^3$. The drainage area controlled by these reservoirs is over 10,000 km$^2$. All these measures have already shown their effect in 22 years on the development of agricultural production and reduction of flood flow and sediment discharge.

## 4. The Effect of Soil and Water Conservation on Runoff and Sediment Discharge

According to the analyses of experimental data of Suide Soil and Water Conservation Experimental Station, Huang He Conservancy Commission, it is known that, under different conditions of rainfall, the runoff and soil loss can be cut down respectively 90—100% and 95—100% by bench-terracing, 65—80% and 75—90% by afforestation, and 50—60% and 60—80% by growing grass (Gong and Jiang, 1977). Usually the detention rate (%) is higher in low intensity rain and lower in heavier rain.

Besides raising the base levels of erosion to check further deepening of the gully beds and prevent sliding of the gully slopes, the function of soil-saving dams is mainly to detain silt to form new land. According to investigations made in-situ, on the average 50,600 tons of silt have been deposited to create one hectare of land and during the past 22 years, $570 \times 10^6$ tons of silt have been deposited in the reservoirs with a capacity larger than $10^6$ m$^3$ each.

In order to make a good comparison of the effect on runoff and sediment discharge through the implementation of soil and water conservation over the entire basin, the hydrological data in the years 1957-78 are presented in two series, namely, 1957-70 and 1971-78. This is because, since 1970, the work on soil conservation has quickened, more dams and reservoirs have been built, the measures formerly established on the slopes have become more effective and the sediment load has markedly decreased, as measured at the gauging stations.

Computation was first directly made by using the index of detention of individual soil conservation measures based on the data obtained from experiments and field investigations (Wubao Central Hydrometric Station, 1975; Wubao Central Hydrometric Station/Yulin Soil and Water Conservation Station, 1976). The total quantities of detention of silt and water were obtained by adding the deposits in the reservoirs and comparing them with observed data of the gauging stations. Then the

| Method of Computation | Direct computation by soil conservation measures | | | Comparison of hydrological data by sequences of years | |
|---|---|---|---|---|---|
| Sequence of years | 1957-70 | 1957-78 | 1971-78 | 1957-70 | 1971-78 |
| Mean annual precipitation over the basin P (mm) | 442 | 423 | 390 | 442 | 390 |
| Mean annual runoff w, observed ($10^6 m^3$) | 1538 | 1407 | 1174 | 1538 | 1174 |
| $w_c$[1], computed ($10^6 m^3$) | 1691 | 1628 | 1519 | 1691 | 1318[2] |
| $\Delta w$, decrease (%) | 9 | 14 | 23 | (24)[3] | 23[4] |
| Mean annual sediment transport | | | | | |
| $w_s$, observed ($10^6$t) | 212 | 173 | 104 | 212 | 104 |
| $w_{sc}$, computed ($10^6$t) | 272 | 240 | 185 | 272 | 136[2] |
| $\Delta w_s$, decrease (%) | 22 | 28 | 44 | (51)[3] | 50[4] |

Remarks:

[1] $w_c = w + \Delta w$ ($10^6 m^3$); $w_{sc} = w_s + \Delta w_s$ ($10^6$t)

where $\Delta w$ and $\Delta w_s$ are respectively the decrease in runoff and sediment transport through measures taken in soil and water conservation

[2] Correction made in accordance with the mean annual precipitation over the watershed in 1957-70

[3] Comparison of data observed before and after 1970

[4] Comparison of computed values before and after 1970

**Table 4. Results of computation of the effect of soil conservation on runoff and sediment discharge in Wuding River Valley (total area = 30,217 km$^2$)**

observed and computed data of the two series of years stated above were compared respectively to check the results computed. The results of computation are shown in Table 4. The difference of effect calculated by the two methods is 6–7%.

In making the comparison between the two sequences of years of hydrological data, as given in Table 4, because the average annual precipitation of 1971-78 was 11.8% less that that of the previous years, the runoff and sediment transport measured in the second period were somewhat smaller than they would otherwise have been. The measured values of runoff and sediment transport of 1957-70 were the results after deduction of the amounts that were detained by conservation measures and hence, they were also smaller than what they would normally have been. In order to eliminate these effects, a correction was made to the annual runoff and sediment transport measured during the first period by adjusting to the values without the effect of conservation, so they are $1691 \times 10^6$ m$^3$ and $272 \times 10^6$ t respectively. Similarly, correction was made to the annual runoff and sediment transport measured in the second period by eliminating the effect of the variation in rainfall, giving values of $1318 \times 10^6$ m$^3$ and $136 \times 10^6$ t respectively.

Altogether, about $1,480 \times 10^6$ tons of sediment have been detained in the Wuding River Valley through measures of soil and water conservation in the past 22 years. This is equivalent to 28% of the total quantity of sediment transported down the entire basin. During 1971-78, the yield of sediment has been reduced 44% (20.1% due to soil-saving dams, 13.2% due to reservoirs, 7.4% due to bench-terracing, afforestation and growing grass, and the remaining 3.3% due to irrigation practice). The mean annual precipitation over the drainage area, the extent of control of slopes and the benefit of reducing sediment load in the years concerned are all shown in Figure 14.

Besides the measures of soil and water conservation in general, there are more reservoirs and dams of larger or medium size built on the various tributaries on the upper reaches of the valley, so that at the Zhaoshiyao gauging station, which controls 15, 325 km$^2$ of drainage area, the effect of reducing sediment is 75%, whereas at the Hengshan gauging station on the tributary Luhe River with a catchment of 2,415 km$^2$, it is as high as 91%.

## 5. Concluding Remarks

(1) The Wuding River Valley is one of the main source areas of the sediment of the Huang He and of coarse sediment in particular. Soil loss from the areas of severe erosion, namely, the gullied hilly loess area and

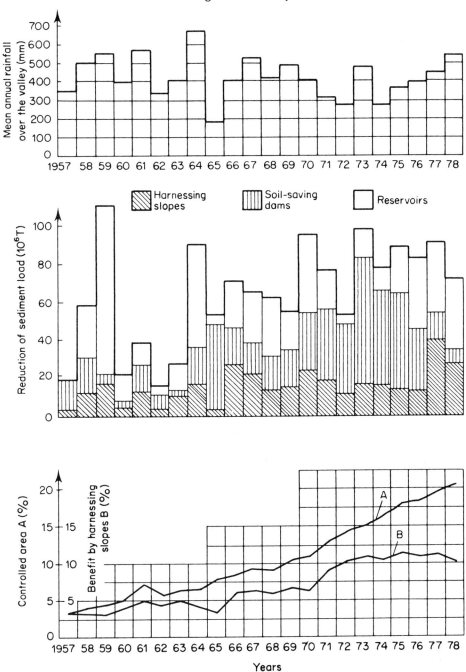

Figure 14. Effect of soil and water conservation measures in reduction of sediment load.

the river sources (upper watersheds) and 'ravine land' area was over 18,000 $t/km^2$. It has been therefore a focal point of soil and water conservation.

(2) Since the founding of new China, with an overall planning appropriate to local conditions and in accordance with the particular features of the areas of different types of soil erosion, soil and water conservation have been practised in a comprehensive way on 20.5% of the area affected by soil erosion. Reservoirs on the tributaries and the main river control a total catchment area of more than 10,000 $km^2$. The total sediment yield has been reduced 44% between 1971 and 1978. Agricultural production has also markedly increased. These results show that soil and water conservation should be the basic means of promoting agricultural production and harnessing the Huang He.

(3) At present, the effect of soil conservation on reducing runoff and sediment discharge is mainly due to measures taken in the gullies. The reservoirs and soil-saving dams, however, have only limited capacities which will further decrease every year as siltation increases. The benefits cannot be guaranteed in the years to come, unless they are to be heightened or more new ones are built or some of the dams are treated to reduce the silt deposition. The effect of measures on the slopes is yet not very pronounced, because the area of control is relatively small and the majority of the trees planted is rather young. But it will certainly increase with time and with the increase of area of control. Hence, it is necessary to properly coordinate the measures taken on the slopes and in the gullies, so that soil erosion can eventually be checked effectively, and the sediment discharge in the river can be steadily reduced.

(4) At present, the degree of control of the Wuding River Valley is still far from being sufficient. It is necessary, therefore, to further mobilize the masses to quicken the pace, to improve the quality of measures being taken and to increase benefits, so as to promote overall development of the production of agriculture, forestry and animal husbandry, to meet the needs of realizing the four modernizations in socialist construction.

## Acknowledgements

The authors wish to acknowledge with thanks the kindness of many of our comrades and Dr P Arens (FAO) who reviewed the manuscript, and made valuable suggestions for revision.

Other participants of this study are: Chen Jiu xian and Wu Zhong of the Shaanxi Provincial Bureau of Soil and Water Conservation; and Zhou Guoping, Lei Sansuo and Zhao Jinhai of the Huang He Conservancy Commission.

## References

Gong Shiyang, and Jiang Deqi, 1977. Soil erosion and its control in small gully water-sheds in the rolling loess area on the middle reaches of the Yellow River. Paper presented at the Paris Symposium on Erosion and Solid Matter Transported in Inland Waters, July 1977. Also published in *Scientia Sinica* No 6, Nov 1978.

Gong Shiyang and Xiong Guishu, 1980. The origin and transport of sediment of the Yellow River. Preprints of International Symposium on River Sedimentation, Chinese Society of Hydraulic Engineering.

Hydrogeological Team no 2, 1976. *A preliminary analysis of the source of coarse sediment in the Wuding River Valley.* Shaanxi Provincial Bureau of Geology (in Chinese).

Jiang Deqi, Zhao Chengxin and Chen Changlin, 1966. A preliminary analysis of the source of runoff and sediment from small watersheds on the middle reaches of the Yellow River. *Acta Geographica Sinica.*

Qian Ning (Ning Chien), Wang Keqin, Yan Linde and Fu Renshou, 1980. The source of coarse sediment in the middle reaches of the Yellow River and its effect on the siltation of the lower Yellow River. Preprints of International Symposium on River Sedimentation, Chinese Society of Hydraulic Engineering.

Wubao Central Hydrometric Station, 1975. A preliminary study on the effect of soil conservation on runoff and sediment transport in Wuding River Valley. Communications on Soil and Water Conservation, Shaanxi Province, Huang He Conservancy Commission (in Chinese).

Wubao Central Hydrometric Station and Yulin Soil and Water Conservation Station, 1976. A preliminary study on the effect of measures of water conservancy and soil conservation on the annual runoff and sediment transport in Wuding River Valley. Symposium on Researches in Soil and Water Conservation in the Huang He Valley, Huang He Conservancy Commission (in Chinese).

# 9

SOIL CONSERVATION PRACTICES AND EROSION CONTROL
IN INDIA - A CASE STUDY

by

D.C. Das
Ministry of Agriculture and Irrigation
New Delhi, India

ABSTRACT

Erosion problems and conservation hazards through floods and droughts have occurred frequently over the centuries. Their effect on the national economy has been more acute in recent years. Owing to increasing human and livestock populations, and other developmental activities, competition for the land has become keen. By reviewing various factors, such as the land use distribution, hydrological and other soil group zones, floods, droughts and existing delineations of the country for various conservation purposes, India has been divided into ten compact Soil Conservation Regions. Considering the latest estimates of different types of erosion prevalent in various parts of the country, a composite soil erosion map has also been prepared.

The erosion problem has been examined for various soil conservation regions in conjunction with the flood and/or drought hazards, and critical socio-economic conditions. Rainfall intensity and the erosion ratio of the soil along with elevation and total rainfall have been collated and interpreted in order to study the soil conservation practices currently in vogue. Problems such as sheet erosion, gully and ravine erosion; torrent, stream/channel erosion; landslides and roadside erosion; coastal erosion; erosion due to shifting cultivation; and wind erosion are specifically discussed. Experimental results and field observations are used to illustrate the extent of hazards under different types of erosion in various soil conservation regions. Various practices now being advocated to treat different types of erosion, national efforts to preserve the life of multi-purpose reservoirs and to solve the problem of ravinous areas through an integrated programme of soil conservation on a watershed basis, are summarized.

## SOIL CONSERVATION PRACTICES AND EROSION CONTROL
## IN INDIA - A CASE STUDY

### INTRODUCTION

Soil erosion in India has been severe for centuries, but its impact on regional and national economy has only been acutely felt during recent decades. The high rate of population growth, both human (Anon, 1975) and livestock, (Anon, 1976 a), has resulted in over exploitation of natural resources to meet the ever-increasing demand for food, fodder and fuel. The increased competition on the same lands for these three essentials is resulting in widespread damage to our forests, pastures and wastelands as well as to fallow and agricultural lands. Thus increasing misuse of natural resources poses one of the greatest threats to better land use management, and the problem is compounded by competing demands from different sectors for expanding development and industrial activities on similar types of land.

Erosion problems in India, therefore, cannot be seen in isolation from natural calamities, viz. flood and drought, or without any reference to socio-economic conditions. An attempt has been made to present the erosion problems and the current efforts being made to solve them on the basis of broad soil conservation regions.

### BACKGROUND INFORMATION

#### Land Use Distribution

Erosion problems vary according to the land use practices. The distribution of areas under various land uses is given in Table 1. Out of the total geographical area of 328 million ha, about 43.20% is cultivated and 20.02% is forest. The vast stretch of the Indo-Gangetic Plain, Deccan Plateau and Rajasthan Desert have very little forest. This poses not only an ecological and erosion problem but also aggravates the shortage of fuel and fodder. Barren and uncultivable wasteland constitute 9.07%, cultivable waste 4.81%, fallow land 6.23% and permanent grassland occupies 3.97% of the total area. The grasslands by and large are poor and over-grazed and so are the wastelands. Both suffer from high erosion hazard.

#### Hydrological Soil Group Zones

Hydrological soil groups give the relative ability of a soil to produce runoff and, therefore, to a considerable extent its susceptibility to erosion. These ratings have been determined for a large number of Indian soil series (Bali, Y.P., 1969) and based on them, India has been delineated into four hydrological soil group zones (Vandersypen et al, 1972). Areas having soils of "B" group constitute 53.60%, "D" 18.49%, "C" 16.82% and "A" 11.09% only (Fig. 1).

Table 1    DISTRIBUTION OF AREAS UNDER DIFFERENT LAND USES IN INDIA   (1972-1973)

| Classification | Area in thousand hectares Break-up | Area in thousand hectares Total | Percentage of total area Break-up | Percentage of total area Total |
|---|---|---|---|---|
| 1. Area under forest | | 65 708 | | 20.02 |
| 2. Area not available for cultivation: | | 46 444 | | 14.10 |
| i) Area under non-agricultural uses | 16 571 | | 5.03 | |
| ii) Barren and uncultivable land | 29 873 | | 9.07 | |
| 3. Other uncultivable land including fallow land: | | 33 728 | | 10.11 |
| i) Permanent pastures and other grazing land | 13 079 | | 3.97 | |
| ii) Land under miscellaneous tree crops and groves not included in net sown areas | 4 358 | | 1.33 | |
| iii) Cultivable waste | 15 841 | | 4.81 | |
| 4. Fallow land: | | 20 565 | | 6.23 |
| i) Fallow land other than current follows | 8 173 | | 2.47 | |
| ii) Current fallows | 12 392 | | 3.76 | |
| 5. Net area sown | | 140 233 | | 43.20 |
| 6. Total reporting area | | 306 223 | | 93.66 |
| 7. Area for which no return exists | | 22 035 | | |
| 8. Total Geographical area | | 328 258 | | 100% |

120

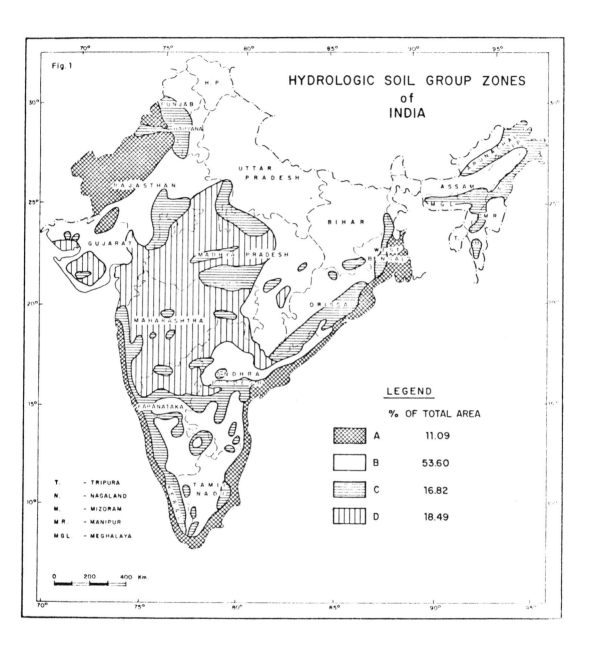

Fig. 1

HYDROLOGIC SOIL GROUP ZONES
of
INDIA

LEGEND

% OF TOTAL AREA

| | | |
|---|---|---|
| A | 11.09 |
| B | 53.60 |
| C | 16.82 |
| D | 18.49 |

T.    – TRIPURA
N.    – NAGALAND
M.    – MIZORAM
M R.  – MANIPUR
M G L. – MEGHALAYA

0    200    400 Km.

**121**

## Flood Hazard

Floods visit India every year and cause untold misery to an average area of 6.7 million ha, out of which 2.2 M ha are cropped. The total affected area is estimated to be 20 M ha. Annual estimated damage due to floods has been put at Rs. 1 260 million (One US $ = Rs. 9.15) out of which 70% relates to agricultural crops (Anon, 1971a and 1974a). Areas affected by flood are shown in Fig. 2.

## Drought Hazard

Like floods, the country experiences drought every year. About 3 million people with their livestock migrate in search of new pastures (Anon, 1973). This migration has often resulted in excessive grazing and the consequent destruction of the protective vegetative cover, leaving the soil bare to the erosive action of both wind and water.

The role of drought in conservation land use management in India is very great. About 75% of the cultivated land is rainfed and produces 42% of the country's foodgrain. It is estimated that even after all irrigation potential is achieved, at least 55% of the cultivated land will remain rainfed (Anon, 1974b). Vast stretches of non-agricultural land in these tracts have scanty rainfall and are suited to grassland development and livestock farming. After reviewing many definitions and classifications, the climatic crop growth indices, limiting rainfall amounts for different growth levels and expressed as a ratio of rainfall to different evapotranspiration demands, have been used to delineate the country into four drought classes, as shown in Fig. 3 (Das et al, 1974).

A comparison of Fig. 2 and 3 and a study of Table 2 reveals that intriguingly the drought has not been a curse to arid and semi-arid (i.e. low rainfall) regions alone. Many areas, in spite of fair annual rainfall, face drought right within the main rainy season (George et al, 1973). Again, in spite of evergreen forest and grasslands, many parts, e.g., the Nilgiris, are subject to regular droughts (Shri Niwas et al, 1967; Das et al, 1971a).

SOIL CONSERVATION REGIONS

Many attempts have been made to delineate a number of zones or regions and sub-zones of areas for the formulation of plans for development and utilization of natural resources. Physiographically, India has been divided into four major divisions (Singh, 1971) and into six water resource regions (Khosla, 1949). It has also been divided into seven main hydrological zones with the problems caused by sedimentation in the multipurpose reservoirs to the soil conservation programme mainly in mind (Bali J.S. 1969). At the Soil Conservation Centre, Dehra Dun, considering major soil groups, vegetation classes, physiography, rainfall and availability of irrigation, the country has been divided into twenty land resource regions and 186 land resource areas (Gupta et al, 1970). Again, for studying soil erosion with a distinct bias toward geology, the country has been divided into five first order physiographic regions (Ahmad, 1973).

After reviewing all these existing delineations and keeping in mind the integrated approach for analysing soil erosion problems as well as for appreciating the current erosion control practices, the delineation made by Gupta et al, (1970) has been regrouped into ten major Soil Conservation Regions as shown in Fig. 4. The distribution of areas under various soil conservation regions along with the annual rainfall and temperature ranges as well as available data on intensity ratio (a measure of erosiveness of rain), and rilling hazard ratio (a measure of erodibility of soil) are presented in Table 2.

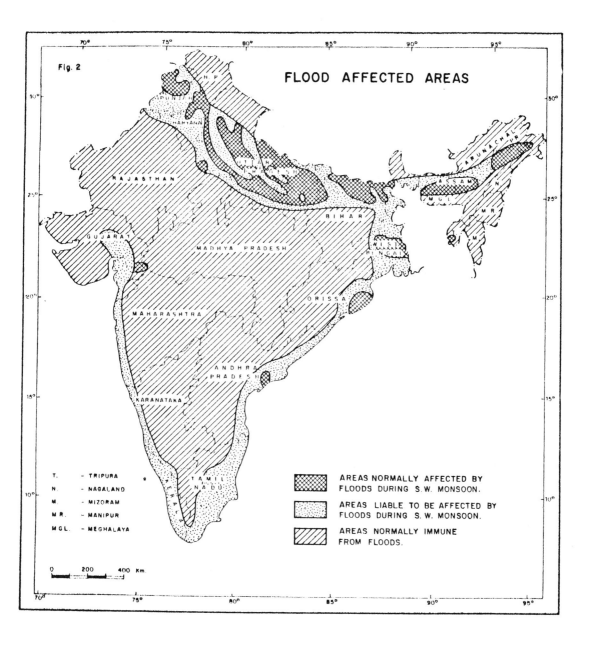

Fig. 2

FLOOD AFFECTED AREAS

T.    – TRIPURA
N.    – NAGALAND
M.    – MIZORAM
M R.  – MANIPUR
M G L. – MEGHALAYA

AREAS NORMALLY AFFECTED BY
FLOODS DURING S.W. MONSOON.

AREAS LIABLE TO BE AFFECTED BY
FLOODS DURING S.W. MONSOON.

AREAS NORMALLY IMMUNE
FROM FLOODS.

0    200    400 Km.

123

Table 2

SOIL CONSERVATION REGIONS OF INDIA

(particulars relate to erosion and other conservation problems)

| S No. | Soil Conservation Region | Area 1/ M ha | Area as percent of total | Rainfall 2/ mm | Mean annual temp. 3/ C. | Intensity ratio * 4/ | Rilling hazard ratio ** 4/ | Hydrological soil group Fig. 1 5/ | Drought hazards (after climatic crop growth index) Fig. 3 | Flood hazard Fig. 2 |
|---|---|---|---|---|---|---|---|---|---|---|
| 2. | 3. | 4. | 5. | 6. | 7. | 8. | 9. | 10. | 11. | |
| 1. | Northern Himalayan Region (A,B,C) 1/ | 34.60 | 10.66 | 200-2500 | 10-20 | 2.05 Dehra-Dun | 2.13 | B | None | None Except a pocket in Jhelum Valley |
| 2. | North Eastern Himalayan Region (G,H) | 17.70 | 5.47 | 1500-2500 | 10-20 | – | – | B and C | None | – |
| 3. | Indo-Gangetic Alluvial Region (D,E,F,L) | 50.90 | 15.60 | 700-1100 Some areas in the West get less than 400 | 20-28 | 2.04 Agra 2.16 Boroda | 7.0 reclaimed ravine Agra 5.0 Baroda 3.46 Ahmedabad | North West-C Rest-B | Eastern part subject to moderate drought; Central part to large; rest has severe drought | Normally affected |
| 4. | Assam Valley and Gangetic Delta Region (I,P) | 11.18 | 3.50 | 1500-2500 | 20-25 | – | – | Assam Valley-C Gangetic Delta-A | None | Normally affected |
| 5. | Rajasthan Desert, Runn of Kutch and contiguous semi-arid region (J and K) | 23.85 | 7.31 | 150-500 | 25-28 and greater than 28 in extreme West | – | Dispersion ration 6/ 33-83% Rajasthan 31.4% Kutch 1.3-6.5 Kota 7/ | Western part – A Rest – B | Rajasthan desert and Runn of Kutch sub-ject to disastrous drought; Rest to severe | – |

124

Table 2 (cont'd)

| 1. | 2. | 3. | 4. | 5. | 6. | 7. | 8. | 9. | 10. | 11. |
|---|---|---|---|---|---|---|---|---|---|---|
| 6. | Mixed Red, Black and Yellow Soil Region (M) | 11.57 | 3.58 | 600–700 | 20–28 | 1.83 Kota | 1.31 Kota 4.57 Sawai, Madhopur | B | Centrally located part affected by large drought and eastern part by severe drought | – |
| 7. | Black Soil Erosion (N) | 67.45 | 20.63 | Outer part between 600–750 Inner part 500–600 | 20–28 | 2.36 Bellary | 4.49–4.79[8] Jabalpur 1.80 Bellary 6.14 Dharwar | D | North East part large; Rest severe hazards | – |
| 8. | Eastern Red Soil Region (O) | 57.45 | 17.55 | 1000–1500 | 20–25 | 1.73 Midnapur | 1.78–4.10 Raipur [2] 2.5–21.60 Gullied area, Midnapur [2] 1.09 Midnapur; 4.23 Santhal Pargana | B Mostly a belt behind East Coast C | Moderate hazard | Damodar River |
| 9. | Southern Red Soil Region (R) | 34.77 | 10.62 | around 750 | 20–25 (10–15 in the Nilgiris) | 1.73 in the Nilgiris | 1.00 Minimum (Hyderabad, the Nilgiris) | Mostly B A belt behind West coast and patches C; small pockets D | A belt along the Western Ghat and also East part exposed to moderate one. Rest exposed to severe drought | Normally free |
| 10. | East–West Coastal and Islands Region (Q,S,T) | 16.20 | 12.60 | East coast about 1000 | 20–25 | – | – | A | West coast and most islands free from drought; central liable to large | Subject to flood hazard |
| | | 325.67 | 100.00 | | | | | | | |

\* Intensity Ratio 4/ = Intensity for 5 minutes for 10 years recurrence / Qualifying intensity for 5 minutes = 7.63 cm/ha

\*\* Rilling Hazard Ratio 4/ = Erosion ratio of soil of an area / Limiting safe value of erosion ratio – 10

1/ after Gupta et al, 1970    6/ Gupta, R.S. 1958
2/ Anon, 1972b              7/ Mehta et al, 1958
3/ after Swaminathan, 1973  8/ Ballal, 1954
4/ Das et al, 1971; Das, 1976
5/ Vandersypen et al, 1972  9/ Bhattecharya, 1957

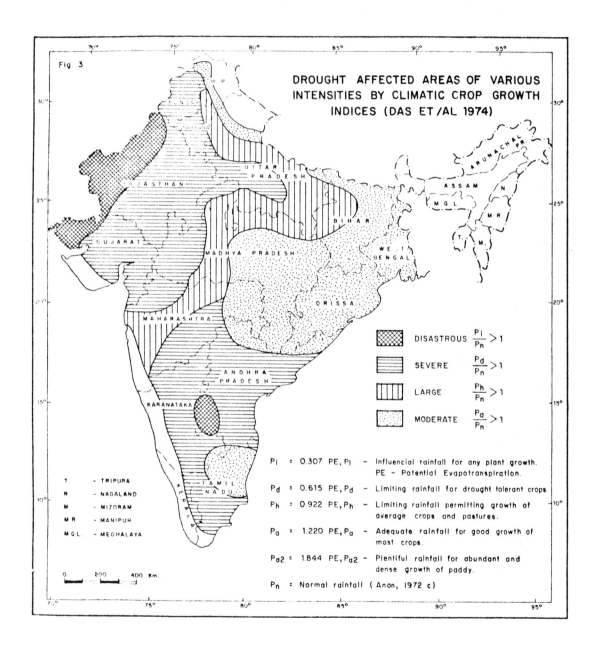

Fig. 3

**DROUGHT AFFECTED AREAS OF VARIOUS INTENSITIES BY CLIMATIC CROP GROWTH INDICES (DAS ET /AL 1974)**

| | | |
|---|---|---|
| ▨ | DISASTROUS | $\frac{Pi}{Pn} > 1$ |
| ▤ | SEVERE | $\frac{Pd}{Pn} > 1$ |
| ▥ | LARGE | $\frac{Ph}{Pn} > 1$ |
| ⋯ | MODERATE | $\frac{Pa}{Pn} > 1$ |

$Pi = 0.307\ PE,\ Pi$ – Influencial rainfall for any plant growth. PE – Potential Evapotranspiration.

$Pd = 0.615\ PE,\ Pd$ – Limiting rainfall for drought tolerant crops

$Ph = 0.922\ PE,\ Ph$ – Limiting rainfall permitting growth of average crops and pastures.

$Pa = 1.220\ PE,\ Pa$ – Adequate rainfall for good growth of most crops.

$Pa2 = 1.844\ PE,\ Pa2$ – Plentiful rainfall for abundant and dense growth of paddy.

$Pn$ = Normal rainfall ( Anon, 1972 c)

T. – TRIPURA
N. – NAGALAND
M – MIZORAM
M R – MANIPUR
M G L. – MEGHALAYA

0   200   400 Km.

**126**

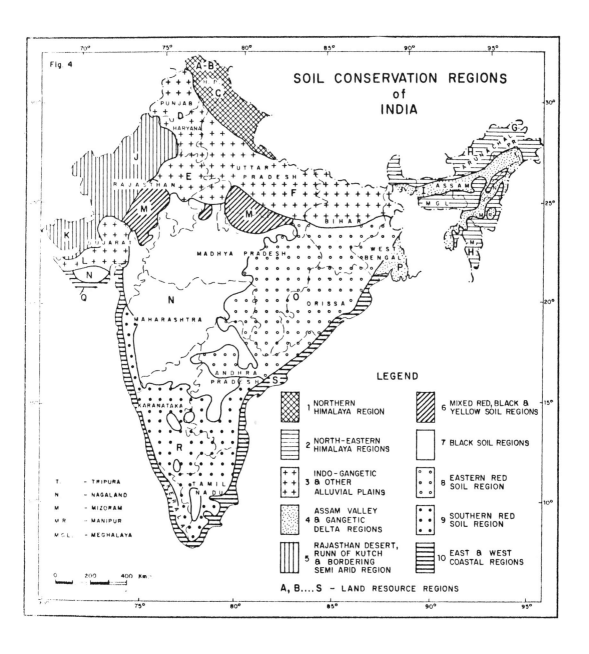

Fig. 4

SOIL CONSERVATION REGIONS
of
INDIA

LEGEND

1 NORTHERN HIMALAYA REGION

2 NORTH-EASTERN HIMALAYA REGIONS

3 INDO-GANGETIC & OTHER ALLUVIAL PLAINS

4 ASSAM VALLEY & GANGETIC DELTA REGIONS

5 RAJASTHAN DESERT, RUNN OF KUTCH & BORDERING SEMI ARID REGION

6 MIXED RED, BLACK & YELLOW SOIL REGIONS

7 BLACK SOIL REGIONS

8 EASTERN RED SOIL REGION

9 SOUTHERN RED SOIL REGION

10 EAST & WEST COASTAL REGIONS

T.   - TRIPURA
N    - NAGALAND
M    - MIZORAM
M R  - MANIPUR
M.G.L. - MEGHALAYA

0    200    400 Km.

A, B....S - LAND RESOURCE REGIONS

127

SOIL EROSION

Past Estimates

There seems to have been no coordinated attempt to document the incidence of various types of erosion and their extent and distribution throughout the country. In the late forties Kaith et al, (1948) submitted a brief report to this effect which urged consideration of the problems and their solution on the basis of natural units. Later on the Planning Commission (Anon, 1964) made a quick estimate of agricultural land subject to erosion, while the Working Group on Soil Conservation of the Fourth Plan (Anon, 1966) attempted to estimate the extent of erosion and other conservation hazards on all types of land. Bali 1974 outlined the distribution of areas under various conservation hazards with specific stress on the erosion problem and sedimentation of multi-purpose reservoirs. According to his estimate, a total of 175 million ha can be considered susceptible to erosion. In another attempt, while giving the macro view of land and water management problems in India, Vohra (1975) indicated that 140 million ha are seriously affected by water and wind erosion. Ahmad (1973) gives some detailed estimates of areas subject to various types of erosion.

Extent of Different Types of Erosion

The National Commission on Agriculture (1975b) provided the latest estimate on the basis of available information from various sources. Many parts of the country are subject to more than one type of erosion. The extent of erosion has been estimated as follows:

|  | (Million ha) |
|---|---|
| Total Geographical area | 328 |
| Total area subject to serious water and wind erosion | 150 |
| Area at critical stage of deterioration due to erosion | 69 |
| Area subject to wind erosion | 32 |
| Area effected by gullies and ravines, about | 4 |
| Area effected by shifting cultivation, about | 3 |
| Area under rainfed farming (non-paddy) | 70 |

Erosion Map

On the basis of various erosion maps prepared by Ahmad (1973) and with the necessary modifications called for by the latest figures available, a composite map of India showing the extent of various types of erosion is presented in Fig. 5.

SHEET EROSION

Sheet erosion exists almost throughout the country. Areas subject to sheet as well as gully erosion and areas subject to notable sheet erosion only are indicated specifically in Fig. 5. It has been estimated that about 6 000 million tonnes of soils are eroded every year from about 80 million ha of cultivated lands; about 6.2 million tonnes of nutrients (2.5 of N, 1.5 million t of P and 2.2 million t of K) are carried away. The nutrients lost in this way are much greater than the quantity that we are using at present (Swaminathan, 1973). The different rates of soil loss under various land uses in different regions are given in a recent compilation (Tejwani et al, 1975), and the figures were obtained from ICAR research stations.

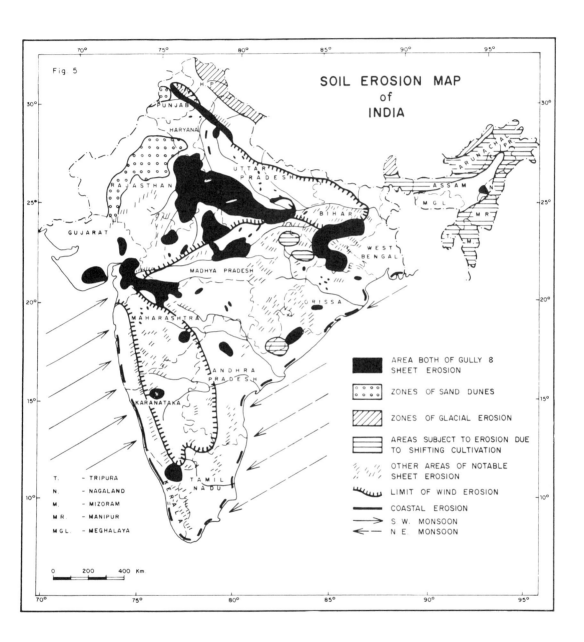

Fig 5

SOIL EROSION MAP
of
INDIA

AREA BOTH OF GULLY &
SHEET EROSION

ZONES OF SAND DUNES

ZONES OF GLACIAL EROSION

AREAS SUBJECT TO EROSION DUE
TO SHIFTING CULTIVATION

OTHER AREAS OF NOTABLE
SHEET EROSION

LIMIT OF WIND EROSION

COASTAL EROSION

S. W. MONSOON

N. E. MONSOON

T.    - TRIPURA
N.    - NAGALAND
M.    - MIZORAM
M R.  - MANIPUR
M G L. - MEGHALAYA

0    200    400 Km.

**129**

Performance of Conservation Practices

The conservation treatments meant to reduce or prevent sheet erosion while achieving the desired moisture conservation and/or runoff disposal, range from contour farming to bench terracing. The performance of different conservation practices for controlling sheet erosion has been evaluated at a number of research stations (Tejwani et al, 1975) and also reviewed elsewhere (Das, 1974). To illustrate the relative effectiveness of a few practices, the results from the Nilgiri hills in the southern red soil region are given. On these hill slopes, that have lateritic soils, up and down cultivation of potato caused a soil loss of 39 t/ha/year whereas nutrient losses were Rs 333/ha (at the market price of 1967). Simple contour cultivation reduced the soil loss to 15 t/ha/year and nutrient losses to Rs 130/ha. When the cultivation was on non-paddy benches the soil loss was reduced to about 1 t/ha/year and the nutrient losses to Rs. 10/ha/year (Raghunath et al. 1967).

Terracing is by far the most effective and widely practised field measure for controlling or preventing erosion in different soil conservation regions. Terracing has also been adapted in different ways to meet varied physiographic and climatic conditions. In a general way, it can be defined as a series of mechanical barriers across the land slope to break the slope length and also to reduce the slope degree wherever necessary. On the steeper hill slopes there are graded (non-paddy) or levelled (paddy) benches constructed by restricted levelling and with provision for disposal. The system also applies to lower slopes where it is developed by raising bunds and then achieving the levelling through puddling. For rolling and flatter lands, with high rainfall, the system means graded bunds or banks (narrow base graded bunding) and for lands with scanty or erratic rainfall, narrow base (level) terracing (contour bunding) is practised to conserve moisture as well as reduce soil erosion. For better moisture conservation on larger areas as well as for control of soil erosion, it was observed that land treatment between the bunds, e.g. partial levelling (Saha and Patel, 1970) or cultivation along the contour, offers better protection as well as moisture conditions for higher production. For rolling lands and hill slopes with inadequate soil depth, graded trenches or contour trenches are used for forestation or plantation crops like tea or coffee. Field as well as experimental data show that terracing controls runoff, checks soil erosion, helps to improve moisture conservation, groundwater storage and crop yields (Satpute, 1972; Das, 1974).

Treatments vary considerably from region to region, for instance, the high slopes of the Northern Himalayas are being put under good forest cover through afforestation and better forest management.

Narrow base terracing (bunding)

The Indo-Gangetic alluvial plains have an erosion intensity ratio value of about 2, whereas the rilling hazard ratio varies from 3 to 7. A sizable part of the region is liable to regular flooding as well as moderate to severe drought (Table 2, Fig. 2 and 3). The problem is, therefore, one of erosion control, disposal of runoff and conservation of moisture. Contour bunding, or narrow base terracing on the contour, at a suitable spacing and of a suitable cross section, is widely used to remove excessive runoff resulting from high intensity storms and surplussing arrangements are made wherever necessary.

**130**

The eastern red soil region has an intensity ratio  around 1.73, whereas the rilling hazard ratio ranges from 1.09 to 4.23.  The region is subject to almost recurrent moderate drought in spite of good rainfall (George et al, 1973). The southern red soil region, on the other hand, has a similar intensity ratio but generally a low rilling hazard ratio.  However, the region is by and large subject to severe drough hazards (Fig. 3 and Table 2).  Contour bunding with a surplussing arrangement often proves adequate.  A typical contour bund cross section with a loose rubble wasteweir is shown in Fig. 6.  Small tanks for storing and utilizing water for paddy cultivation on benches are also extensively used.

Fig. 6   Contour bunding (narrow base terrace on contour) with dressed rubble wasteweirs, in black soil region (shallow black soil)

The black soil region has a very high intensity ratio (2.36) and the rilling hazard ratio ranges from 1.3 to 6.14.  The north-east of the region suffers from large droughts whereas the rest to severe ones (Table 2, Fig. 3). The problem is not only to conserve moisture, but also to remove excess rainfall which cannot infiltrate due to low intake rates.  The accumulation of water on land causes damage to the standing crops besides resulting in mechanical failure of the bunds and other structures.  In the northern part of this region, the removal of excess water facilitates a more flexible rainy season crop and gives the possibility of using the excess water to grow a profitable winter crop.  On the rest of this region, the rainfall is low and erratic;  this area remains fallow during the main rainy season and thus subject to heavy erosion.  Contour bunding has not been effective in the region (Das, 1974).

The latest findings in the black soil and the semi-arid southern red soil region indicate that some sort of graded bunding (graded narrow base terracing) with small ponds for storage and reuse of the runoff water for assured cultivation is more effective. Cultivation of crops on broad ridges and furrows seems to hold promise for these regions where stress is laid on saving crop residues, proper tillage and water harvesting besides erosion control (Anon, 1974b and Krishnamoorthy et al, 1974). The different types of small ponds which could be used on existing farm holdings are being tried presently at the All India Coordinated Dry Land Research Centre as well as the International Crop Research Institute for Semi-Arid Regions. Some possible changes in the design details for higher capacity-cost ratio have been developed by Sharma and Kampan (1975).

## Design details for bunding

The spacing of the narrow base terraces can be more rationally determined from the modified Stewart or Ootacamund formula. It provides more flexibility over space and time and involves easily determinable factors as compared to the adjusted empirical formulae in use. The formula is:

$$\text{Vertical Interval} = V.I. = \frac{\sqrt{2s}}{2n}, \text{ m}$$

$$\text{and Horizontal Interval} = \frac{50}{n(\sqrt{s})}, \text{ m}$$

Where,

$s$ = Average slope percentage
$n$ = I.R.P.C.
$I$ = Intensity Ration (defined at the foot of Table 2)
$R$ = Rilling hazard ratio  "    "    "    "  )
$P$ = Proportionate protection offered by supporting conservation practices
$C$ = Proportionate protection offered by cropping management.

The formula is also applicable when trenching.

Concerning the cross-section, the design depends upon the amount of excess rainfall and the depth of impounding desired upstream of the terrace. In the case of graded bunding, it is the channel that is more important and the downstream bund is a spoil bund. For the semi-arid southern red soils 0.2% longitudinal gradient was found to be effective, whereas for the Northern Himalayan foothills the gradients of 0.4 and 0.6% were found to be equally good (cited from Das, 1974).

There are different types of structures; for removing excess rainfall from bunded areas some of the common ones are given in the book, Soil Conservation in India (Rama Rao, 1972). They are:

1. prefabricated escapes
2. dressed/undressed loose stone wasteweirs with end pitching
3. ramp-cum-weir
4. pipe outlets
5. ringwell and pipe outlets
6. channel weir
7. stone/brick masonry low drops/chutes

Paddy and non-paddy bench terracing

In the Northern Himalayas, paddy crops are raised on bench terraces. There are also traditional irrigation systems using the hill streams. In the southern hills, namely, the Nilgiris, lying within the southern red soil zone, a system of non-paddy benches has been developed (Lakshmipathy and Narayanswamy, 1956) see Fig. 7. These benches are provided with a longitudinal gradient to drain into cross disposal drains. The vertical spacing is determined by the depth of the cut available and from the formula $VI = 2 (T - 0.15)$ m, where T is the depth of productive soil in meters; VI, computed as above, is again tested with the minimum width vis-à-vis the land slopes and ease for farm operations (Lakshmipathy and Narayanswamy, 1956). For effective runoff disposal as well as better moisture conditions, these benches should not be longer than 100 m (Das, 1974), whereas they can have a longitudinal gradient upto even 1% under Nilgiri conditions (Kurian et al, 1975). In the Northern Himalayan regions, non-paddy benches of the southern type are being introduced. However, due to heavy rainfall, a more elaborate disposal system is being provided.

On the steep Himalayan slopes, narrow benches are being constructed to raise belts of orchard trees. Such benches are also recommended for raising other plantation crops, and medicinal plants.

Developing low cost bench terracing

Even though there is no substitute for bench terracing when growing crops on hill slopes, it is expensive and initial crop yield is low due to disturbance of the top soil. In order to obviate these disadvantages, benches could be developed over years behind graded barriers of grass, or earthen bunds or stone walls. A typical bench developed behind the graded barrier of Tripsacum laxum is shown in Fig. 8. In Kerala, benches are developed by putting earthen bunds with downstream pitching across the land slope, whereas in the northern Himalayan region, as well as in the low hills of the eastern red soil region, benches are developed by building stone risers (Murthy, 1969).

Vegetative Stabilization of Conservation Measures

All bund channel sections as well as the riser face of benches are stabilized with grass. On locations subject to hot winds Eucalyptus, and Eulaliopsis binata grass or Ricinus communis are grown on bunds and they provide an additional income over years besides serving as wind breaks (Singh and Srivastava, 1973; Tejwani et al, 1975).

GULLY AND RAVINE EROSION

This is possibly one of the most spectacular types of erosion. Ravines annually ravage 8 000 ha throughout the country where approximately 0.5% of the ravine catchment area is being eroded. Annual damage to the tablelands accounts for Rs 40 million (Anon, 1972a). Gullies and ravines damage land, habitations, roads, railways and other public properties, thus adversely affecting rural economy (Anon, 1976b). A typical picture of gully erosion is shown in Fig. 9. They also provide shelter to unsocial elements causing law and order problems. The ravines are mostly found along the rivers of Jamuna, Chambal, Sabarmati, Mahi, Gumoti and in the catchments of the Mayurakshi, Kangasabati in the eastern red soil region. Generally, ravines originate within a short distance of the river and

Fig. 7 Bench terracing (non-paddy) in the southern red soil region of the Nilgiri hills (photo P.K. Thomas)

Fig. 8 Developing non-paddy benches along the graded strips of Tripsacum laxum on the slopes of the Nilgiri hills

near or around the confluences of two channel flows.  At a conservative estimate,
the country is losing a total output of 3 million tonnes of foodgrains annually
due to degradation of land through gullies and ravines (Anon, 1972a).

Fig. 9  Severe gullying in the catchment of the Mayurakshi
on highly erodible soils in eastern red soil region

## Formation of Gullies and Ravines

Opinions vary on the causes leading to gully erosion and ravine formation.
The popular belief is that due to excessive exploitation of the land by continuous
deforestation, over grazing and faulty cultivation, gullies and ravines have
formed.  Ahmad  (1973), however, argues that in the areas where ravines are very
severe there has not been luxurious vegetation in the recent past.  The areas
subject to severe gully erosion are located in western, central and eastern India
and along the foothills of the Himalayas.  The difference in elevation of these
areas ranges from near sea level in the west to about 300 - 1 000 m in the east
In the foothills of the Himalayan region, gullies are being found at elevations

135

as high as 1 000 – 1 300 m. The type of soils, rainfall received and topographical features are also considerably different in these major regions where gully and ravines are found extensively. Table 2 reveals that in the ravines in western India, the intensity ratio is 2.16 whereas the rilling hazard ratio is between 3.46 and 5. The bulk of the ravines in Central India are in the Indo–Gangetic alluvial region and have an intensity ratio of 2.04 and rilling hazard ratio for ravinous areas as high as 7. A part of the ravine system in central India falls in the black soil region as well as in the mixed red, black and yellow soil region. In the latter region, the rainfall intensity ratio is 1.83 whereas the rilling hazard ratio ranges from 1.31 to 6.5. In the red soil region of eastern India, where ravines are very conspicuous, the intensity ratio is 1.73 and the rilling hazard ratio is high, Ahmad (1973) contends that the incidence of gully and ravine erosion is possibly due to geologic uplift. Roy and Mishra (1969), from their field observations, concluded that ravines are a function of depth of the river and are invariably confined to the vicinity of rivers and tributaries. Gully is a function of overland flow and therefore, of catchment characteristics alone, whereas ravine is primarily a function of channel flow and its ingress into the catchment. They hold the view that the ravines proceed from the river or tributaries to the agricultural lands.

## Treatment

Considering the complex physiographic background of the regions where gully and ravine erosion are prevalent and the involvement of socio–economic aspects, attempts have been made to take appropriate action to control this erosion as well as to restore the land to better land use management. The earliest attempts were made in the erstwhile Gwalior State in 1919, there revenue concessions were granted to cultivators who would reclaim the ravines along the Chambal, (Rege and Yash Pal, 1973). In the post–independence years, the Government of Uttar Pradesh and Madhya Pradesh took a number of pioneering steps to control and reclaim ravines along the Jamuna and Chambal (Anon, 1971b; Rege and Yash Pal, 1973).

### Central pilot projects

During the Fourth Plan, Pilot Projects were started in the four major States covering the ravinous areas of western and central India to develope suitable technology for reclamation and control of ravines for agriculture and horticulture and for stabilizing deeper ravines. During the Fifth Plan, the Pilot Projects were extended to more locations to cover the varied physiographic and climatic conditions prevalent in the catchments of different rivers and also to test the feasibility of the technology as well as the viability of the economics of ravine control, reclamation and the stabilization programme. The entire concept for tackling this programme was outlined in the Report of the Working Group on Ravine Reclamation (Anon, 1972); they examined alternatives, e.g. putting the ravines under complete vegetation and reclaiming them for agricultural purposes, and a combined programme on a definite watershed basis. The conclusion was that afforestation alone or reclamation for agricultural purpose alone could not solve the problem. It needs to be solved from the basis of a watershed and from the tablelands to the confluences of ravines forming a natural drainage system. As a result, normal conservation measures, e.g. terracing, bunding, etc. are being provided on the tableland draining into the ravines. By building periphery bunds and drop structures at the gully heads, the ingress of the gullies into the tablelands is being arrested while providing safe ways to take away the runoff through the ravine system. Shallow ravines, up to a depth of about 2 m are being reclaimed for agricultural and horticultural purposes where irrigation could be provided and cash crops grown (Fig. 10). Deeper ravines with steep sides are being stabilized with the help of vegetation and engineering structures of various types (Fig. 11).

Fig. 10    Reclamation of shallow ravines
for agriculture with field
irrigation channels along the
Chambal in the mixed black,
red and yellow soil region
(photo Y.P. Bali)

Fig. 11    Ravines along the Chambal
River, near Kota, in the mixed
black, red and yellow soil
region.  Deep tail gullies and
ravines are being put under
afforestation (photo Y.P. Bali)

Fig. 12    Controlling erosion and restoring gullied lands by creating small
ponds and developing paddy benches in the Mayurakshi catchment in
the eastern red soil region

**137**

In the eastern red soil region gullied areas are being reclaimed by putting earthern checkdams in a series or singly to develop micro storage ponds (Fig. 12). On the downstream side of the storage, paddy benches are developed. Incidentally, this system ensures holding up most of the potential sediment in a zone where soils have a rilling hazard ratio as high as 21.00. The system improves the productive capability of the area and thus encourages considerable cooperation by the people.

Performance of Treatment

The effectiveness of various treatments is being tested at research stations as well as on the Pilot Projects. The combined package of practices on the watershed basis have, by and large, been found effective. In the eastern ravines engineering and biological measures increased rain water retention by 53% (Singh and Dayal cited from Anon, 1974c). In western as well as central India the system of treating ravine watersheds with a combination of biological and engineering measures e.g. raising fuel fodder plantations, constructing peripheral bunds, drop structures, etc. resulted in a reduction of the peak discharge as well as soil loss over the years. More information about different types of structures, construction of bunds, selection of grasses and trees which have been found suitable for the purpose, through experimentation at research stations, are given in detail in an ICAR publication (Tejwani et al, 1975).

Regarding the economics of ravine reclamation, an evaluation report on the U.P. showed that by reclaiming the ravine areas for agriculture, yield could be increased by 16% (Anon, 1971b). The analysis of the integrated project outlined in another report (Anon, 1972a) indicated that the annual direct benefit from the seven year plan for 330 000 ha would be at the rate of 11.2% on the investment. This project would create employment opportunities to the tune of 339 million man days. The economic viability from the areas treated with grass, areas reclaimed for agricultural purposes and areas stabilized by afforestation, have also been analysed at ICAR research stations. It was observed that the investment in treating gullies, including reclaiming the areas, could be recovered within a period of 2-10 years (Tejwani et al, 1975).

Ravines affect habitations, communications and the agricultural productivity of regions where sizable populations are located. Any improvement measures for the area, even on the basis of watershed, therefore require the closest involvement of the people, i.e., beneficiaries, for their maintenance and appropriate follow-up actions not only in respect of agricultural lands, but also in respect of the deep ravines which will be put under fuel-cum-fodder plantations. If the areas are not treated with a view to providing appreciable economic returns to the beneficiaries, any treatment is likely to meet with ultimate failure. On the other hand, Kamnavar et al, 1975) observed that 3% of the initial investment would be required for maintenance and to stabilize the treatment measures during the first two years; in subsequent years it would be nominal. The cost of maintenance could easily be met from the sale of produce, such as grass, fire-wood, etc. from the treated area. Keeping this in mind, the National Commission on Agriculture recommended emphatically that the economic conditions should not be the sole basis in the reclamation of ravine lands. Poverty and the objective of curbing the activities of antisocial elements should receive due consideration. Ravine reclamation as such, should receive national priority and investment should not be denied on account of an unfavourable benefit cost ratio (Anon, 1976b).

[*Editor's Note:* Material has been omitted at this point.]

**138**

EROSION DUE TO SHIFTING CULTIVATION

Shifting cultivation or 'jhuming' is practised mainly in the north eastern Himalayan region and marginally on other non-Himalayan hills (Fig. 5). The practice, evolved over centuries, incorporates a mixed land use pattern appropriate to the physiographic remoteness and lack of communication. This remoteness has forced the people to adopt a system that is self-contained and involves the minimum number of items. The problem of 'jhuming' has engaged the attention of various agencies for quite a few decades; the National Commission on Agriculture (1976c) cites about 19 references and though these reports differ on many points, there is agreement that improvement in land use management in the areas subject to this practice is urgently needed. The latest estimate by Miss Wadia (1975) indicates that about 2.7 million ha are affected by shifting cultivation in the north-eastern Himalayan region alone (including non-Himalayan hill ranges) and the average area cultivated by a family is 0.92 ha.

To meet the exigencies of collective security and to adjust individual needs and responsibilities vis-à-vis community ones, the system seems to have served the people of the area remarkably well. The question does arise as to why it has become urgent to review the effectiveness of this system, which has not only evolved in India under certain conditions, but also in many parts of the world with similar conditions.

Degradation Hazards

Opinions differ on the associated hazards of soil degradation and consequent sediment and flood havoc caused by the practice of 'jhuming' on hill slopes. Some people hold the view that shifting cultivation, which does not involve ploughing and other intensive agricultural operations, does not disturb the soil greatly. It is true that there are no systematic studies available to illustrate the sediment hazards from the areas subject to shifting cultivation. In the past when the 'jhuming' cycle was long, about 20-30 years, it is possible that there was no appreciable degradation of land or ecology; but the cycle has now become as short as 1-3 years which does not permit any natural recuperation by the land. However, from Table 2 it can be seen that the area receives between 1 500 – 2 500 mm of rainfall and the slopes are, by and large, steep. The area is also subject to heavy landslides, accelerated by an intensive road development programme. Floods with a heavy silt load create havoc almost annually in the Assam Valley.

To set aside any doubts, data from the Machkund catchment (No. 6 in Fig. 18) can be cited; it is close to the eastern coast and covers slopes in the eastern Ghat. The entire area is populated by tribes who have practised shifting culti-vation (locally called 'podu') for centuries. Since the Centrally Sponsored Scheme for Soil Conservation was launched in the Machkund catchment during the Third Five Year Plan, about 40% of the critically eroded areas have been treated up to 1975-76, and as a result, the siltation has fallen from 3.38 to 2.51 ha m/100 km$^2$/yr (Table 3). In addition, it can be concluded from field observation that areas subject to shifting cultivation  become depleted of better flora and fauna over the succeeding cycles.

Treatment

To tackle this problem, Dhebar's Commission (cited from Anon, 1976c), suggested that a coordinated approach be taken to provide employment opportunities in forestry, agriculture, animal husbandry and industry for at least 300 days in a year. This needs to be examined within the background of multiplicity of vocations and the rising aspirations of the people of the area to adopt better vocations and a higher standard of living with the opening of the area by a

**139**

Table 3 SEDIMENT PRODUCTION RATES VIS-A-VIS SOIL CONSERVATION WORKS IN A FEW SELECTED
RIVER VALLEY PROJECT CATCHMENTS

| No. Reservoir Catchment | Catchment area 1 000 ha | Sediment Production Rate ha m/100 km$^2$/yr | | |
|---|---|---|---|---|
| | | Assumed | Original (Year) | Reduced (Year) |
| | 56 87 000 Total | | | |
| 1. Bhakra | 37 160 (within India) | 4.28 | 8.38 | 6.14 |
| 2. Maithon DVC | 521 | 1.62 | 15.45 | 13.10 |
| 3. Panchet DVC | 981 | 2.47 | 13.32 | 10.30 |
| 4. Machkund | 195 | 3.57 | 3.38 | 2.51 |

modern communication system and the introduction of currency. Therefore, it
must be carefully considered whether the same family, who may want to send
some if its members for education and other vocations, will be in a position
to cultivate or manage larger areas with the labour available without substantial
support from improved inputs, such as irrigation, fertilizer, better seeds and
appropriate tools, implements and machinery. This integrated development is
being tried on a pilot scale in the north eastern Himalayan region where a
family is being provided with a hectare of irrigated land and another hectare
of dry land with additional support for raising profitable forest, horticulture
or other plantation crops. To improve the ability of families to manage these
enhanced areas, demonstrations are being given on the use of appropriate tools,
implements and machinery. Considering the level of present development, big
tractors or equipment cannot be introduced in these areas to bring about an agricultural revolution within a short time: the process has to be gradual and in stages.
To begin with, the large number of indigenous tools should be studied and,
if found suitable, their designs improved. The storage bin, similar to that
which has proved useful in the Punjab, winnowing and threshing machines may be
of immediate use.

Unless the 'jhumias' or shifting cultivators can be weaned from the
practice by a profitable alternative vocation suiting their socio-cultural
background, the problem cannot be solved. And if this goal cannot be achieved,
the direct objective of reducing erosion and consequent sedimentation accompanied
by heavy floods cannot be fulfilled either.

SOIL CONSERVATION IN THE CATCHMENT OF RIVER VALLEY PROJECTS

India has been building up her irrigation and hydro-electric potential
steadily over the years. These projects are mostly multipurpose. The sites of
the reservoirs are the gift of nature and therefore not profuse in numbers. The
projects provide water for irrigating the downstream areas and thereby increase
agricultural production. The stored water is also used to generate electricity
for developing industries and intensifying agriculture. It is, therefore, essential
that the reservoir capacity created should be preserved as long as possible.
Even though allowance has been made for unavoidable sediment flowing with the
water, the rate of sedimentation in most of the reservoirs has been very high;

a few of examples can be seen in Table 3.  The sediment production rates of some of the multipurpose reservoirs at present range from 2.5 to 18 ha m/100 km$^2$/yr (Gupta, 1975); he also draws attention to the higher silt load of Himalayan rivers than non-Himalayan ones.

## Effect of Sedimentation

The accelerated sedimentation of the reservoirs has, in many cases, already affected the irrigation potential created.  For example, in Nizamsagar reservoir the present capacity of the reservoir is 338 million m$^3$, against the initial 898 million m$^3$.  With this loss of capacity, there is not enough water available to irrigate the designed command area of 110 000 ha for growing sugarcane and paddy.  The lack of supply of sugarcane has already threatened the functioning of sugar factories including that of the Bodhan Sugar Factory, the largest in south east Asia (Suba Rao, 1974).  Furthermore, sediment coming from the catchment is degrading it at a fast rate, thus reducing the all-round productivity.

## Watershed Approach

The problem of preserving the reservoirs is combined with improving productivity in the catchment area and preserving the existing productivity of the command area.  Realizing a complex nature of the problem, a scheme was launched during the Third Five Year Plan to treat a few of the catchments in the river valley projects.  To date, this scheme is operating in 30 such catchments (Fig. 18), which cover an area of about 79 ha each and almost all types of erosion and conservation problems are present in these catchments.  Because of obvious financial constraints and lack of technical personnel, it is impossible to treat all areas immediately and, indeed, it may not be necessary to treat all of them to prevent considerable sedimentation of the reservoirs and degradation of the catchment.  A priority survey is being conducted by the All-India Soil and Land Use Survey of the Government of India to ascertain the critical areas of degradation in these catchments.  The methodology developed takes into consideration the erodibility of soils, topography, vegetation, channel system, status of erosion, erosiveness of the climate as well as the proximity of the watershed to the reservoir or main tributaries draining into the reservoir (Bali and Karale, 1973).  From the area surveyed so far in various catchments, it appears that about 15% of the land can be considered as highly critical and should be treated immediately.  Such areas are being identified as small watersheds having a national treatable area ranging from 2 000 to 4 000 ha.

The macro plans are drawn up first on the basis of sub-catchment/sub-basins and then more detailed micro plans are prepared for small watersheds or sub-watersheds within the sub-catchment/sub-basin, although, the actual implementation of works should proceed from the micro level, i.e. on the basis of watershed/sub-watersheds.  This helps utilize the available resources to the maximum, and accommodates inevitable changes occurring while planning details or even while implementing the plan.

## Implementation of the Programme

During the Fifth Plan, about 240 such watersheds are receiving saturation treatment, i.e. remedial measures for all types of land and associated drainage systems.  The measures for treating agricultural land comprise bunding, bench terracing and grassed waterways and sometimes ponds, non-agricultural lands are provided with closures for restricting grazing, afforestation and grassland development.  Gullies and stream bank erosion are treated with structures, periphery bunds and vegetative measures.  In extremely degraded land, contour dyking and ditching is done to improve soil and moisture conservation, also enclosures are built for regeneration of grass and shrubs.  Sediment trap basins

**141**

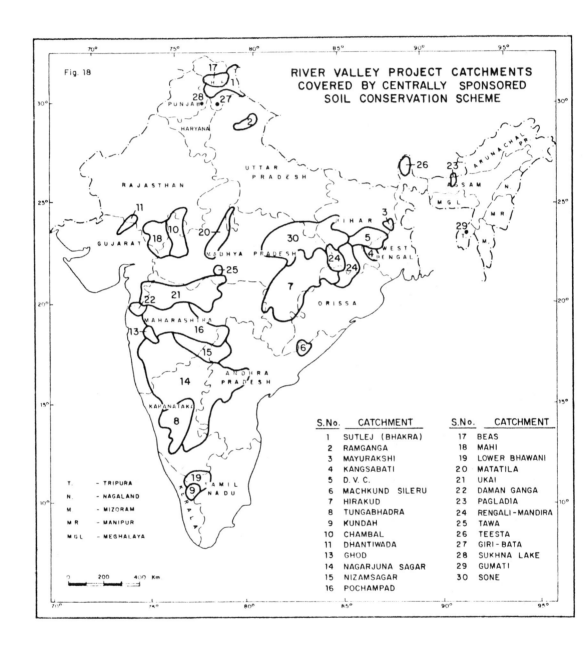

Fig. 18

RIVER VALLEY PROJECT CATCHMENTS
COVERED BY CENTRALLY SPONSORED
SOIL CONSERVATION SCHEME

| S.No. | CATCHMENT | S.No. | CATCHMENT |
|---|---|---|---|
| 1 | SUTLEJ (BHAKRA) | 17 | BEAS |
| 2 | RAMGANGA | 18 | MAHI |
| 3 | MAYURAKSHI | 19 | LOWER BHAWANI |
| 4 | KANGSABATI | 20 | MATATILA |
| 5 | D.V.C. | 21 | UKAI |
| 6 | MACHKUND SILERU | 22 | DAMAN GANGA |
| 7 | HIRAKUD | 23 | PAGLADIA |
| 8 | TUNGABHADRA | 24 | RENGALI-MANDIRA |
| 9 | KUNDAH | 25 | TAWA |
| 10 | CHAMBAL | 26 | TEESTA |
| 11 | DHANTIWADA | 27 | GIRI-BATA |
| 13 | GHOD | 28 | SUKHNA LAKE |
| 14 | NAGARJUNA SAGAR | 29 | GUMATI |
| 15 | NIZAMSAGAR | 30 | SONE |
| 16 | POCHAMPAD | | |

T.   - TRIPURA
N.   - NAGALAND
M    - MIZORAM
MR   - MANIPUR
MGL  - MEGHALAYA

0   200   400 Km

**142**

Fig. 19   Sediment detention tanks in a watershed of the Damodar Valley
Corporation, eastern red soil region

Fig. 20   Good vegetation restores the eroded landscape owing to better soil
and moisture conditions created by sediment detention tanks – Damodar
Valley Corporation, eastern red soil region

on upstream watersheds are constructed to hold sediment, reclaim badly eroded land and improve the all round moisture condition (Fig. 19); they also moderate floods and recharge groundwater (Bhumbla, 1976). These small ponds help in reclaiming downstream land for paddy cultivation and can be used for irrigation too; they also provide conditions conducive to the growth of better trees and other vegetation (Fig. 20).

The effectiveness of the scheme is reviewed from time to time and in a number of cases where it has been in operation for a considerable time there is an indication that sediment production rates are falling in spite of the fact that the areas treated were not very large (Das and Kaul, 1976); this is also evident from Table 3. However, the pace of the conservation programme needs to be accelerated considerably in order to reduce sedimentation rates below the permissible ones quickly, thus increasing the useful life of the reservoirs.

### Catchment Area Authority

It has been well established that, whatever the type of erosion or conservation hazard, the source area as well as the affected problem area must be treated together. Thus, irrespective of the type of erosion, the planning and implementation has to be on the basis of a natural unit, i.e. a watershed or catchment. The solution does not only cover protection and stabilization, but also its continuing effectiveness through appropriate maintenance and follow-up practices. This is possible when the beneficiaries are involved intimately in managing the treated land by a judicious land use plan which gives sizable direct and indirect benefits (Bhumbla, 1976).

Considering this vital need, catchment area development authorities are being constituted for different catchments in different States embarking on developmental activities in an integrated manner on a catchment basis.

### COASTAL EROSION

India has a coastline of 5 700 km, along the east and west (Singh, 1971) where sea and wind erosion cause problems in certain places, especially during the monsoon. The various stretches subject to erosion are shown in Fig. 5. The problem is acute in the north of the east coast and the southern portion of the west. Littoral drift at about 1 million tonnes has been observed to be north-east during the south-west monsoon, and south-west during the north-east monsoon at about 0.25 million tonnes; this leaves a net northern drift of 0.75 million tonnes/year. This information relates to Pondicherry only (Deivasingamany, 1975). The area affected is roughly between 50 and 80 thousand $km^2$ (Anon, 1976b).

### Treatment

About 50 years ago, French engineers, for the first time, built a vertical coastal wall to safeguard the territory in Pondicherry, but even two decades ago, the sea in Pondicherry was about 400 m from its present position. Over the years, the following measures have been taken to check the problem of sea and wind erosion along the coast:

1.  creation of protective sea walls whereever essential;

2.  plantations of casuarina on the sandy foreshores, buttressed by inland plantations of cashewnut, eucalyptus and coconut.

The National Commission on Agriculture (1976) recommended a similar technique.

In certain pockets along the west coast, fertile lands are inundated by sea water; special attention should be paid to the prevention of salt water incursion through appropriates drainage and proper follow-up and maintenance works to enable the growing of profitable crops. A composite treatment for these lands has yet to be evolved.

WIND EROSION

A very sizable portion of India comprising semi-arid and arid regions is subject to varying degrees of wind erosion; the area is shown in Fig. 5 and it includes the desert of Rajasthan, the Runn of Kutch, coastal areas in the south and contiguous areas of Rajasthan and Gujarat. It also extends deep into the Indo-Gangetic alluvial, black soil and southern red soil regions. In the early 1950s, it was estimated that the desert had been encroaching upon adjoining lands in Uttar Pradesh and Haryana at the rate of 0.8 km/year and that for the last 50 years the encroachment was estimated to be 13 000 ha. There may be differences of opinion regarding the marching of the desert, but the damage done by blowing sand-laden winds to the adjoining lands cannot be ignored. The great Indian desert in reality does not fulfil the usual idea of a desert – a great stretch of sand, a tract entirely devoid of streams with few rocks and a large number of sand hills. The Indian desert contains a substantial number of flora; its population, though thin compared to many parts of India, is considerable and it has large herds of camel, sheep, goats, etc. (Blanford cited from Misra 1967). The air over this region contains a quantity of precipitable moisture comparable with that of many other locations in the country with a less severe environment. It could therefore, be said that, apart from the continuous interaction of meteorological, geological and hydrological changes over the centuries, for which evidence is available, the desert condition has definitely been accentuated by the activities of man. The arid land of Gujarat, particularly the Runn of Kutch, on the other hand, has the twin problems of excess salt and an acute shortage of fresh water.

The dispersion ratio for desert soils ranges from 33 to 83 and that for Kutch is 65 (Table 2), which indicates their high erodibility.

Approach to the problem

Considering the inherent limitation of precipitation and high temperature, it may not be possible to develop the greater part of the region to the extent that it can provide a sizable area of intensive agriculture, in spite of the fact that the Rajasthan Canal scheme is bringing a total change in environment to a portion of the desert. For the rest of the area, in the future, the solution needs to be based on principles applied to arid zone development. Any programme planning should take the following into consideration:

    i.    resources in the great Indian desert area must be managed under arid and semi-arid condition and not under the extreme limitation of desert conditions;

    ii.   a livestock-cum-agricultural economy prevailed in the region for ages and still, in spite of the advancement of technology, will hold the key for any sustained solution; therefore, grassland or pasture development will play a very important role;

    iii. the delicate environmental balance needs to be restored and ways and means found to prevent frequent disturbance of this delicate balance.

**145**

Treatment

An integrated desert development technology is being developed by the Central Arid Zone Research Institute at Jodhpur to tackle the problem. The main thrust is on stabilization of sand dunes, afforestation, raising of wind breaks and shelterbelts, and contour bunds and stubble mulch farming. Work on development of water harvesting techniques is also in progress. Among the plants becoming effective in the Rajasthan desert and adjoining areas are Acacia tortilis (Kaul, 1970), Eucaluptus species, Acacia nilotica, Dalbergia sissoo and grasses like Cenchrus ciliaris, C. setigerus Lasiurus sindicus and Dichanthium annulatum (Bhimaya et al, 1968).

The technique of stabilizing shifting sand dunes by afforestation consists of: (i) protection against biotic interference, (ii) treatment of sand dunes by fixing materials in parallel strips or chess board design using local shrub materials and starting from the crest down to the heel of the dune to protect seedlings from exposure or burial, and (iii) afforestation of such treated dunes by direct seeding and planting (Kaul, 1970).

Along the coast in the southern red soil region, there are the problems of coastal sand dunes and considerable wind erosion. Here, sand dune fixation and provision of wind breaks are the common practices. The species usually used are Prosopis juliflora, Azadirachta melia, Acacia nilotica, Casuarina equisetifolia, Jatropha species, etc. (Sivasundaram, 1967).

**146**

BIBLIOGRAPHY

Anon.   Study of soil conservation programme for agricultural land.  Programme Evaluation
1964      Organization, Planning Commission Publication No. 41, New Delhi.

Anon.   Report of the working group on soil conservation for formulation of Fourth Plan.
1966      Ministry of Food and Agriculture, Government of India Pub., New Delhi.

Anon.   Floods and their control in India, Central Water and Power Commission, Ministry
1971a     of Irrigation and Power, New Delhi.

Anon.   An evaluation of ravine reclamation programmes in Uttar Pradesh P.R.A.I., U.P.,
1971b     Lucknow.

Anon.   Report of the Kosi technical committee, River Valley Projects Dept. Patna.
1971c

Anon.   Ravine reclamation programme for dacoit infested areas of U.P., Uttar Pradesh,
1972a     Madhya Pradesh and Rajasthan.  Report of Working Group on Ravine Reclamation,
          Ministry of Home Affairs, New Delhi.

Anon.   Rainfall Atlas of India.  Indian Meteorological Department, New Delhi.
1972b

Anon.   Draft Fifth Plan, Vol. II. Planning Commission, New Delhi.
1974a

Anon.   Crop life saving research procedures, Travelling Seminar, ICAR, New Delhi.
1974b

Anon.   Annual Report.  Central Soil Water Conservation Research and Training Institute,
1974c     Dehra Dun.

Anon.   Interim Report on Desert Development, National Commission on Agriculture,
1974d     New Delhi.

Anon.   Food Statistics, Twenty-fifth Bulletin, Directorate of Economics and Statistics,
1975      Ministry of Agriculture and Irrigation, New Delhi.

Anon.   Eleventh Census, 1972.  Livestock (Information updated to 18.1.1976).
1976a     Directorate of Economics and Statistics, New Delhi.

Anon.   Report of the National Commission on Agriculture, Part V.  Chapter 17 and 18.
1976b     Ministry of Agriculture and Irrigation, New Delhi.

Anon.   Report of the National Commission on Agriculture, Part IX Chapter 42.  Ministry
1976c     of Agriculture and Irrigation, New Delhi.

Anon.   Kosi project and its achievement.  Note prepared by Irrigation and Electricity
1976d     Department, Government of Bihar, Patna.

Ahmad, Enayat.  Soil Erosion in India.  Asia Publishing House, Bombay.
1973

Bhattacharya, J.C.  Erosion studies in lateric areas of W.B.  J. Ind. Soc. Soil
1957      Sci. 5 (2): 103-108.

Bhimaya, C.P., Kaul, R.N. and Ahuja, L.D.  Forage, production and utilization in arid and
1968      semi-arid range lands of Rajasthan.  Proc. Symp. Reclamation and Use of Wastelands
          in India, National Institute of Science, New Delhi, 216–223.

Bhumbla, D.R.  Soil conservation in siwaliks.  Indian Farming;  March, New Delhi.
1976

Bali, J.S.  Soil conservation for the large dam projects in India.  The Harvestor, Vol. II
1969      (i) 31–39.  Kharagpur.

Bali, J.S.  Soil degradation and conservation problems in India.  Presented at the
1974      FAO/UNEP Expert Consultation on Soil Degradation, June 1974.  FAO Rome.

Bali, Y.P.  Hydrologic soil groups of India.  Resource Inventory Centre, Department of
1969      Agriculture, New Delhi.

Bali, Y.P. and Karale, R.L.  Priority delineation of river valley project catchments.  All
1973      India Soil and Land Use Survey, Ministry of Agriculture and Irrigation (Dept. of
          Agric).  New Delhi (Official material).

Ballal, D.K.  A preliminary investigation into some of the physical properties affecting
1956      erosion in M.P.  J. Ind. Soc. Soil Sci. 2 (1): 37–41.

Banerjee, S. and Lal, V.B.  Flood control in India – basic approach to the problem.  Indian
1972      J. Power and River Valley Dev., Oct. 410–416, New Delhi.

Das, D.C., Raghunath, B. and Sreenathan A.  Bench terracing and soil and water conservation
1967      in the Nilgiris.  J. Ind. Soc. Agril. Engrs.  IV(2): 49–59.

Das, D.C., Chandrasekhar, K., Kurian, K. and Thomas, P.K.  Investigations on drought
1971a     from various climatic parameters at Ootacamund, J. Ind. Soc. Agril. Engrs. VIII(4):
          13–23.

Das D.C., Raghunath, B. and Poornachandran, G.  Comparison of the estimated spacings of
1971b     bunds and terraces by modified Stewart formula (Ootacamund) with a few existing
          formulae for few places.  The Harvester XIII:   33–39 I.I.T., Kharagpur.

Das, D.C., Tejwani, K.G. and Ram Babu.  Hydrological investigations at ICAR Soil Conservation
1973      Reserach Demonstration and Training Centres (International Hydrologic Decade),
          No. 22, October.  New Delhi.

Das, D.C., Bali, J.S. and Hupta, R.P.  Delineation of India by climatic crop growth indices
1974      into various drought prone areas for conservation planning.  Presented at XIII
          Annual Convention of ISAE, Allahabad.

Das, D.C.  Terracing – definition, classification and suitability as observed in India.
1974      Soil Cons. Digest 2(2): 30.  Association for Soil and Water Conservationists and
          Trainees (ASWCT) Dehra Dun.

Das, D.C.  Design of narrow base terracing (bunding).  Soil Cons. Digest 4 (2):
1976      ASWCT, Dehra Dun (in press).

Das, D.C. and Kaul, R.N.  Centrally sponsored schemes of soil conservation in river valley
1976      catchments.  J. Inf. Doc. Agril. Engrs. 'Today' I(ii).

Deivasigamnym J.  Problem of sea erosion, flooding and salt water intrusion in Pondicherry.
1975      Presented at Technical Discussion Watershed Management, Bangalore.

George, C.J., Rama Sastri, K.S. and Rentala, G.S.  Incidence of droughts in India.
          Agreement No. 5 IMD Pune.

Gupta, G.P. Sediment production – States report on data collection and utilization. Soil
1975    Cons. Digest 3(2): 10–21 ASWCT Dehra Dun.

Gupta, R.S. Investigation on the Desert soils of Rajasthan: fertility and mineralogical
1958    studies. J. Ind. Soc. Soil Sci. 5: 115–120.

Gupta, S.K. and Dalal, S.S. Gabion structures for soil conservation. Indian Forester 93
1970    (6): 383–392, FRI. Dehra Dun.

Gupta, S.K., Tejwani, K.G., Mathur, H.N. and Arivastava, M.M. Land resource regions and
1970    areas in India. J. Ind. Soc. Soil Sci. 18 (2): 187–198.

Kaith, D.C., Khan, M.H., Kalamkara, R.J., Riaz, A.G., Gandhy, D.J., Sen A.T. and Roy
1948    Chaudhry, S.P. A soil conservation and land utilization programme for
        India (Report).

Kamnavar, H.K., Singh, A. and Dayal, R.B. Maintenance of soil conservation measures in
1975    ravine control works. Soil Conservation Digest 3 (2): 53–58, ASWCT Dehra Dun.

Kaul, R.N. Afforestation in Arid Zones. Indo–Pakistan. Dr. W. Junk (Ed). N.V. Publishers,
1970    The Hague.

Khosla, A.M. Appraisal of water resources: analysis and utilization of data. Paper
1949    presented at the United Nations Scientific Conference on the Conservation and
        Utilization of Resources.

Krishnamoorthy, CH., Choudhury, S.L. and Spratt, E.D. New horizons for dry farming in
1974    the rainfed tropics – Results of AICRP on dry land agriculture. Presentation
        FAO/UNDP Expert Consultation Meetings, Hyderabad.

Kurian, K., Poorna Chandran, G., Das, D.C. and Lakshmanan, V. Longitudial grade for
        bench   terraces at Ootacamund J. ISAE. XII (3 and 4): 27–28.

Lakshmipathy, B.M. and Narayanswamy. S:1 Bench Terracing. J. Soil and Water Cons.
1956    in India. 4(4): 161–168.

Mehta, K.M., Mathur, C.M., and Shankarnarayana, H.S. Investigation on the physical
1958    properties of Kota soils in relation to their erodibility. J. Ind. Soc. Soil
        Sci. 6: 227–253.

Misra, V.C. Geography of Rajasthan. National Book Trust, India, New Delhi.
1967

Murthy, V.V.N. Stone terracing of hill slopes. J. Soil and Water Cons. in India.
1969    17 (3 and 4): 35–38.

Palit, B.K. and Kapur, O.P. Erosion control along railway lines. Presented at XIII
1975  ·  Annual Convention ISAE, Allahabad.

Pathak, S. Gully and stream erosion control. Presented at Technical Discussion,
1975    Watershed Management, Simla.

Raghunath, B., Sreenathan, A., Das, D.C., and Thomas, P.K. Conservation evaluation of
1967    various land use practices on steep to moderately steep land in the Nilgiris.
        Part I and II. Presented at VI Annual Convention of ISAE, Bangalore.

Raghunath, B., Das, D.C. and Thomas, P.K. Some results of investigations on hydrology of
1970    the sub-watershed in the Nilgiris (India). Symp. Results of Research on
        Representative and Experimental Basins, Wellington (NZ) IASI – UNESCO Pub. 96: 3.118.

**149**

Rama Rao, N.S.V. Soil conservation in India. ICAR, New Delhi.
1972

Rege, N.D. and Yash Pal. Ravine reclamation – Review of approach. Ministry of Agriculture
1973    (Dept. of Agri.), New Delhi.

Roy, K. and Mishra, P.R. Formation of Chambal ravines. Indian Forester. 95(3): 160–164.
1969    FRI, Dehra Dun.

Satpute, R.V. Analysis of the types of soil conservation works so far done in the country
1972    under different soil and agro–climatic conditions including hilly areas.
        (Unpublished report).

Shah, R.L. and Patel G.A. Measuring benefit due to bunding in Panch Mahal Dist. of
1970    Gujarat State of India. J. Agro. 15(4): 410–414.

Shamra. P.N. and Kampen, J. Small runoff storage facilities for supplemental irrigation –
1975    some preliminary observation. ICRISAT, Hyderabad.

Shri Niwas, Das, D.C. and Kakshmanan, V. Studies on rainfall behaviour in the Nilgiris.
1967    Presented at VI Annual Convention ISAE, Bangalore.

Singh, B., Mathur, H.N. and Gupta, S.K. Vegetative and engineering measures for torrent
1971    training and stream bank protection in Doon Valley. Indian Forestor 97(1):
        47–54.

Singh, R.L. (Ed). India: A Regional Geography. National Geographical Society of India,
1971    Varanasi.

Singh, D.P. and Srivastava. R.P. Cumulative time deposit by planting trees and grass on
1973    contour bunds. Soil Cons. Digest 1(i): 56–58. ASWCT Dehra Dun.

Sivasundaram, T. Soil erosion by wind in Madurai District. J. Soil and Water Cons.
1967    India 15(3 and 4): 43–47.

Subba Rao, K.V. Silt studies of Wizam Sagar Project. PWD, Government of Andhra Pradesh,
1974    Hyderabad.

Swaminathan, M.S. Agriculture on space ship earth. Coromandel Lecture (3).
1973a    26 February 1973.

Swaminathan, M.S. Our agricultural future. Sardar Patel Memorial Lecture. AIR. India
1973b    International Centre, New Delhi.

Tejwani, K.G., Gupta, S.K. and Mathur, H.N. Soil and water conservation research
1975    (1958–71). ICAR, New Delhi.

Vandersypen, D.R., Bali, J.S. and Yadav, Y.P. Handbook of Hydrology. Central Unit
1972    (Soil Cons.) Ministry of Agriculture, New Delhi.

Vohra, B.B. Land water management problems in India. Training Div., Dept. of Personnel
1975    and Admn. Reforms, Cab. Sect. New Delhi.

Wadia, F.K. Director, Economics and Statistics, North Eastern Council Secretariat,
1975    Shillong.

Reprinted from GeoJournal **4**:279–287 (1980)

## On the Soil Erosion Problem in the Temperate Humid Area of Central Europe

Richter, G., Prof. Dr., Geogr. Inst., Universität Trier.
POB 3825, D-5500 Trier, FR Germany

### Introduction

Soil erosion in Central Europe is generally not considered to be a serious problem. In comparison to the severely eroded areas in more continental climatic regions or in tropical highlands this viewpoint is correct, especially at the macro level. However, one must take into account that the Central European cultural landscape has been densely settled for a long time and thus subject to anthropologically related soil erosion processes and superficial changes for hundreds if not thousands of years. In this sense soil erosion in Central Europe must be seen as an amalgam of present and historical factors.

### The Role of Soil Erosion in the Evolution of the Cultural Landscape of Central Europe

In the last decades a number of studies from various areas of Central Europe have shown that soil erosion has played an important role in history. This evidence is obtained from archival and geomorphological field work. The oldest cultural example of water erosion is provided by Scheffer & Meyer (1958). In a loess slope near Göttingen they discovered neolithic diggings from several time periods which had been filled up through erosion. Similar findings from the same area for the Roman time period are reported by Meyer & Willerding (1961).

Correlative fluvial deposit were searched for by Wildhagen & Meyer (1972) in a 20 km long section of the Leine Valley near Göttingen. The dating of the fluvial loam indicates that soil erosion in the neolithic times was limited. It is in the iron age (about 800 BC to 600 AD) that the fluvial accumulation begins. In the last 2000–2500 years, this phase of fluvial loam accumulation has raised the floodplain an average of 124 cm with a maximum of 210 cm. Such an increase in the plain level would require a sediment load of circa 25 mio m$^3$ for the 20 km section studied.

It is evident that such large quantities of fluvial deposits could only be the result of large scale erosion in the cleared areas of the river basin. Bork & Rohdenburg's (1979) work in the Untereichsfeld area makes this point clear. In the presently forested areas which arose after the abandonment of settlement at Tiedershusen (Fig 1) in 1465, the following series of events took place: after the land clearing process whose exact time period is unknown, the old and middle Holocene para-brown earth was largely removed and replaced by Colluvium. Potsherds discovered in such fields were dated as belonging to the time period (13. to the middle of the 15. century). Then the settlement was abandoned and the fields left to forest with the Colluvium changing into para-brown earth. The furrows between the formerly cultivated plots were in part further eroded and filled a new with Colluvium.

Thus there occurred not only reduction of the natural soil profiles of earlier times, but there was also a change in profiles types and profile renewal due to the influence of man. Therefore soil profiles especially in loess and red sandstone regions, even in forested areas, have little in common with precultivation soils.

One should, however, not conclude that the expansion of erosion in earlier times is only correlated with the expansion and/or more intensive utilization of agricultural lands. A series of studies have shown that it is often in the less intensively cultivated fields and in plots between cultivation periods that intensive gullying occurs (for example Richter & Sperling, 1967). A dramatic increase in soil erosion damage arose in previous eras when under population pressure, land cultivation was extended to steep slopes, plot rotation was changed for the worse through new agrarian systems and hilling practices or when new crops or crop rotations were introduced.

Thus one can speak of soil erosion cycles which was initiated through one of the above mentioned practices and stopped when land use was better related to local soil conditions. The memory of such soil erosion cycles usually disappeared after a few generations. It was therefore possible that in Central Europe agricultural mispractices consistently arose with consequent effects. Evidence of such cycles can be found in the works of Hard (1965, 1968, 1970), Hempel, Lena (1954, 1957), Herz (1964), Käubler (1937, 1938, 1952), Klug (1964), Linke (1963), Richter & Sperling (1967), Schultze (1965), Vogt (1953, 1958 a–b) and Wagner (1961, 1965).

With regard to wind related soil erosion similar findings are known. Examples of wind erosion from as early as the first centuries AD have been documented (Richter 1965, pp. 25–27, 1967). In the old moraine regions of Northern Germany, it was the heath areas which were subjected the most to wind erosion. The natural climax vegetation was devastated through the removal of ground cover for fertilization purposes, sheep grazing and forest clearing. Consequently the periglacian sands were blown to dunes or 'wailing sands'. The recultivation of such loose sandy areas occured mostly only through state projects. From archival sources Iwersen (1953), v. Gehren (1952) give evidence of such state undertakings. In 1792 the Danish King set out to reduce the wind blown sands through a series of counter-measures. The revegetation of blowout areas and dunes was ordered as well as the establishment of wind breaks. Edicts were ordered against grazing, burning and the removal of groundcover.

In the case of winderosion one can also identify destructive cycles. One such destructive period set in at the end of the 18. century as the result of overpopulation and a wool based economy. An overgrazing and overutilization of the heatherlands occurred. A second destructive wave set in with the cultivation of light poor sandy soils after WW I. The same process was repeated after WW II in Niedersachsen and Schleswig-Holstein. Here large areas were deforested for agricultural purposes concomitant with the recultivation of much pastureland and the introduction of row crops. Only after the checking of such practices and the successful establishment of hedges were wind erosion damages reduced.

The study of present soil erosion in Central Europe should only be carried out when local variations as well as the historical factors are included.

### The Regional Aspect

The regional differentiation is examined on the basis of five landscape types between the North Sea and the Alps, Fig 2. The soil erosion maps done for the German Democratic Republic by Flegel (1958) and Kugler (1976) and the Czechoslovakian erosion map prepared by Stehlik (1975) agree with Richter's work in principle.

Type I stretches from the North Sea to the Baltic and covers Schleswig-Holstein. This area can be subdivided into: marsh areas and moors along the west coast, the Elbe valley and Eider-Treene lowlands, old glacial moraine area into which Würm sanders were deposited, and the young (Würm) moraine area along the Baltic. The largest region affected is the old moraine landscape with the Würmian sanders. Large superficial areas were covered with glacio-

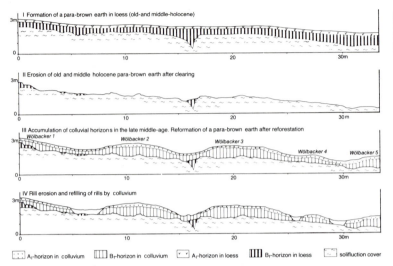

Fig 1    Profiles Tiedershusen near Göttingen: Soil- and relief-formation phases under the influence of soil erosion (Bork & Rohdenburg p. 124, 1979)

fluvial sands. In other old moraines large amounts of windblown sands are found, some of which are of periglacial origin, other portions are wind eroded during the last 2000 years.

In March, April and May this area is characterized by low rainfall, large amounts of sunshine and strong winds. As the fields have little vegetation cover at this time, winds exceeding 5 Beaufort create extensive blowout damage. Areas affected by such conditions in the 1950's have been estimated to cover 37.000 ha (Iwersen 1953, 1955). Hassenpflug (1971, 1972) has analyzed this wind erosion problem through airphoto interpretation and site analysis (Fig 3). The air photo displays the sandfans in the lee of eroded fields. Hassenpflug reports that the damages has been increased in the 1970s. Since the program "Nord" has been implemented, large areas have been recultivated by drainage. Windbreaks were still not thick and high enough to prevent heavy damages. Since then the situation has improved.

The young (Würm) moraine areas are composed of normal to sandy loam deposited on moraines. On large fields of this hilly land water erosion is widely established. Short and moderate inclined slopes (under 6°) hinder extensive soil erosion. Hassenpflug (1971) has estimated that over the last 200 years some 4 cm of the soil profile have been removed. In the old moraine areas of Holstein similiar rates of erosion have occurred as in type 2. The marshy regions along the west coast are not endangered.

Landscape type 2 (Fig 2) includes the old Riss moraines of NW Germany. This landscape is subdivided into the sea marches along the coast, in the flat moraine plains (Geest) between rivers Ems, Weser and Elbe, and further east the moraine area of the Lüneburg Heath which belongs to a younger glacial period. In the entire region the problem of wind erosion dominates, although to a lesser extent than in Schleswig-Holstein. However, the higher sandy moraine plains near the German Bight and the Hümmling Hills are threatened by blowouts. Here in the 1940's and 50's extensive damage resulted from deforestation and expansion of cultivated areas. v. Gehren (1954) has computed on the basis of erosion reports that in 1949 some 11 % of the cultivated land in the districts of Stade, Oldenburg, Aurich and Osnabrück suffered from wind erosion. This was verified by Grosse (1953, 1955) through map analysis and ground surveys of large blowout areas. Reforestation and the planting of hedges seem to improve the situation.

The Lüneburg Heath with its more hilly relief, higher share of loamy soils and larger forest cover has had little damage. In landscape type 2 are, however, widespread sandy loess islands near Cloppenburg, Syke and Lüneburg. In spite of the low relief energy considerable water erosion has been noticed by Grosse (1955) and Flohr (1952).

Landscape type 3 (Fig 2) includes the N portion of the Central European Hills (Mittelgebirge). It stretches from the Münster flats, through to the Weser Uplands, and the Harz. With the exception of the Harz, the Uplands are composed of a variety of mesozoic sediments. They form ridges. Faulting, folding, lifting and sinking have modified the relief. There are widespread deposits of loess covering the rolling landscape along the N rim of the Mittelgebirge, the "Börden". The loess has mainly accumulated in river valleys and tectonic basins. All loess areas are intensively utilized for sugar beets and grain. Relief, soil structure and landuse have made this region one of the most susceptible to water erosion all over Germany (Fig 4). (Grosse 1955, Hempel 1954, 1958, 1963; Hempel, Lena 1957, Richter 1963, 1972).

Landscape type 4 is comprised in part of the Rhenish Slate Plateau (from 400 to 700 m asl). These tertiary peneplains are deeply dissected by the Rhine, Mosel and Lahn river systems which have formed steep slopes and notched valleys. The tributary valleys gradually penetrate into the tertiary peneplains. With the exception of the loess filled basins in the E portion of the region, the soils on greywacke- and slate are sceletal brown earth. The region falls into two climatic zones. The plateaus and ridges are influenced by maritime factors while the valleys are more continental. With increasing elevation landuse shifts from cultivation to grazing and forestry. In the loess basins, cultivation was intensified instead, and thereby soil wastage has increased.

On the slopes, the traditional grape cultivation has increased especially in the Mosel-region. By enlarging vineyards and in order to mechanize grape production, many terraces, rocks and hedges were destroyed, and thereby erosion was extended. After summer thunderstorms channel and gully erosion has noticeably increased. In winter soil sliding and creeping occurs and landslides become more frequent (Fig 5) (Richter, 1978).

The above erosion process has been studied and measured at the University of Trier's soil erosion research station for more than

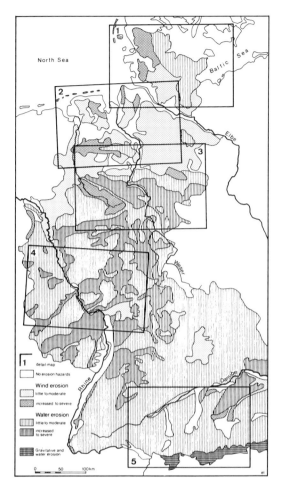

**Fig 2** Federal Republic of Germany: Average Hazards of Soil Erosion (Richter 1978, p. 148)

five years. The results indicate that even in the skeletal shale soils on slopes which are relatively resistant to erosion remarkable soil losses of 0.5 – 2 t/ha/year occur. Above all it is the fine material that is eroded (Richter & Negendank, 1977). At present soil erosion on the steep slopes of the Mosel valley is a serious problem.

Similarly the primarily loess covered Rhine-Hessian hills in the south of the Rhenish Slate Mountains are severely eroded (Gegenwart 1952, Ruppert 1952, Leser 1965 and Richter 1972).

The last landscape type (Fig 2) in its W section comprises the foreland of the Bavarian Alps and the limestone alps. The lower Bavarian hills (Niederbayern) only suffer from erosion to any great extent. This latter section due to a composition of tertiary sands and loess, is susceptible to washouts and furthermore is widely cultivated. The remaining portion of the foothills is composed of glaciofluvial remnants (Schotterplatten) which are more resistant to water erosion. The young moraine areas are due to high rainfall,

used for grazing and, therefore, protected. With regard to wind erosion, well drained cultivated moor fields along the Danube and Isar are susceptible.

In the Alps arable land is negligable as compared to grazing and woodland. Erosion is the joint effect of gravity and water on over-grazed slopes. Rain and snow melt-water seeps into the soil. Clay and marl soils may become waterlogged. This results in mudslides and slope breakaways (Grottentaler & Laatsch 1973, Karl & Danz 1969, Laatsch 1971, 1974, Laatsch & Grottentaler 1973, Schauer 1975 and Wendl 1969). In general, because of the above conditions many of the Alpen slopes are unstable. Any economic activity must try to maintain a permanent forest and grass cover. Otherwise gullying cannot be kept under control.

In summary, this brief traverse from the North Sea to the Alps demonstrates that a substantial variation in soil erosion exists in Central Europe.

Jung (1956) has studied the relationship between crops and water erosion in selected landscapes within Germany. His studies indicate that slopes have, due to erosion, a 10–50 % lower yield as compared to plains.

Soil erosion in Central Europe can be summarized as follows: The erosion cannot be classified as massive wastage. However, conservation effort should relax, otherwise serious erosion problems would arise. Modern agriculture practices are not carried out without any risk. One should, however, be able to calculate the extent of such risks. For that purpose more knowledge on quantifying soil erosion is required.

## Quantification Problems

At the qualitative level, it is possible to estimate the influence of various factors on soil erosion. It is difficult, however, to measure precisely and compare the impact of such factors. Although both simple and complex erosion models do exist, such as those developed in the USA (Wischmeier & Smith, 1961), to measure major erosion factors and provide estimates of annual rates of erosion, such models are not completely transferable to Central Europe. Before such models can be applied here, one requires model calibrations especially with regard to rainfall (R-factor) and soil (K-factor) based on Central European conditions.

The erosion index of rainfall (EI), whose annual value is called R-factor, is defined as the product of the kinetic energy revealed through the maximum thirty minute intensity of precipitation per hour, divided by one hundred. This index may be calculated from the quantity and intensity of precipitation for every period of rainfall. Such measurements can only be obtained through a system of automatic registrations by recording precipitation gages. The lack of such a net of recorders has prevented the application of such models to Central Europe. To overcome this data handicap other measures have been applied. A common procedure is the use of precipitation data available and to map the frequency with which daily averages are exceeded (Gegenwart 1952, Hartke & Ruppert 1959, Kern 1961, Masuch 1970). Richter (1965), for example, has calculated the number of days which exceed 10 mm precipitation of the annual average on the basis of an empirical formula developed by Johannsen (1952). Bermanakusumah (1975) attempted to estimate the R-factor with another formula, using the average annual precipitation including snowfall. So far all attempts to calculate the R-factor were made without intensity measurements.

To the author's best knowledge, the first attempt to calculate the R-factor according to Wischmeier based on European data over many years is the work by Laurant & Bollinne (1976, 1978). They utilized the following formula adapted from Wischmeier and transformed into the metric system:

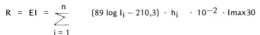

Fig 4    Water erosion caused by a heavy thunderstorm near Hildes-
heim in the "Börde"-loess region

Fig 5    Creeping and sliding of topsoil in vineyards near Trier caused
by snow-melting

$$R = EI = \sum_{i=1}^{n} (89 \log l_i - 210,3) \cdot h_i \cdot 10^{-2} \cdot I_{max30}$$

$l_i$ = intensity of rainperiod expressed in cm/hr.

$h_i$ = amount of precipitation in the period given in cm

$I_{max30}$ = maximum intensity during a 30 minute period expressed
in cm/hr.

Calculation of the average R-factors was based on a period of
20 to 40 years for three Belgian stations Uccle, Saint-Hubert and
Spa (Laurant & Bollinne, 1978). Results from all three stations

indicate that during the summer months, the EI value is greatly increased, This can be attributed to the greater amounts of precipitation ($\geq 12,7$ mm) in summer (Fig 6). The comparison of average
rainfall with the mean R-value also indicates that the rain-sum is by
itself not sufficient to estimate the R-factors.

| station | mean annual precipitation (mm) | mean value EI |
|---|---|---|
| Uccle | 835 | 64,9 |
| Saint-Hubert | 1071 | 123,1 |
| Spa | 1114 | 142,7 |

(adapted from Laurant & Bolline, 1978)

**154**

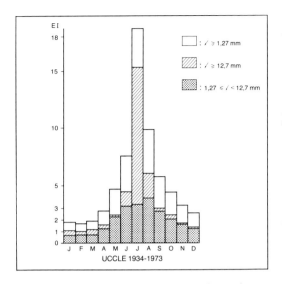

**Fig 6** Average monthly values of EI-units at Uccle (Belgium) during thirty years (Laurant & Bollinne 1978, p. 226)

Measurements taken at the University of Trier's research station at Mertesdorf agree with the observation that the EI values vary throughout the year. Although rainfall frequency and quantity as well as runoff are relatively equal throughout the year, soil losses are high only in summer (Fig 7). The question of soil wastage will be examined more closely as more data are forthcoming. The EI values from Mertesdorf will be correlated with surface runoff and soil losses for 13 measurement plots. Now only a few data are available:

Station Mertesdorf, plot 132 (2 x 32 mm, ∢ 21°), vineyard

| half year | $P_{mm}$ Precipitation | EI after Wischmeier (metric system) | runoff (l) | soil loss kg/ha |
|-----------|------------|------------|--------|---------|
| winter 74/75 | 316 | 9 | 203 | 44 |
| summer 75 | 290 | 45 | 210 | 474 |
| winter 75/76 | 143 | 2 | 61 | 5 |
| summer 76 | 165 | 6 | 77 | 13 |
| winter 76/77 | 312 | 13 | 65 | 9 |
| summer 77 | 358 | 60 | 182 | 50 |
| 3 years | 1584 | 134 | 798 | 595 |

Similar problems arise when the soil factor K is used. A simple procedure to estimate K-factors based on five soil characteristics was developed by Wischmeier, Johnson & Cross (1971). This procedure is most appropriate for the American Midwest (Wischmeier, 1976).

In Central Europe calculation of soil erodibility was attempted by analysis of soil erosion forms, and by the examination of soil profiles on slopes. A qualitatively derived rank order of erodibility was thus achieved. These rank orders have been summarized by Schmidt (1979).

Richter (1965) on the basis of 98 slope profiles has gained similar results. The average soil resistance to water erosion is related to the average decrease in profile depth per degree of slope. The

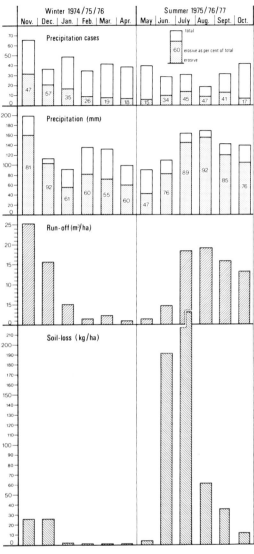

**Fig 7** Soil erosion measurement station Mertesdorf, plot 132: Added monthly sums of precipitation, run-off and soil-loss, Nov. 1974 — Oct. 1977 (Richter, in print)

above relationship only produced a ranking of soil resistance and not a measurement for soil loss.

Becher, Schäfer, Schwertmann, Wittmann & Schmidt (1977) have calculated the soil K-factor for Hallertau using Wischmeier's method. They were, however, unable to relate these K-estimates accurately to the susceptibility to erosion. Plot-measurements are needed to calibrate the K-values. This is the crux of the problem. It is relatively easy to adapt American soil loss values to Central

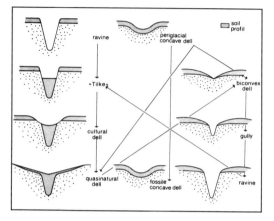

**Fig 8**   Dell-forming in cultural landscapes
(after Richter 1976. Fig 4)

European research. It is, however, impossible to know the significance of such data without parallel empirical measurements.

The testing and establishment of R- and K-factors on sites in Central Europe is therfore and indispensible prerequisite for the application of soil loss models, while the LS-factor (slope lenght and steepness), the C-factor (landuse) and the P-factor (conservation measures) are directly transferable. Such tests are among the major future investigations in quantifying the washing away process in Central Europe.

A second task is the typing together of results from plot measurements with soil profiles and forms of erosion in the field. Soil profiles are hardly appropriate for direct comparison with soil loss from sample plots, as they mirror the overlapping results of centuries of erosion often under different land use conditions. Unfortunately one still does not understand fully the weathering intensity and the profile structures at the beginning of any given land use.

The possibilities of relating plot measurements with actual soil loss nearby are somewhat better. Schmidt (1979) has suggested a complex three stage system of measurement and mapping whereby the following can be done:

1)   plot measurements
2)   simultaneous measurements carried out in nearby fields with sediment traps (with "Feldblechen")
3)   at intervals repeated mapping and measurement of the erosion forms according to the calculation procedures set out by Flohr (1962).

Furthermore one should also measure the pebbles, suspended matter and matter in solution, that are transported from the watershed. Thus the problem. of comparing runoff load in the valley streams with data from plot and field measurements in the same basin is resolved, and leads one step closer to the extrapolation of data from research plots.

## Soil Erosion and the Natural Equilibrium of Landscape

Soil erosion should not be seen as an isolated phenomenon but rather be understood as an integral part of the cultural landscape. The question arises for the relationship between normal erosion or geologic erosion and soil erosion. To what extent has man through

his impact on vegetation and through land use affected the cycle of weathering and removal, and thus affected the soil and water balance? This question is also related to man's impact on geomorphologic features. The second aspect deals with the relationship between soil erosion and total denudation (including solution rates). This means to what extent in Central Europe has the impact of man affected the environment of a drainage basin. This includes the growing problem of water eutrophication due to soil erosion.

The question of the relationship between soil erosion and geologic erosion has been theoretically formulated by Mortensen (1954/55). Mortensen understood soil erosion as a quasi-natural process. Although man controlls the extent to which erosion occurs, this process is subdued to natural laws. Man may intensify such natural processes indirectly. It is possible, however, that new removal mechanisms may arise which could not have occurred under natural conditions. Soil erosion is therefore the difference in soil removal between natural and quasinatural conditions and processes. A sensitive indicator of the fact that soil erosion is more than a natural removal process is the soil profile and soil type. This is especially evident in culturlandscapes with loess soils. Fig 1, based on Bork & Rohdenburg (1979) demonstrates how para borwn earth is replaced by colluvial accumulations. In extreme cases thin soil profiles are developed with a light humous A horizon on top of unweathered loess. This is a backward development in soil type to "Pararendzina". Generally brown, parabrown and black soils are converted to cultivated brown earth types, however under strong soil removal conditions the above types are converted often to rankers. The process whereby rendzina evolves into Braunerde-Rendzina or Terra Fusca may also be reversed by soil erosion. On the other hand at the foot of slopes the buildup of large colluvial and often pseudogleyed profiles of macadamized colluvium is frequent.

The extent of such profile transformation has been shown by Richter, Müller & Negendank (in press). Their mapping of a 32 km2 area of the Rhenish Slate Plateau near Trier 1:10,000 produced the following findings:

18 % of the mapped area show none to light evidence of erosion
35 % of the mapped area show slight to medium profile shortining
23 % of the mapped area have strongly eroded profiles (20–40 cm)
12 % of the mapped area have extremely eroded profiled (<20 cm)
13 % of the mapped area show colluvial and pseudogleyic profiles

In total some 50 % of the area have been changed through soil erosion.

Evidence that soil erosion caused by land use overwhelmes the natural process is also provided by various landform patterns. That is, especially in the border regions between areas which have been in pasture or woods for a long time and arable fields. For example, a slope which is forested on the upper reaches and arable downhill displays the arable surface several meters lower than that of the woods. This difference in elevation is the result of a more rapid erosion process due to cultural factors. If the pattern of land use is reversed a zone of earth accumulation can be observed at the borders of cultivated and non-cultivated land. The accumulation zone may also count for meters. Investigation of such accumulation or erosion zones permits to estimate the volume of soil removal per a certain period of time. Wandel & Mückenhausen (1949) have estimated that at the borders of different land use some 50–70 cm of the soil profile being eroded within 50–55 years. This corresponds to 1 m / 100 yr. Likewise hedges accumulate soil and contribute to the formation of agricultural terraces.

Soil erosion process and the volume eroded over a long period can also be calculated from fluvial deposits in valleys. On this basis Wildhagen & Meyer (1972, p. 122) have calculated that over 300 years some 4 mio m3 of soil were deposited in the Leine valley. To attain this volume some 1.1 cm must have been removed through-

**156**

out the drainage basin. It was further estimated that in this time (600—900 AD) some 25 % of the study area was cultivated. Thus the overall rate of erosion for cultivated areas may be 4 cm/300 yr. In comparison with normal erosion processes which erode the landscape at the rate of 4—8 cm/1000 yr., the Leine valley represents a substantial increase as the result of agriculture. Such volumes of depositional matter over time have modified, to some extent, relief in culturlandscapes. Linke (1963) attempted to categorize such anthropogenic land forms. Sperling (1978) has systematized such land forms and pointed out, that not all of them were formed through semi-natural processes. Some of them were primarily the result of man's activities especially of plowing methods.

Another common feature in this context is the turn about the drainage pattern. Initially the runoff was directed to depressions in the preexisting periglacial surface like dells. Additional channels were then created by field roads. The development of drainage patterns is demonstrated by Richter (1976), Fig 8. Gullies if overgrown were at times filled up again. The stage when boxlike valleys are formed by soil accumulation in gullies is called Tilke. Additional soil accumulation changes the Tilke into dells which strongly resemble their genuine periglacial forms. Man has played an important role in filling up the Tilkens.

Soil removal by water happens by suspension and by solution. Verstappen (1977, 1978) on the basis of measurements in a small forested drainage basin in the Oesling (Northern Luxemburg) found that some 62 kg/ha was carried away annually in solution and this was the most significant type of denudation. Completely different results were obtained from studies carried out at Mertesdorf near Trier. The origin and soil profiles of the two areas are similar with the exception that the slopes are somewhat steeper at Mertesdorf (20°) and covered with vineyards. At Mertesdorf the findings indicate that annually some 520 kg/ha of soil were removed mechanically and some 18 kg/ha by solution.

Another significant component of soil erosion has been mentioned by Schwille (1973) in his report on chemical pollution of surface and ground waters in the Mosel valley: "The high nitrogen values are directly attributable to intensive application of chemical fertilizers to vineyards". Similar works by Bernhardt, Such & Wilhelms (1969), Wiechmann (1973), Wohlrab, Wenzel & Mollenhauer (1974), Süssmann & Wohlrab (1975), Walther (1976) and Sokollek (1976/77) indicate that at present one must calculate at least temporarily with significant amounts of fertilizers in rivers. However, the role of fertilizers carried away from arable land in the eutrophication of water networks may be smaller than that resulting from urban and industrial wastes. At this point not a sufficient amount of research has been made in order to understand more fully the role of erosion in removing fertilizers.

The role of soil erosion in the complex household of a landscape has not been exactly quantified in Central Europe.

## Conclusion

Previous research on soil erosion in Central Europe was mainly focused on historical aspects, with soil erosion forms and their regional setting. It is now time to utilize a more quantitative approach. Furthermore in the future, soil erosion should be seen in the context of the entire equilibrium of landscape systems, that means as part of economic-ecological research.

## Acknowledgement

Thanks are given to Prof. Dr. W. Hassenpflug (Kiel), to Prof. Dr. H. Rohdenburg (Braunschweig), to Dr. R. G. Schmidt (Basel) and to Dr. J. Slupik (Krakov). They supported my work by sending figures and photographs and by giving additional information. Furtheron it is a pleasure to acknowledge the help of Prof. Dr. G. Romsa (Windsor) especially for translation and improving in style.

## References

Becher, H.H., Schäfer, R., Schwertmann, U., Wittmann, O. & Schmidt, F.: Bestimmung der Erosionsanfälligkeit nach Wischmeier an hopfengenutzten Böden der Hallertau. Zeitschr. f. Kulturtechnik und Flurbereinigung 18, 339—349 (1977)

Bermanakusumah, R.: Untersuchungen über Bodenverlagerung und Erodierbarkeit einiger Mittelgebirgsböden Hessens. Dissertation Gießen 1975.

Bernhardt, H., Such, A. & Wilhelms, A.: Untersuchungen über die Nährstofffrachten aus vorwiegend landwirtschaftlich genutzten Einzugsgebieten mit ländlicher Besiedlung. Münchener Beiträge zur Abwasserwirtschaft, Fisch- u. Flußbiologie 16, 60—118 (1969)

Bork, H.-R. & Rohdenburg, H.: Beispiele für jungholozäne Bodenerosion und Bodenbildung im Unter-Eichsfeld und Randgebieten. Landschaftsgenese und Landschaftsökologie 3, 115—134 (1979)

Flegel, R.: Die Verbreitung der Bodenerosion in der Deutschen Demokratischen Republik. Bodenkunde und Bodenkultur 6, Leipzig (1958)

Flohr, E.F.: Bodenzerstörungen durch Frühjahrsstarkregen im nordöstlichen Niedersachsen. Göttinger Geographische Abh. 28, Göttingen 1962.

Fournier, F.: Climat et erosion. Paris 1960.

Gegenwart, W.: Die ergiebigen Stark- und Dauerregen im Rhein-Main-Gebiet und die Gefährdung der landwirtschaftlichen Nutzflächen durch Bodenzerstörung. Rhein-Mainische Forschungen 36, Frankfurt 1952.

v. Gehren, R.: Das Auftreten von Bodenverwehungen in früheren Jahrhunderten und ihre Bekämpfung. Umschaudienst des Forschungsausschusses Landschaftspflege und Landschaftsgestaltung d. Akad. f. Raumforschung u. Landesplanung 213—218, Hannover (1952)

v. Gehren, R.: Die Bodenverwehungen in Niedersachsen 1947—1951. Veröff. d. Niedersächsischen Amtes f. Landesplanung u, Statistik, Reihe D 6, Hannover (1954)

Grosse, B.: Untersuchungen über die Winderosion in Niedersachsen unter besonderer Berücksichtigung der Cloppenburger Geest und des Hümmlings. Forschungs- u. Sitzungsberichte der Akademie für Raumforschung und Landesplanung Hannover 2, 180—250 (1953)

Grosse, B.: Die Bodenerosion in Westdeutschland. Ergebnisse einiger Kartierungen. Mitteil. aus dem Institut für Raumforschung Bonn 11, Bad Godesberg (1955)

Grottenthaler, W. & Laatsch, W.: Hangerosion im Laintal bei Benediktbeuren. Forstwiss. Centralblatt 92, 1, 1—19 (1973)

Hard, G.: Zur historischen Bodenerosion. Ztschr. f. d. Geschichte der Saargegend 15, 209—219 (1965)

Hard, G.: Grabenreißen im Vogesensandstein. Rezente und fossile Formen der Bodenerosion im mittelsaarländischen Waldland. Berichte zur dt. Landeskunde 40, 1, 91—91 (1968)

Hard, G.: Exzessive Bodenerosion um und nach 1800. Erdkunde 24, 291—308 (1970)

Hartke, W. & Ruppert, K.: Die ergiebigen Stark- und Dauerregen in Süddeutschland nördlich der Alpen. Forschungen zur dt. Landeskunde 115, Remagen 1959.

Hassenpflug, W.: Studien zur rezenten Hangüberformung in der Knicklandschaft Schleswig-Holsteins. Forschungen zur dt. Landeskunde 198, Bonn-Bad Godesberg 1971.

Hassenpflug, W.: Formen und Wirkungen der Bodenverwehung im Luftbild. In: Hassenpflug, W. & Richter, G.: Formen und Wirkungen der Bodenabspülung und Verwehung im Luftbild. Landeskundliche Luftbildauswertung im mitteleuropäischen Raum 10, 43–82 (1972)

Hempel, L.: Beispiele von Bodenerosionskarten im Niedersächsischen Bergland sowie Bemerkungen über Berücksichtigung der Erosionsschäden bei der Bodenschätzung. Neues Archiv für Niedersachsen 4/6, 140–143 (1954)

Hempel, L.: Soil erosion and water run-off on open ground and underneath wood. Int. Ass. of Scientific Hydrology General Assembly of Toronto, Vol. 1, 108–114, Gentbrugge 1958.

Hempel, L.: Bodenerosion in Nordwestdeutschland. Erläuterungen zu Karten über Schleswig-Holstein, Hamburg, Niedersachsen, Bremen und Nordrhein-Westfalen. Forschungen zur dt. Landeskunde 144, Bad Godesberg 1963.

Hempel, Lena: Flurzerstörung und Bodenerosion in früheren Jahrhunderten. Ztschr. f. Agrargeschichte und Agrarsoziologie 2, 114–122 (1954)

Hempel, Lena: Das morphologische Landschaftsbild des Unter-Eichsfeldes unter Berücksichtigung der Bodenerosion und ihrer Kleinformen. Forschungen zur dt. Landeskunde 98, Remagen 1957.

Herz, K.: Die Ackerflächen Mittelsachsens im 18. und 19. Jahrhundert. Reprint from: Sächsische Heimatblätter, Dresden 1964.

Iwersen, J.: Windschutz in Schleswig-Holstein, aufgezeichnet am Beispiel der schleswigschen Geest. Gottorfer Schriften zur Landeskunde Schleswig-Holsteins 2, Schleswig 1953.

Iwersen, I.: Windschutz auf der schleswigschen Geest. Jahrbuch f. d. Schlewigische Geest 3, 6–11 (1955)

Johannsen, H.H.: Niederschlagstage — Niederschlagssummen. Berichte des dt. Wetterdienstes in der US-Zone 42, 221–223 (1952)

Jung, L.: Untersuchungen über den Einfluß der Bodenerosion auf die Erträge in hängigem Gelände. Schriftenreihe für Flurbereinigung 9, Stuttgart 1956.

Karl, J. & Danz, W.: Der Einfluß des Menschen auf die Erosion im Bergland, dargestellt an Beispielen im bayerischen Alpengebiet. Mit einem Beitrag von J. Mangelsdorf. Schriftenreihe der Bayer. Landesstelle für Gewässerkunde 1, München 1969.

Käubler, R.: Die Tilke als junge Form des Kulturlandes. Geographischer Anzeiger 38, 361–372 (1937)

Käubler, R.: Junggeschichtliche Veränderungen des Landschaftsbildes im mittelsächsischen Lößgebiet. Wiss. Veröff. d. dt. Museums für Länderkunde Leipzig, NF, 5, 71–90 (1938)

Käubler, R.: Beiträge zur Altlandschaftsforschung in Ostmitteldeutschland. Petermanns Geograpische Mitteilungen 96, 245–249 (1952)

Kern, H.: Große Tagessummen des Niederschlages in Bayern. Münchener Geogr. Hefte 21, Regensburg 1961.

Klug, H.: "Reche" und "Rosseln" in Rheinhessen. Anthropogene Kleinformen in der morphologischen Hanggestaltung einer Agrarlandschaft. Mitteilungsblatt zur rheinischen Landeskunde 13, 131–133 (1964)

Kugler, H.: Geomorphologische Erkundung und agrarische Landnutzung. Geographische Berichte 80, 190–204 (1976)

Laatsch, W.: Bodenschutz im Bergland des bayrischen Alpengebietes. Forstwiss. Centralblatt 90, 159–174 (1971)

Laatsch, W.: Hangabtragung durch Schnee in den oberbayrischen Alpen und seine Begünstigung durch unpflegliche Almwirtschaft. Forstwiss. Centralblatt 93, 1, 23–34 (1974)

Laatsch, W. & Grottenthaler, W.: Labilität und Sanierung der Hänge in der Alpenregion des Landkreises Miesbach. Bayer. Staatsmin. f. Ern., Ldw. u. Forsten, München 1973.

Laurant, A. & Bollinne, A.: Characterisation des pluies en Belgique du point de vue de leur intensité et de leur érosivité. Pedologie 28, 2, 214–232 (1978)

Leser, H.: Die Unwetter vom 4. und 5. Juli 1963 im Zeller Tal (Pfrimmgebiet, südliches Rheinhessen) und ihre Schäden. Berichte zur dt. Landeskunde 35, 74–89 (1965)

Linke, M.: Ein Beitrag zur Erklärung des Kleinreliefs unserer Kulturlandschaft. Wiss. Zeitschr. d. Martin-Luther-Universität Halle-Wittenberg, Math.-Klasse 12/10, 735–752 (1963)

Masuch, K.: Häufigkeit und Verteilung bodengefährdender sommerlicher Niederschläge in Westdeutschland nördlich des Mains zwischen Weser und Rhein. Forschungen zur dt. Landeskunde 181, Bad Godesberg 1970.

Meyer, B. & Willerding, U.: Bodenprofile, Pflanzenreste und Fundmaterial von neuerschlossenen neolithischen und eisenzeitlichen Siedlungsstellen im Göttinger Stadtgebiet. Göttinger Jahrbuch 21–38 (1961)

Mortensen, H.: Die "quasinatürliche" Oberflächenformung als Forschungsproblem. Wiss. Zeitschr. d. Ernst-Moritz-Arndt-Universität Greifswald 4, Math.-Nat.-Reihe, 625–628 (1954/55)

Richter, G.: Die Hilfe des Luftbildes für die praktische Bodenerosionsbekämpfung. International Archives of Photogrammetrie 14, 327–332 (1963)

Richter, G.: Bodenerosion. Schäden und gefährdete Gebiete in der Bundesrepublik Deutschland. Forschungen zur dt. Landeskunde 152, Bad Godesberg 1965.

Richter, G.: Kaiserzeitliche Waldverwüstung in der schleswigschen Geest. Ein Beitrag zum Heideproblem. Mitteil. d. floristischsoziologischen Arbeitsgemeinschaft Stolzenau/Weser 11/12, 223–229 (1967)

Richter, G.: Formen und Wirkungen der Abspülung im Luftbild. In: Hassenpflug, W. & Richter, G., Formen und Wirkungen der Bodenabspülung und -verwehung im Luftbild. Landeskdl. Luftbildauswertung im mitteleuropäischen Raum 10, 11–41 (1972)

Richter, G.: Einleitung. In: Richter, G. & Sperling, W., Ed., Bodenerosion in Mitteleuropa. Darmstadt 1976.

Richter, G.: Soil erosion in Central Europe. Pedologie 28, 145–160 (1978)

Richter, G.: Bodenerosion in den Reblagen an Mosel-Saar-Ruwer. Formen, Abtragungsmengen, Wirkungen. 41 Dt. Geographentag Mainz, Tagungsbericht u. Wiss. Abhandlungen 371–389, Wiesbaden 1978.

Richter, G., Müller, M.J. & Negendank, J.F.W.: Landschaftsökologische Untersuchungen im Gebiet zwischen Mosel und unterer Ruwer. Forschungen zur dt. Landeskunde 214 (in press)

Richter, G. & Negendank, J.F.W.: Soil erosion processes and their measurement in the German area of the Moselle river. Earth Surface Processes 2, 261–278 (1977)

Richter, G. & Sperling, W.: Anthropogen bedingte Dellen und Schluchten in der Lößlandschaft. Untersuchungen im nördlichen Odenwald. Mainzer Naturwiss. Archiv 5/6, 136–176 (1967)

Ruppert, K.: Die Leistung des Menschen zur Erhaltung der Kulturböden im Weinbaugebiet des südlichen Rheinhessen. Rhein-Mainische Forschungen 34, Frankfurt 1952.

Sokollek, V.: Untersuchungen über den Einfluß der Landnutzung auf Oberflächenabfluß und Versickerung. Jahresbericht des ökologischen Forschungsstation der Justus-Liebig-Univ. Gießen, Waldeck 3, 119–136 (1976/77)

Sperling, W.: Anthropogene Oberflächenformung: Bilanz und Perspektiven in Mitteleuropa. 41. Dt. Geographentag Mainz, Tagungsber. u. wiss. Abh. 363–370, Wiesbaden 1978.

Süssmann, W. & Wohlrab, B.: Wirkungen verschiedener Bodennutzung auf das Abflußregime und den Stoffeintrag in Gewässern. Teil 2. Jahresber. d. Ökologischen Forschungsstation d. Justus-Liebig-Univ. Gießen, Waldeck 2, 45–76 (1975)

Schauer, T.: Die Blaikenbildung in den Alpen. Schriftenr. d. Bayer. Landesamtes f. Wasserwirtschaft 1, 8, München 1975.

Scheffer, F. & Meyer, B.: Bodenkundliche Untersuchungen an neolithischen Siedlungsprofilen des Göttinger Leinetalgrabens. Göttinger Jahrbuch 3—19 (1958)

Schmidt, R.G.: Probleme der Erfassung und Quantifizierung von Ausmaß und Prozessen der aktuellen Bodenerosion (Abspülung auf Ackerflächen). Physiogeographika 1, Basel 1979.

Schultze, J.H.: Bodenerosion im 18. und 19. Jahrhundert. Forschungs- und Sitzungsberichte der Akademie für Raumforschung und Landesplanung 30, 1—16 (1965)

Schwille, F.: Die chemischen Zusammenhänge zwischen Oberflächenwasser und Grundwasser im Moseltal zwischen Trier und Koblenz. Bes. Mitt. zum dt. Gewässerkdl. Jahrbuch 38, Koblenz 1973.

Stehlik, O.: Potential soil erosion by running water on territory of the Czech Socialist Republic. Studia Geographica 42, Brno 1975.

Verstraten, J.M.: Chemical erosion in a forested watershed in the Oesling, Luxembourg. Earth Surface Processes 2, 2/3, 175—184 (1977)

Verstraten, J.M.: Water-rock interactions in (very) lowgrade metamorphic shales. A case study in a catchment in the Oesling, Luxembourg. Diss. Amsterdam 1978.

Vogt, J.: Erosion des sols et technique de culture en climat tempéré de transition (France et Allemagne). Revue de Géomorphologie Dynamique 4, 157—183 (1953)

Vogt, J.: Zur historischen Bodenerosion in Mitteldeutschland. Petermanns Geographische Mitt. 102, 199—203 (1958a)

Vogt, J.: Zur Bodenerosion in Lippe. Erdkunde 12, 132—135 (1958b)

Vorndran, G.: Hangabtragsbilanzen. Zeitschr. f. Geomorphologie, Supplementband 28, 124—133, Berlin/Stuttgart 1977.

Wagner, G.: Die historische Entwicklung von Bodenabtrag und Kleinformenschatz im Gebiet des Taubertals. Mitt. d. Geogr. Ges. München 46, 99—149 (1961)

Wagner, G.: Die Bodenabtragung im Wandlungsprozeß der Kulturlandschaft. Berichte zur dt. Landeskunde 35, 91—111 (1965)

Walther, W.: Der Stoffaustrag bei kleinen Einzugsgebieten mit ackerbaulicher Nutzung. Veröff. d. Inst. f. Stadtbauwesen der TU Braunschweig 19, 133—152 (1976)

Wandel, G. & Mückenhausen, E.: Neue vergleichende Untersuchungen über den Bodenabtrag an bewaldeten und unbewaldeten Hangflächen im Nordrheinland. Geol. Jahrb. 65, 507-550 (1949)

Wendl, U.: Die Sanierung des Halblechgebietes. Wasser u. Boden 1, 14—17 (1969)

Wiechmann, H.: Beeinflussung der Gewässereutrophierung durch erodiertes Bodenmaterial. Landwirtschaft u. Forschung 26, 1, 37—46 (1973)

Wildhagen, H. & Meyer, B.: Holozäne Boden-Entwicklung, Sediment-Bildung und Geomorphogenese im Flußauen-Bereich des Göttinger Leinetal-Grabens. Göttinger Bodenkundliche Berichte 21, 1—158 (1972)

Wischmeier, W.H.: Use and misuse of the universal soil loss equation. Journal Soil and Water Conservation 31, 1, 5—9 (1976)

Wischmeier, W.H., Johnson, C.B. & Cross, B.V.: A soil erodibility nomograph for farmland and construction sites. Journal Soil and Water Conservation 189—192 (1971)

Wischmeier, W.H. & Smith, D.D.: A universal soil-loss equation to guide conservation farm planning. Transactions of the 7. Internat. Congress of Soil Science, Madison Wisconsin/USA 1, 418—425, Amsterdam 1961.

Wohlrab, B., Wenzel, V. & Mollenhauer, K.: Wirkungen verschiedener Bodennutzung auf das Abflußregime und den Stoffeintrag in Gewässern. Teil 1. Jahresber. d. Ökologischen Forschungsstation d. Justus-Liebig-Univ. Gießen 1, 1—40 (1974)

**159**

# 11

Reprinted from *Prog. Phys. Geogr.* **4**:24–47 (1980)

# Soil erosion and conservation in Britain

## by R. P. C. Morgan

Despite the overwhelming evidence of the universal dangers and universally widespread character of soil erosion, it is still the generally held belief that it is unimportant in Britain . . . . There is . . . visual evidence in ploughed fields on hill slopes after rain of the finer material being washed downhill—this can be seen especially after sudden downpours on arable land in the drier eastern counties of England. It is only a small step from this to real 'gully' erosion of which examples are by no means unknown (Stamp, 1962, 295–96).

Soil erosion by water is generally thought to be unimportant on agricultural land in Britain. . . . But erosion does occur . . . on soils with an inherently poor structure such as sandy soils which are cropped in such a way as to leave the soil bare at some seasons. In the West Midlands near Kidderminster and Ross-on-Wye . . . soils have suffered bad erosion . . . and the soils of the Lower Greensand are also known to be vulnerable (Warren, 1974, 41).

There is no risk of erosion except on steep slopes. . . . At present only very infrequent gully erosion is likely to occur on sloping land in sandy areas (Davies *et al.*, 1972, 226).

Soil erosion is the removal of soil by wind, water and mass movements at a faster rate than that at which new soil forms. In soil erosion and conservation texts (Bennett, 1939; Stallings, 1957; Hudson, 1971; Morgan, 1979) a fundamental distinction is made between erosion at natural rates and that at accelerated rates. It is generally implicit that for natural rates of erosion a balance exists between the rates of soil loss and soil formation and that man is the chief cause of upsetting this balance and, therefore, of accelerated erosion. On a global scale soil erosion is associated with misuse of the land where the soil is inadequately protected by a plant cover. It occurs with continuous arable cropping, overgrazing and intensive recreational use. It results in thinning of the top soil, loss of plant nutrient, reduced crop yield and lower grassland productivity, so that farming costs are increased and incomes reduced. Eroded soil is washed into rivers, resulting in pollution and, through sedimentation, reduced channel capacity and flooding. Soil erosion thus brings about a deterioration in land quality and, ultimately, abandonment of the land.

There are clearly considerable differences of opinion on the importance of soil erosion in Britain. Why should these differences exist and what is the true situation? This paper examines the evidence for soil erosion occurring in Britain, assesses its importance both nationally and locally, and comments on the adequacy or otherwise of the current programme of research into erosion and methods of erosion control with particular reference to British conditions.

## I Historical accounts

There is ample evidence, albeit somewhat qualitative, in both historical and geographical sources, for erosion in Britain. Defoe, in the eighteenth century, writes of Bagshot Heath:

> Here is a vast tract of land . . . much of it is a sandy desert and one may frequently be put in mind here of Arabia Deserta . . . for in passing this heath on a windy day, I was so far in danger of smothering with the clouds of sand . . . that I cou'd neither keep it out of my mouth, nose or eyes (quoted in Hoskins, 1970, 140).

Gilbert White describes the effects of a cloud-burst at Selborne on 5 June 1784:

> There fell . . . prodigious torrents of rain on the farms . . . which occasioned a flood as violent as it was sudden; doing great damage to the meadows and fallows, by deluging the one and washing away the soil of the other. The hollow lane towards Alton was so torn and disordered as not to be passable till mended, rocks being removed that weighed 200 weight (White, 1788; edn 1977, 267).

References to similar events are frequently found in local and not widely circulated journals. Oakley (1945) provides a detailed account of a cloud-burst near West Wycombe in the Chiltern Hills, illustrated by maps and photographs of the eroded lands and the areas of deposition. In attempting to place this event in its historical and spatial perspective, he cites articles on similar storms, mostly in the Yorkshire Wolds. A major literature search would undoubtedly identify much evidence of this type. Its value, however, would be somewhat limited because, in these accounts, there is insufficient information to even estimate how much soil has been lost. In most cases, it is the rapid, short duration and probably extreme event which is described. Between these events must occur more moderate ones with small amounts of soil loss which go unreported but which, cumulatively, are equally if not more important in their contribution to total soil loss.

## II Press accounts

> Tornado on Sunday, May 21, in south Bedfordshire, followed by hail and rain, tore carrots out of the ground, battered brassicas and defoliated chrysanthemums. More

serious, however, was the soil erosion caused in many fields. This picture shows a field of cabbage in which the plants had been completely covered by sand from other fields. Then the rain-floods swept round the contours of the land, carrying away the soil, leaving half the plants exposed again and half still buried (caption from *The Grower*, 3 June 1950).

The south-westerly gale at the week-end which lifted clouds of valuable top-soil from newly cultivated fields and blanketed the town in dust on Sunday was the most notable feature of the past week's weather. The Cawdor road was blocked near the Howford Bridge and the Forres road was also affected (*The Nairnshire Telegraph*, 30 March 1976).

Individual instances of erosion are often reported in the local press and warrant a paragraph or two of coverage. Within the archives of local newspapers throughout the country must lie sufficient information to provide a base for analysing the frequency and spatial extent of erosion. The task of assembling the information from such widely distributed sources is enormous, however, and it is not surprising that no study of this kind has yet been attempted. In some cases it is possible to trace the occurrence of erosion and its effects over a period of time. The problems of Shepton Beauchamp in Somerset have resulted in a series of reports and letters in *The Western Gazette* since the mud deluges of June 1966 and July 1968. The more recent correspondence emphasizes a change in the drainage system which, as a result of housing development, has cut off a former flood overflow channel. Since the problem seems to predate this change, this cannot be the sole cause of erosion. Other letters refer to enlargement of fields on the steeply rolling land around the village, changes associated with the removal of hedges and ditches, an increase in root-crop cultivation at the expense of grass and the cost of erosion in road clearing operations and blocked road drains.

Only occasionally do instances of erosion command wider publicity in the national press. Major flood disasters and the erosion associated with them are generally well-covered but these are strictly extreme natural events and should not be interpreted as examples of erosion accelerated by man. National press coverage tends to be restricted to the more spectacular examples of erosion resulting from wind (*The Sunday Times*, 7 April 1968).

## III  Local archives

Local authority records, normally kept in the Surveyor's Department, detail the dates when road clearance is carried out and these provide further evidence on the frequency and location of erosion. Records from Somerset County Council show that clearance work was required at Shepton Beauchamp 14 times between 16 October 1976 and 18 February 1977. It is small wonder that the inhabitants of the village have blocked up the lower parts of their front doors and keep sandbags at the ready.

The accounts of reservoir siltation found in the unpublished records of water authorities are another source of evidence. One such report was made in the early 1960s to the then Manchester Corporation Waterworks, now part of the Pennine Division of the North West Water Authority. The report states that in 1959, 2 286 630 dm³ of peat and sand was removed from Bottoms Reservoir in Longdendale at a cost then of £4500. Erosion has clearly always been a problem in the surrounding catchment because settling tanks were provided in the main river, upstream of the entrance to the reservoir, at the time of construction in 1850. In 1960 it was estimated that routine clearance of the tanks yielded 254 700 dm³/y of sediment at an annual cost of £800. Surprisingly, no detailed study of the problem appears to have been commissioned and no serious attempt made to find a solution. This type of evidence suggests that masses of data on sedimentation rates remain to be researched in water authority archives.

## IV  The scientific evidence

The evidence presented in scientific journals and texts appears to contradict that found in the local press. It is invariably assumed that the absence of intense rainfall and the presence of hedgerows and trees in the landscape mean that soil erosion is unimportant and rarely occurs in Britain. Rodda (1970) puts it even more strongly by suggesting that these conditions have enabled British farmers to follow cultivation practices which would have proved disastrous if followed in other parts of the world. Where soil erosion is mentioned, it is only to stress its insignificance, as in Hudson (1967) and Davies *et al.* (1972). Both these references, however, contain impressive pictures of rill erosion in Britain to illustrate their point. In the FAO review of legislation for soil conservation Britain is not even mentioned (Christy, 1971) and Fournier (1972), in a review of erosion in Europe, writes with reference to the maritime region extending from central Portugal to the north of Ireland and England:

> The climatic conditions of the maritime fringe of moderate temperate Europe encourage the vegetation. They allow it to form an effective protective covering, even when it has deteriorated. . . . The traditional agricultural methods favour conservation, especially the separation of fields by hedges (Fournier, 1972, 91–92).

With a clear appreciation of erosion problems elsewhere in Europe and an awareness of changing agricultural practices in France, however, Fournier adds a cautionary note:

> The systematic elimination of the hedges, favoured by certain supporters of regrouping, may give rise to a dangerous situation. On the windy plateaux of western Brittany, the action of wind is beginning to erode the soil. . . . On the schist ridges, in Vendée for example, runoff develops freely and is already beginning to scrape away the soils. . . . It is necessary to act with caution and avoid changing the traditional systems without careful consideration (Fournier, 1972, 92).

**163**

It is somewhat surprising that geomorphologists have not paid more attention to areas of accelerated erosion, given the opportunity that they afford as field laboratories for the study of erosion processes actually at work. Yet, the *Bibliography of British geomorphology* (Clayton, 1964), covering contributions up to and including 1962, lists only three references on soil erosion: one (Richardson, 1939) covers erosion and soil conservation generally and the others (Spence, 1955; 1957) deal with wind erosion in the Fens. Greater interest has been shown more recently and Richter (1977) cites 17 articles on soil erosion in Britain in his *Bibliographie zur Bodenerosion und Bodenerhaltung 1965–75*. These exclude three papers presented at a symposium of the British Geomorphological Research Group in 1965 on current rates of erosion in the British Isles which clearly deal with accelerated erosion in the North Yorkshire Moors (Curtis, 1965), mid-Wales (Thomas, 1965) and the New Forest (Tuckfield, 1965).

Douglas (1970) gives a comprehensive review of papers dating back to the 1860s which emphasize man's role in contributing to erosion. The review is biased towards descriptions of gully erosion which warrants nine citations compared with three for sheet erosion and four for wind erosion. Although this reflects the more spectacular nature of gullying and the greater visual evidence for its impact, it is by no means certain that this is an adequate indicator of its importance (Jacks and Whyte, 1939, 43).

Only three of the papers mentioned by Douglas (1970) have an agricultural context, the remainder being concerned with heathlands and upland moorlands where the rate of natural erosion is often high anyway. In these areas it is difficult to distinguish between natural and accelerated erosion because the latter often develops spontaneously at points which are highly susceptible to erosion because of natural instability, areas of slumping being an example.

Evans (1971) examines the evidence for soil erosion under three main headings: erosion in the uplands; wind erosion in eastern England; and sheet and rill erosion in lowland England generally. It is instructive to investigate this evidence afresh because of the light it throws on our understanding of the erosion problem.

The work on upland Britain concentrates on erosion of the peat moors. Through the studies of Bower (1960; 1961; 1962), which combine aerial photograph interpretation and field observations, a clear picture is obtained of the different patterns and processes of erosion and the extent of the area affected. The results of water erosion and mass movement are distinguished and two types of dissection of the peat are recognized, one consisting of freely and intricately branching gullies and the other of widely spaced, rarely branching, individual gullies. Apart from the work of Johnson (1958), examining headward extension of gullies since 1911, and Conway and Millar (1960), who give data on rates of silting in reservoirs in the Tees and Troutbeck valleys, little information is available on rates of erosion. Without detailed measurements, discussion on possible causes and dates of the initiation of erosion relies on personal interpretations of often equivocal evidence. Bower (1961) considers that much of the erosion now taking place is natural and reserves judgement on the importance of human activity. She is taken to task by Radley (1962; 1965), however, for failing to emphasize the role of man, particularly through fire. Yet, it should be pointed out that Radley was writing following the effects of abnormally fierce and extensive fires on the Pennine Moors in 1959 which resulted in the cutting of braided rill networks and gullies up to 1 m deep. Barnes (1962), reviewing the argument, ends with a plea for more evidence.

A similar situation, in which details of the extent and effects of erosion but not the rates are known, exists for wind erosion. The work of Radley and Simms (1967), Robinson (1968) and Wilkinson *et al.* (1968) shows that wind-erodible soils cover much of Yorkshire, Lincolnshire and the east Midlands. It also shows that, in these areas, it is only particular fields or, in many cases, parts of fields, as on the crests in gently undulating terrain, that are susceptible to erosion. All workers agree, however, that the cause of the problem lies in farming practices such as arable farming of land which is not suitable for that purpose and the removal of hedgerows.

The published evidence for rill erosion is sparse. Apart from mentions in Stamp (1962), Brade-Birks (1944) and similar texts, little notice was taken of it until Evans (1971) cited many instances of its occurrence on arable land. Ample evidence for it exists in the form of thinner soils on convexities, where subsoils are turned up on ploughing and crop growth is often stunted, and soil wash deposits found as valley infils. In a later article, Evans (1978) presents data on 103 sites in lowland England affected by sheet and rill erosion and shows that the conditions required for its occurrence are far from extreme. A typical eroded site has slope lengths of 50 to 400 m, a relief of less than 25 m and a mean slope steepness of less than 6 per cent. Again, however, no information is presented in rates of soil loss.

A fourth area of concern has attracted much interest recently and this is erosion associated with excessive recreational use of the land, particularly that related to footpaths and trackways. Yet again, few measurements of rates of erosion have been made. Studies have concentrated on the deterioration of the plant cover; increases in the area of bare ground; rates of path widening, particularly in relation to wetness and roughness; the intensity of use of paths by walkers; and the effects of trampling on compaction, infiltration rates and bulk density of the soil (Bayfield, 1971; 1973; Bayfield and Lloyd, 1973; Liddle, 1973; 1975; Liddle and Moore, 1974; Liddle and Greig-Smith, 1975; Coleman, 1977).

## V    Measurements of soil erosion

### 1    *The story so far*

Although there is clearly a great deal of scientific evidence for soil erosion, it is essentially qualitative in nature and is not likely to attract much attention. At present, deciding whether or not erosion is a problem relies too heavily on an intuitive interpretation of the evidence. These descriptive studies must be supported by data on measured rates of erosion.

Most geomorphological studies show that recorded erosion rates in Britain are low (Table 1) but it has been suggested elsewhere (Morgan, 1977) that one of the reasons for this is that most measurements have been carried out in areas with the minimum of interference by man. They are therefore representative of natural rates of erosion. The role of man in influencing erosion is becoming increasingly recognized and, in some studies, extremely high rates of erosion have been recorded. Imeson's (1971; 1974) work in the North Yorkshire Moors relates to heather burning; Evans's (1971; 1977) figures in Hey Clough in the Peak District are for soil removed by sheet wash from sheep scars; and the data of Al-Ansari *et al.* (1977) on the Almond Valley, eastern Scotland, refer to

**Table 1**    Rates of water erosion in Great Britain

| Method of measurement | Region | Rate (t/ha/y) | Source |
|---|---|---|---|
| 1 Sampling of sediment in river water | R. Nith, Dumfries/ Galloway | 0.68 | Geikie, 1868 |
| | R. Clyth, Highland | 0.25 | Geikie, 1868 |
| | R. Tyne, Northumberland | 0.68 | Hall, 1967 |
| | R. Derwent, Derbyshire | 1.17 | Hall, 1967 |
| | Hodge Beck, Yorkshire | 4.80 | Imeson, 1971 |
| | Lower Swansea Valley, West Glamorgan | 0.60 | Bridges and Harding, 1971 |
| | Grains Gill, Cumbria | 400.00 | Harvey, 1974 |
| | Exeter, Devon | 1.92 | Walling, 1974b |
| | R. Almond, Tayside | 2.70–9.40 | Al-Ansari *et al.*, 1977 |
| 2 Slope measurements; sediment traps; nails | Yorkshire Pennines | 0.01 | Young, 1960 |
| | Water of Deugh, Dumfries/Galloway | 0.01 | Kirkby, 1967 |
| | Mid-Wales | 1.50 | Slaymaker, 1972 |
| | Hey Clough, Derbyshire | 34.40 | Evans, 1971; 1977 |
| | Hodge Beck, Yorkshire | 0.94 | Imeson, 1974 |
| 3 Reservoir surveys | Strines, Yorkshire | 1.27 | Young, 1958 |
| | North Esk, Midlothian | 0.26 | Ledger *et al.*, 1974 |
| | Hopes, East Lothian | 0.25 | Ledger *et al.*, 1974 |

Based on table in Morgan (1977). The rates quoted are either those given in the original sources or are conversions based on a bulk density of 1.0 g/cm$^3$.

**Table 2**  Soil loss by water erosion in individual storms in Great Britain

| Date | Region | Rate (t/ha) | Source |
|---|---|---|---|
| 21 May 1973 | Balsham, Cambridgeshire; rills following storm of 7.4 mm of rain. | 3.3 | Evans and Morgan, 1974 |
| March–June 1975 | Cromer Ridge, north Norfolk; gullies; storms of 21.2 mm on 7 May and 19.3 mm on 15 June; dates of erosion uncertain and several storms may be involved. | 195.0 | Evans and Nortcliff, 1978 |
| 24–26 September 1976 | Worfield, east Shropshire; rills and gullies after 83 mm of rain. | 156.0 | Reed, 1979 |

soil loss from agricultural land under turnips. The high rates recorded by Harvey (1974) in the Howgill Fells relate to gullying.

These somewhat limited but growing number of assessments of annual rates of accelerated erosion can be further supplemented by data for individual storms (Table 2). Very high rates of soil loss have been recorded, especially from rill erosion on agricultural land in the English Midlands (Evans and Morgan, 1974; Reed, 1979) and gullying along farm tracks in north Norfolk (Evans and Nortcliff, 1978).

## 2  The Silsoe study

It was to provide basic data on soil erosion on agricultural land in Britain that measurements of soil loss were started in May 1973 in the Silsoe area of Bedfordshire. The object of the study was to measure the rate of erosion by rainsplash, overland flow and rill wash, separately and in combination, on the sandy loam soils of the Cottenham Series, derived from the Lower Greensand. Since February 1977 additional measurements have been made on boulder clay and chalk soils. Data are now being obtained for a variety of ground cover conditions ranging from bare soil to grass, cereals and woodland.

The experimental set-up is described in detail elsewhere (Morgan, 1977) but briefly the soil eroded by rainsplash is collected in field splash cups (Morgan, 1978) and overland flow and its associated sediment are collected in 0.5 m wide sediment traps (Morgan, 1979). Rill erosion is assessed volumetrically. Data are obtained after every rainstorm on one of the bare soil plots on the Cottenham Series and at monthly and 100-day intervals on all plots. For reasons of space and because they are the most comprehensive only the results for erosion by overland flow are considered here.

**Table 3**    Mean annual soil loss on mid-slope sites in Silsoe area

| Location | Ground cover | Soil | Slope (°) | Soil loss (t/ha) |
|---|---|---|---|---|
| Silsoe | Bare ground | Sandy loam | 11 | 10.80 |
| | Bare ground | Sandy loam | 11 | 10.90 |
| | Grass | Sandy loam | 7 | 0.24 |
| | Grass | Sandy loam | 11 | 0.98 |
| Maulden | Woodland | Sandy loam | 11 | 0.05 |
| Ashwell | Winter wheat | Chalk | 10 | 0.30 |
| Meppershall | Winter wheat/spring barley | Boulder clay | 10 | 0.36 |
| Pulloxhill | Spring barley | Boulder clay | 10 | 0.23 |
| Woburn | Winter oats/winter wheat/ winter beans | Sandy loam | 7 | 0.43 |

Data are derived from soil collected in 0.5 m wide sediment traps. To convert data from a unit width to a unit area basis, a contributing slope length of 10 m is assumed, giving a catchment area of 5 m$^2$. A 10 m value for slope length coincides with the spacing of depositional splays and micro-fans in the field after an erosive event. It is also a convenient basis for comparison with erosion rates measured in the United States from standard plots of 1.83 m width and 22.2 m length, allowing for appropriate scaling. Data are for two adjacent traps, giving a catchment area of 10 m$^2$, 1 m wide

The mean annual soil loss on the mid-slope sites (Table 3) shows an expected ranking of woodland, grassland, cereals and bare soil for land use with increasing soil loss. The data for the bare soil sites provide an indication of soil loss under market gardening where the land has little or no cover for much of the year. The data also show a ranking of clay, chalk and sandy loam for soils with increasing soil loss.

In order to assess the severity of erosion the measured values of soil loss must be compared with a maximum acceptable value which, theoretically, corresponds to the rate at which loss of soil by erosion is balanced by that of new soil formation. The latter is virtually impossible to determine with any degree of acceptable accuracy and no data are available for agricultural soils in Britain. The assumption is sometimes made that the rates of erosion and soil formation are in equilibrium in undisturbed areas and, therefore, that a measurement of the natural rate of erosion in such areas can be used as a crude indicator of the rate of new soil formation. Based on the figures quoted in Table 1, 0.2 t/ha/y might be selected as a suitable maximum acceptable value. In practice, much higher rates are used. In the United States a rate of 11 t/ha/y is the generally acceptable standard (Hudson, 1971). A lower value would seem to be more appropriate for the Silsoe area, however, given the shallow top soils on both the Lower Greensand and the Chalk. Using the guidelines of Arnoldus (1977) for determining acceptable soil losses, a value of 2 t/ha/y seems to be more suitable. Setting the results of the study against this standard, it is clear that the sandy loam soils of the area are at risk unless kept under grass or cereal cover.

**Table 4**  Soil loss from erosive storms at Silsoe

| Date | Soil loss (t/ha) | Storm rainfall (mm) | KE > 10 (J/m$^2$) | Duration (h) | Cumulative % of total soil loss |
|---|---|---|---|---|---|
| 7.8.78 | 8.61 | 11.9 | 258 | 5.5 | 14 |
| 6.7.73 | 7.07 | 34.9 | 444 | 8.0 | 26 |
| 5.5.73 | 6.17 | 6.7 | 0 | 4.16 | 37 |
| 21.5.73 | 6.00 | 7.1 | 73 | 2.0 | 46 |
| 27.8.73 | 5.09 | 16.8 | 87 | 6.5 | 55 |
| 6.5.73 | 4.00 | 2.2 | 17 | 0.33 | 61 |
| 8.8.74 | 3.68 | 17.0 | 190 | 13.5 | 67 |
| 16.7.74 | 2.87 | 7.6 | 139 | 13.0 | 72 |
| 2.8.78 | 2.50 | 8.3 | 22 | 14.0 | 76 |
| 13.7.74 | 2.47 | 16.5 | 209 | 10.5 | 80 |
| 16.7.76 | 2.26 | 27.4 | 382 | 8.67 | 84 |
| 19.6.73 | 2.04 | 39.6 | 0 | 23.83 | 88 |

Data are for the mid-slope site of 11° with bare ground. Soil loss figures are conversions from data collected on a unit width basis (Table 3). The KE > 10 index refers to the kinetic energy of all rains falling at intensities equal to or greater than 10 mm/h for ten minutes

At the site monitored on a storm-by-storm basis, 88 per cent of the soil loss since 2 May 1973 has taken place in 12 storms (Table 4). As can be seen from the characteristics of the storms, erosion is associated with events of only moderate magnitude. The maximum storm rainfall recorded during the study period so far is 39.6 mm which compares with a daily total with a five-year return period of 37.0 mm (Natural Environment Research Council, 1975). The frequency of erosion, however, is surprisingly high. In the six years of study only 1975 and 1977 produced no major erosion event whereas soil losses over 2 t/ha were recorded on the mid-slope for individual storms on two or more occasions in 1973, 1974 and 1978.

It can be seen from Table 3 that, in the study area, land use and soil type account for differences in erosion rates between sites. At a particular site, however, as shown by data for 30 storms, the major control over variations in soil loss is runoff (r = 0.79). The addition of a rainfall parameter, in the form of the KE > 10 index, defined in Table 4, improves the correlation coefficient to 0.82. The correlation coefficient between soil loss and KE > 10, although significant at the 5 per cent level, is low at 0.39. Even so, the KE > 10 index yields better correlations with soil loss than does total rainfall.

## 3  Silsoe data in a British context

The erosion rates on the bare sandy loam soils of the Silsoe area are very high when compared with rates measured elsewhere in Britain (Tables 1 and 2). Only in upland Britain on sheep scars in the Peak District and in gullies in the Howgill Fells have higher rates been recorded. Such comparisons are of limited value, however, because of the paucity of data on erosion rates in Britain referred to earlier. It is more helpful to compare the erosion in individual events in the Silsoe area with soil loss from individual storms elsewhere in Britain (Table 2). It can then be seen that the erosion rates considered severe at Silsoe are, in fact, at worst equal to and often much lower than those recorded under similar conditions of land use and soil in other parts of the country.

Even with data from the Silsoe study there is an insufficient base for extrapolating measured rates of accelerated erosion over the whole of Britain with a view to identifying areas of greatest erosion risk. An alternative approach to a national evaluation is therefore required, based on analyses of rainfall and soils data (Morgan, 1979). The problem is to select suitable rainfall and soil parameters and the results of the Silsoe study are of assistance here.

Whatever parameter is chosen for rainfall is worthless unless it actually correlates with soil loss. To use commonly accepted indices such as $EI_{30}$, defined for an individual storm as the product of the kinetic energy of the rainfall and the maximum rainfall intensity over a 30-minute period (Wischmeier and Smith, 1965), and KE > 25, defined as the kinetic energy of all rainfall at intensities greater than 25 mm/h (Hudson, 1971), is of little value until a relationship is shown to exist between them and soil loss for Britain. In fact, various attempts to correlate rainfall parameters with erosion have not proved particularly successful in Britain. Bridges and Harding (1971) obtained correlation coefficients of 0.71 and 0.85 between total rainfall and bank erosion for unspecified time periods in the Lower Swansea Valley but found no significant relationship between erosion and rainfall intensity. Harvey (1974), however, found that rainfall intensity correlated well with sediment yield of gullies (r = 0.67 for rainfall intensities greater than 2.5 mm/h for a duration of one hour). Because kinetic energy and rainfall intensity are so closely correlated (Hudson, 1971), Harvey's results should hold for a straight kinetic energy index, such as KE > 10. Surprisingly, therefore, Walling (1974a) found no significant correlation between sediment yield and kinetic energy in the Exeter area where the relationship between sediment yield and total rainfall was, however, equally poor. Evans (1978) has suggested that the most useful parameter may prove to be the total rainfall over the previous two days but, as yet, this parameter has not been widely tested.

That poor correlations exist between soil loss and rainfall in Britain is not unexpected. With rates of splash erosion being low and most erosion occurring by overland flow and rill wash, resulting in high correlations between soil loss and runoff, the effectiveness of the rainfall is dependent on how much runoff is produced. This, in turn, depends on the infiltration and soil moisture storage characteristics of the soil and the type of rainfall event. The Silsoe study shows that erosion results from two types of event:

the short-lived intense storm, frequently of convectional origin; and the prolonged storm of low intensity, normally of frontal origin (Table 4). Notwithstanding the contradictory findings of other workers, it also appears from the Silsoe study that an index of rainfall energy is the most useful of the possible rainfall parameters as a guide to erosion risk. Although, as indicated earlier, the correlation coefficient between soil loss and the KE > 10 index is poor, it is statistically significant at the 5 per cent level, is better than that achieved with any of the other rainfall parameters tried and, unlike some of the other indices used by other workers, is reasonably stable, ranging from 0.39, the value given above for individual storms, to 0.41 for data aggregated in 100-day periods. The KE > 10 index is therefore selected as a basis for a national assessment of erosion risk.

## VI   Erosion risk in Britain

Values for the KE > 10 index are estimated from data contained in the Flood Studies Report (Natural Environment Research Council, 1975) using the following procedure. Taking the two-day rainfall total with a five-year return period and employing the conversion factors contained in the report in tables 2.6, 2.7, 2.8, 2.9 and 3.6, the amount of rain expected in ten minutes is estimated for return periods ranging from twice in one year to once in a hundred years. These rainfall values are expressed as intensities. The annual frequency of rainfall at 10 mm/h for ten minutes is also calculated. Values of kinetic energy are determined for all these intensity values using the relationship between kinetic energy and rainfall intensity established by Hudson (1971). Multiplying the kinetic energy values by their annual frequency of occurrence and integrating over a complete year yields a value for mean annual KE > 10. Mean annual values have been determined for each 10 × 10 km grid square of the Ordnance Survey grid referencing system and an isoline map produced (Figure 1).

These rainfall energy values are measures of erosivity which is defined as the potential of rain to cause erosion (Hudson, 1971). Values in Britain are very low, rising above 1400 J/m$^2$/y only in parts of the Pennines, the Welsh mountains, Exmoor and Dartmoor. Along most of the west coast erosivity is less than 900 J/m$^2$/y and over much of the Outer Hebrides, Orkneys, Shetlands and the north coast of Scotland falls below 700 J/m$^2$/y. It is not surprising that few instances of soil erosion have been recorded in these areas. In contrast, in eastern and southern England, where erosivity is over 1100 J/m$^2$/y, sheet and rill erosion are more commonly observed, particularly on sandy soils. In Scotland, high erosivity values occur in the Almond Valley and the Lammermuir Hills where, as indicated earlier, there is much evidence for high rates of soil erosion. Although higher values of erosivity are expected in mountainous

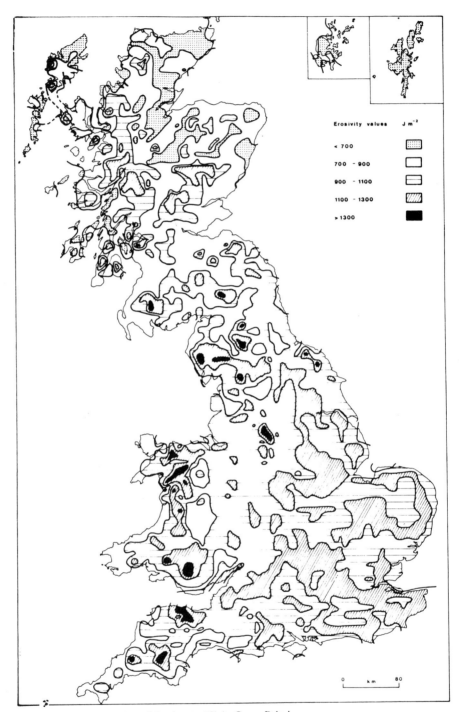

**Figure 1** Mean annual erosivity (KE > 10) in Great Britain

areas, in fact, over much of the Lake District, Pennines, Welsh Mountains and Scottish Highlands, erosivity is relatively low and values are comparable with those of lowland England. The reasons for this pattern of erosivity have not been investigated but since erosivity relates more to the intensity rather than the quantity of rain, it is presumably influenced by the origin of the rainfall, i.e. convectional, relief or frontal.

Erosivity is only an expression of the potential for erosion. How this potential is converted into actual erosion depends on the local conditions of soils, slope and land use. As a first attempt at analysing the effect of soil type and the interaction between soils and erosivity, 504 sites where water erosion has been recorded in the last few years (Evans, personal communication) were located on the 1:1 million Soil Map of England and Wales, produced by the Soil Survey. The occurrence of these sites was examined by soil type and erosivity group. The results (Table 5) show that certain brown calcareous earths (Mapping Group 15) are liable to erode even where erosivity is less than 900 $J/m^2/y$; that most of the erodible soils such as rendzinas, other brown calcareous earths, many types of brown earths and stagnogleys do not normally erode unless erosivity exceeds 900 $J/m^2/y$; and that certain rendzinas (Mapping Group

**Table 5**   Frequeucy of eroded sites in England and Wales by soil type and erosivity group

| Soil type | Mapping group[1] | Erosivity group[2] | | | Total observations |
|---|---|---|---|---|---|
| | | <900 | 900–1100 | >1100 | |
| Rendzinas | 10 | 1 | 33 | 15 | 49 |
| | 11 | 0 | 1 | 9 | 10 |
| Brown sands | 12, 13 | 0 | 11 | 27 | 38 |
| Brown calcareous earths | 14 | 0 | 7 | 1 | 8 |
| | 15 | 10 | 17 | 16 | 43 |
| | 16, 17 | 0 | 6 | 9 | 15 |
| Brown earths | 18, 20, 21, 23, 25, 27, 29, 30, 31, 62, 63 | 5 | 35 | 10 | 50 |
| | 22, 24 | 0 | 4 | 9 | 13 |
| Argillic brown earths | 32, 33, 37 | 0 | 16 | 8 | 24 |
| | 36 | 2 | 1 | 7 | 10 |
| Palaeo-argillic earths | 38, 39 | 0 | 9 | 4 | 13 |
| Calcareous pelosols | 45 | 0 | 1 | 4 | 5 |
| | 46 | 0 | 46 | 46 | 92 |
| Stagnogleys | 51, 53, 54, 55, 56, 58, 69 | 1 | 78 | 22 | 101 |
| | 52, 57 | 0 | 11 | 22 | 33 |
| | | | | | 504 |

[1]As on the 1 : 1 million map, Soils of England and Wales published by the Soil Survey
[2]$_{KE} > 10$ index. Mean annual values ($J/m^2$) taken from Figure 1

11), brown sands, certain brown earths (Mapping Groups 22 and 24), some calcareous pelosols (Mapping Group 45) and some stagnogleys (Mapping Groups 52 and 57) are not generally erodible unless erosivity is greater than 1100 J/m²/y.

Although it is tempting to produce a map combining the erosivity and erodible soil groups as an indicator of the areas of greatest erosion risk, it is undoubtedly misleading to do so. As emphasized earlier, soil erosion is very localized in its occurrence, being restricted to particular fields or even parts of fields rather than to broad areas. Since the erodible soil groups would cover about a third of the map, any calculations of the area at risk would clearly overestimate the real nature of the soil erosion problem. To predict where erosion is going to occur requires a knowledge of local slope, plant cover and land management conditions, as well as information on rainfall and soils. Erosion is all too often associated with bad management of the land. It is likely that under good management with a good plant cover, none of the soils identified as erodible would actually erode. Nevertheless, the national study has value in showing which areas, because of particular combinations of soil and erosivity, are at risk and where greatest attention must therefore be paid to the way in which the land is used.

## VII   The cost of erosion

Virtually nothing is known of the cost of soil erosion in Britain beyond the occasional estimates found in local authority records or press reports. Although it would obviously be difficult to isolate the costs related to erosion from those of farming practice generally, no one seems to be sufficiently motivated even to try. Until an attempt is made, however, it will be difficult to make a case for spending more money on soil erosion and conservation research, except from a purely academic point of view and from the practical applications it may have to overseas conditions. Evidence from France suggests that an economic survey would show that greater expenditure on research is justified, particularly for agricultural soil erosion. Schwing (1978) estimates the cost of erosion to an individual farmer in the vineyards of Alsace as FF 2300/ha in a moderate event, rising to FF 10 000/ha for an extreme event. No evidence is given, however, on what proportion of total farm costs these figures represent.

## VIII   Why is conservation not applied?

The approaches to erosion control are well established and described in basic texts (Hudson, 1971). A strategy for erosion control is first developed by using land capability classification to evaluate land according to its limitations for cultivation, particularly those associated with erosion. The details of the conservation system are then determined in terms of agronomic techniques, mechanical measures and a suitable range of soil management practices. Agronomic techniques centre on the use of plant covers to control erosion by reducing raindrop impact on the soil surface through rainfall interception; reducing water and wind velocities through imparting roughness to the flow; increasing infiltration as plant roots open up the soil, thereby reducing runoff; and increasing the organic content of the soil. Mechanical measures are aimed at controlling the volume of runoff, reducing the velocity of water and wind, and increasing depression storage. They include ridging, terracing, shelterbelts and the installation of artificial waterways. Soil management is concerned with the application of fertilizers and manures to increase organic content and retain the aggregate structure of the soil and with tillage practices.

Although land capability classifications have been developed by the Soil Survey (Bibby and Mackney, 1969) and the Agricultural Land Service (1974) for general assessments of the physical attributes of the land, they have not been used in Britain for soil conservation design. Provision exists for erosion to be included in these classifications as a limiting factor but only in the cases of soils susceptible to wind erosion is it regularly applied. In Europe too the land capability classification is rarely used and conservation strategies are generally formulated in relation to erosion surveys carried out using various systems of geomorphological mapping (Verstappen and van Zuidam, 1968; Demek, 1971). The advantages of this geomorphological approach over the land capability classification have been emphasized by Tricart (1973) and Tricart and Kilian (1979). Geomorphological mapping is used in the only comparable survey carried out in Britain which is that undertaken jointly by the National College of Agricultural Engineering and Clwyd County Council to provide a basis for developing a management strategy, incorporating erosion control, for the Moel Famau Country Park (Baker *et al.*, 1979). The only other local authority report on soil erosion in Britain known to the author is by Clwyd's predecessor at Mold, Flintshire County Council, on the sand dune coast between Prestatyn and the Point of Air (Brown, 1973).

The question of determining the details of a conservation system rarely arises in Britain because conservation measures are not employed. The reasons for this in British farming are not hard to find. The use of agronomic techniques requires introducing new crops, often in a rotation system, and, in many areas, the crops which would provide suitable erosion control are not economic. The results of the Silsoe study, for example, show that erosion could be easily controlled on the sandy soils of Bedfordshire by putting the land down to grass. Yet, where farming is dominated by cereals and market-gardening this is clearly an impractical solution. Many of the mechanical measures which would be highly recommended in the United States are unsuitable for Britain. They are too expensive and, all too frequently, cannot be accommodated in existing field layouts. It is not uncommon in lowland England for fields in areas of irregular relief to have several depressions or dry valleys within them and local slopes up to 11 or 12 degrees. Fields on spur ends often slope in two directions, making contouring virtually impossible.

Given the limited awareness of soil erosion as a problem and the lack of incentives for introducing conservation measures, it is hardly surprising that very little research has been carried out in this country on conservation techniques. The work on the agricultural side has focused on the use of shelterbelts to control wind erosion. However, with the exception of fundamental studies by Caborn (1957) on the spacing and porosities of belts and the microclimatic effects of shelter, which paralleled or were in advance of studies in the United States, most work, at least until recently, has only repeated American experiments. This work is summarized by Davies and Harrod (1970) in their review of wind erosion in Britain and the various control measures available. Although little information is provided on costs, they conclude that only horticultural crops can justify much expenditure and that most farmers are forced to rely on tillage practices, inter-row cropping and the protection afforded by natural shelter to control erosion. With the continued removal of trees and hedgerows in eastern England and the poor maintenance of many of the shelterbelts which remain, researchers have looked to other means of controlling wind erosion. Experiments have been carried out in intercropping the maincrop with barley and mustard (Williams, 1972). As a result of these and other field trials information is now available on the advantages and disadvantages of the following methods of wind erosion control: planting straw; barley inter-row cropping; direct drilling; spraying the soil with polyvinyl acetate emulsion; erecting hessian windbreaks; planting willow hedges; and mixing peaty and sandy top soils with a clay subsoil. Although the last method is the cheapest, there is not everywhere a clay subsoil to permit its use. The next cheapest and more widely applicable technique is barley inter-row cropping. At £23.20/ha, including the cost of barley seed, drilling and subsequent destruction by herbicides before the main crop is fully mature (Rickard, 1978), it is easily the most attractive method of wind erosion control.

Research on soil conservation in upland areas has been largely carried out by the Countryside Commission and is essentially of a very practical nature. As a result of various field trials, methods of revegetating eroded areas and renovating footpaths are well-established, as exemplified by the restoration project at Tarn Hows in the Lake District (Barrow *et al.*, 1973). Probably because of the lack of a theoretical base to the work, which would be provided by research into erosion processes under visitor pressure, many of the solutions adopted by the Commission seem unnecessarily expensive. The problem of gully erosion on Box Hill was dealt with by large-scale landforming operations to fill in the gullies before the land was replanted with grasses, shrubs and trees (Streeter, 1977). Preliminary surveys investigated the extent of ground lowering but by techniques which did not allow the effects of erosion and compaction by trampling to be separated (Streeter, 1975). Thus, the cause and mechanics of the erosion were never properly established and without this information the Commission was not in a position to consider alternative strategies. In fact, for an organization devoted to conservation generally, soil erosion and erosion control form a surprisingly small part of the work of the Countryside Commission; only 13 of the 216 projects listed in the Commission's research register (Countryside Commission, 1976) deal with the topic. Practical low-cost conservation measures have been adopted by other bodies, for example, the work of the National Trust at Dundgeon Ghyll in the Lake District and that of the Malvern Conservators.

## IX What needs to be done?

There is an obvious need for basic research into conservation techniques to determine which are most relevant to British requirements. This is particularly important for the control of agricultural soil erosion because, at present, only very limited advice can be given to farmers and the conservation measures which are suggested are rarely attractive. The most profitable topics to investigate would appear to be alternative crops to grow; tillage practices; and the timing of cultivations.

### 1 *Alternative crops*

Research under this heading needs to examine the interception capacities of different crops; the changes in the kinetic energy and drop-size distribution of the rainfall reaching the ground surface as a result of interception by the crop and the coalescence of water drops on and their fall from the canopy; and the roughness imparted to water and air flow with different planting densities and at various stages of crop growth. Little fundamental work of this type has been carried out even in the United States. Yet, only through detailed studies of the relationships between crop morphology and soil erosion will a thorough understanding be obtained of why different crops result in different amounts of soil loss. Although it is logical to begin with investigations of the major crops such as wheat, barley, maize and grass, it is important, given the nature of the farming economy in the areas where most erosion occurs, to look at cabbages, leeks, carrots, rhubarb and strawberries.

### 2 *Tillage practices*

In spite of much fundamental research on tillage practices in Britain, examining the ideal soil conditions for seed-bed preparation and such effects as smearing and the breakdown of aggregate structure of the soil (Davies *et al.*, 1972; Spoor and Muckle, 1974), no attempt has been made to relate this work to erosion control. Where practicable some farmers have taken to contour cultivation, using a reversible plough so that soil is always turned upslope but this is only a partial solution to the problem and does not tackle the cause of erosion. Although the need for special tillage practices on erodible soils is recognized in the United States (Schwab *et al.*, 1966), only direct drilling has been recommended in Britain as a means of erosion control (Pidgeon and Ragg, 1979) and no studies have been made to determine the hydrological changes in the soil or the reduction in soil loss which may occur with its use.

Before recommending special practices which are often expensive and inconvenient to adopt, it would seem appropriate to investigate the effects of tillage on soil loss. The two main relevant effects are changes in surface roughness and increases in depression storage. Evans (in press) shows how roughness varies during the year from the rough surfaces produced by ploughing to the smoother ones resulting from disc and tine harrowing. No studies have related soil loss to these changes or attempted to define a critical roughness condition which will reduce erosion to an acceptable level. Basic work of this kind is being carried out by Martin (1978) which will eventually lead, it is hoped, to better recommendations to farmers on tillage practices for erosion control.

Soil conservation cannot be considered, however, in isolation from other aspects of land management. This is particularly true on sandy soils where water shortage is a common cause of crop failure. It would seem logical on these soils to look at soil and water conservation together and examine ways of increasing depression storage of water, thereby reducing runoff, possibly through tied ridging. Research into soil erosion control cannot be easily separated from the whole question of the more efficient use of water in agriculture.

## 3  *Timing of cultivations*

Several studies emphasize the importance of timing tillage operations to avoid structural damage to the soil (Spoor and Muckle, 1974; Spoor, 1975) but no one seems to have considered whether this work might not have an application for erosion control. The most likely reason for this is that in their search to derive a suitable index to describe the erodibility of a soil, researchers have tended to view soil erodibility as a constant for a particular soil. The results of the Silsoe study, however, show that soil erodibility varies throughout the year. The sandy soils in Bedfordshire are ten times more erodible in summer than in winter (Martin and Morgan, 1979). Further investigations of these seasonal variations are required to determine whether erosion can be reduced by restricting tillage operations to particular periods of the year.

## X Conclusions

Sufficient evidence exists to show that soil erosion in Britain is not insignificant. It may not be frequently observed by the general public, particularly on agricultural land where the evidence is either removed by cultivation or obscured by crop growth, but it certainly occurs. Its incidence is, however, localized in both time and space. Only certain storms give rise to erosion and only certain fields are affected. Thus, erosion is far from being a national problem and any attempt to make it an environmental issue, as a recent article (Reed, 1979) seems to try to do, is unwarranted.

Even so, the local importance of the problem must be recognized both on agricultural land and in areas under recreation pressure. Research is urgently needed to establish the true cost of the problem and to find solutions. Most of the conservation measures in current use are too expensive and inflexible to be attractive. Some very fundamental studies relating soil conservation methods to the mechanics of soil erosion are required so that realistic measures can be developed for the future.

*National College of Agricultural Engineering, Silsoe, Bedford*

*Acknowledgements*

The Silsoe study was carried out with a grant from the Natural Environment Research Council. Dr R. Evans provided information on the distribution of eroded sites. Messrs L. Martin and I. Bain drew attention to the local reports of soil erosion.

## XI References

**Agricultural Land Service** 1974: *Agricultural land classification of England and Wales*. London: Ministry of Agriculture, Fisheries and Food. (13 pp.)

**Al-Ansari, N. A., Al-Jabbari, M.** and **McManus, J.** 1977: The effect of farming upon solid transport in the River Almond, Scotland. *Symposium on erosion and solid matter transport in inland waters, International Association of Scientific Hydrologists Publication no.* 122, 118–25.

**Arnoldus, H. M. J.** 1977: Predicting soil losses due to sheet and rill erosion. *FAO Conservation Guide* 1, 99–124.

**Baker, C. F., Morgan, R. P. C., Brown, I. W., Hawkes, D. E.** and **Ratcliffe, J. B.** 1979. Soil erosion survey of the Moel Famau Country Park. *Clwyd County Council Planning Department, Country Park Research Report no. 2/National College of Agricultural Engineering Occasional Paper no. 7*, Mold and Silsoe (38 pp.).

**Barnes, F. A.** 1962: Peat erosion in the Southern Pennines: problems of interpretation. *East Midland Geographer* 2(20), 216–22.

**Barrow, G., Brotherton, D. I.** and **Maurice, O. C.** 1973: Tarn Hows experimental restoration project. *Countryside Recreation News Supplement* 9, 13–18.

**Bayfield, N. G.** 1971: Some effects of walking and skiing on vegetation at Cairngorm. In Duffey, E. and Watt, A. S., editors, *The scientific management of animal and plant communities for conservation, eleventh Symposium of British Ecological Society*, Oxford: Blackwell, 469–85.

**Bayfield, N. G.** 1973: Use and deterioration of some Scottish hill paths. *Journal of Applied Ecology* 10, 635–44.

**Bayfield, N. G.** and **Lloyd, R. J.** 1973: An approach to assessing the impact of use on a long distance footpath—The Pennine Way. *Countryside Recreation News Supplement* 8, 11–17.

**Bennett, H. H.** 1939: *Soil conservation*. New York: McGraw-Hill. (993 pp.)

**Bibby, J. S.** and **Mackney, D.** 1969: Land use capability classification. *Soil Survey, Technical Monograph no. 1*, Harpenden. (12 pp.)

**Bower, M. M.** 1960: The erosion of blanket peat in the Southern Pennines. *East Midland Geographer* 2(13), 22–33.

1961: The distribution of erosion in blanket peat bogs in the Pennines. *Transactions of the Institute of British Geographers* 29, 17–30.

1962: The cause of erosion in blanket peat bogs. A review of evidence in the light of recent work in the Pennines. *Scottish Geographical Magazine* 78, 33–43.

**Brade-Birks, S. G.** 1944: *Good soil*. London: English Universities Press. (304 pp.)

**Bridges, E. M.** and **Harding, D. M.** 1971: Micro-erosion processes and factors affecting slope development in the Lower Swansea Valley. In Brunsden, D. editor, *Slopes: form and process*, Institute of British Geographers Special Publication 3, 65–79.

**Brown, I. W.** 1973: *Reconnaissance survey of sand dune erosion: Prestatyn to The Point of Air*. Mold: County Planning Department, Flintshire County Council.

**Caborn, J. M.** 1957: Shelterbelts and microclimate. *Forestry Commission Bulletin no. 29*, Edinburgh: Her Majesty's Stationery Office.

**Christy, L. C.** 1971: Legislative principles of soil conservation. *FAO Soils Bulletin no. 15*. (68 pp.)

**Clayton, K. M.** 1964: *A bibliography of British geomorphology*. London: George Philip. (211 pp.)

**Coleman, R. A.** 1977: Simple techniques for monitoring footpath erosion in mountain areas of north-west England. *Environmental Conservation* 4, 145–48.

**Conway, V. M.** and **Millar, A.** 1960: The hydrology of some small peat-covered catchments in the Northern Pennines. *Journal of the Institute of Water Engineers* 14, 415–24.

**Countryside Commission** 1976: *Research Register no. 8*, Cheltenham: Countryside Commission.

**Curtis, L. F.** 1965: Soil erosion on Levisham Moor, North York Moors. In *Rates of erosion and weathering in the British Isles*, Symposium of the British Geomorphological Research Group.

**Davies, D. B., Eagle, D. J.** and **Finney, J. B.** 1972: *Soil management*. Ipswich: Farming Press. (254 pp.)

**Davies, D. B.** and **Harrod, M. F.** 1970: The processes and control of wind erosion. *NAAS Quarterly Review* 88, 139–50.

**180**

**Demek, J.** 1971: *Manual of detailed geomorphological mapping*. Brno: Czechoslovak Academy of Science Institute.

**Douglas, I.** 1970: Sediment yields from forested and agricultural lands. In Taylor, J. A., editor, *The role of water in agriculture*, Oxford: Pergamon, 57–88.

**Evans, R.** 1971: The need for soil conservation. *Area* 3, 20–23.

1977: Overgrazing and soil erosion on hill pastures with particular reference to the Peak District. *Journal of the British Grassland Society* 32, 65–76.

1978: Distribution of water eroded fields in lowland England and their characteristics. Paper presented to *Workshop on assessment of erosion in USA and Europe*, Rijksuniversitet, Gent, Belgium.

in press. Mechanics of water erosion. In Kirkby, M. J. and Morgan, R. P. C., editors, *Soil erosion*, Chichester: Wiley.

**Evans, R.** and **Morgan, R. P. C.** 1974: Water erosion of arable land. *Area* 6, 221–25.

**Evans, R.** and **Nortcliff, S.** 1978: Soil erosion in north Norfolk. *Journal of Agriculture Science Cambridge* 90, 185–92.

**Fournier, F.** 1972: *Soil conservation*. Nature and Environment Series, Council of Europe. (194 pp.)

**Geikie, A.** 1868: On denudation now in progress. *Geological Magazine* 5, 249–54.

**Hall, D. G.** 1967: The pattern of sediment movement in the River Tyne. *Institute of the Association of Scientific Hydrologists Publication no. 75*, 117–42.

**Harvey, A. M.** 1974: Gully erosion and sediment yield in the Howgill Fells, Westmorland. In Gregory, K. J. and Walling, D. E., editors, *Fluvial processes in instrumented watersheds*. Institute of British Geographers Special Publications 6, 45–58.

**Hoskins, W. G.** 1970: *The making of the English landscape*. Harmondsworth: Penguin. (326 pp.)

**Hudson, N. W.** 1967: Why don't we have soil erosion in England? In Gibb, J. A. C., editor, *Proceedings of Agricultural Engineering Symposium*, London: Institute of Agricultural Engineers Paper 5/B/42.

1971: *Soil conservation*. London: Batsford. (320 pp.)

**Imeson, A. C.** 1971: Heather burning and soil erosion on the North Yorkshire Moors. *Journal of Applied Ecology* 8, 537–42.

1974: The origin of sediment in a moorland catchment with particular reference to the role of vegetation. In Gregory, K. J. and Walling, D. E., editors, *Fluvial processes in instrumented watersheds*, Institute of British Geographers Special Publication 6, 59–72.

**Jacks, G. V.** and **Whyte, R. O.** 1939: *The rape of the earth. A world survey of soil erosion*. London: Faber. (313 pp.)

**Johnson, R. H.** 1958: Observations on the stream patterns of some peat moorlands in the Southern Pennines. *Memoirs and Proceedings of the Manchester Literary and Philosophical Society* 99, 1–18.

**Kirkby, M. J.** 1967: Measurement and theory of soil creep. *Journal of Geology* 75, 359–78.

**Ledger, D. C., Lovell, J. P. B.** and **McDonald, A. T.** 1974: Sediment yield studies in upland catchment areas in south-east Scotland. *Journal of Applied Ecology* 11, 201–206.

**Liddle, M. J.** 1973: *The effects of trampling and vehicles on natural vegetation*. Unpublished PhD thesis, University College of North Wales, Bangor.

1975: A selective review of the ecological effects of human trampling on natural ecosystems. *Biological Conservation* 7, 17–36.

**Liddle, M. J.** and **Greig-Smith, P.** 1975: A survey of tracks and paths in a sand dune ecosystem. *Journal of Applied Ecology* 12, 893–930.

**Liddle, M. J.** and **Moore, K. G.** 1974: The microclimate of sand dune tracks: the relative contribution of vegetation removal and soil compression. *Journal of Applied Ecology* 11, 1057–1068.

**Martin, L.** 1978: Accelerated soil erosion from tractor wheelings: a case study in mid-Bedfordshire. Paper presented to *Seminar on agricultural soil erosion in temperate non-Mediterranean climates*, Université Louis Pasteur, Strasbourg, France.

**Martin, L.** and **Morgan, R. P. C.** 1979: The impact of drought and post drought on soil erosion in mid-Bedfordshire. In *Atlas of drought in Britain, 1975–76*, London: Institute of British Geographers.

**Morgan, R. P. C.** 1977: Soil erosion in the United Kingdom: field studies in the Silsoe area, 1973–75. *National College of Agricultural Engineering, Occasional Paper no. 4.* (41 pp.)

1978: Field studies of rainsplash erosion. *Earth Surface Processes* 3, 95–99.

1979: *Soil erosion*. London: Longman. (114 pp.)

**Natural Environment Research Council** 1975: *Flood studies report. II Meteorological studies*. London: NERC.

**Oakley, K. P.** 1945: Some geological effects of a cloud burst in the Chilterns. *Records of Buckinghamshire* 15, 265–80.

**Pidgeon, J. D.** and **Ragg, J. M.** 1979: Soil, climatic and management options for direct drilling cereals in Scotland. *Outlook on Agriculture* 10, 49–55.

**Radley, J.** 1962: Peat erosion on the high moors of Derbyshire and west Yorkshire. *East Midland Geographer* 3(17), 40–50.

1965: Significance of major moorland fires. *Nature* 205, 1254–59.

**Radley, J.** and **Simms, C.** 1967: Wind erosion in east Yorkshire. *Nature* 216, 20–22.

**Reed, A. H.** 1979: Accelerated erosion of arable soils in the UK by rainfall and runoff. *Outlook on Agriculture* 10, 41–8.

**Richardson, E. G.** 1939: Erosion and soil conservation in theory and practice. *Science Progress* 34, 63–75.

**Richter, G.** 1977: *Bibliographie zur Bodenerosion und Bodenerhaltung 1965–75*. Universität Trier. (97 pp.)

**Rickard, P. C.** 1978: *Systems of wind protection for field crops*. London: Ministry of Agriculture, Fisheries and Food Note. (12 pp.)

**Robinson, D. N.** 1968: Soil erosion by wind in Lincolnshire, March 1968. *East Midland Geographer* 4(30), 351–62.

**Rodda, J. C.** 1970: Rainfall excesses in the United Kingdom. *Transactions of the Institute of British Geographers* 49, 49–60.

**Schwab, G. O., Frevert, R. K., Edminster, T. W.** and **Barnes, K. K.** 1966: *Soil and water conservation engineering*. New York: Wiley. (683 pp.)

**Schwing, J. F.** 1978: Les manifestations graves et les évènements catastrophiques d'érosion des sols dans le vignoble Alsacien. In *Erosion agricoles des sols. Problèmes de méthode. Applications en Alsace*. Strasbourg: UER de Géographie, Université Louis Pasteur, Recherches Géographiques no. 9, 29–46.

**Slaymaker, H. O.** 1972: Patterns of present subaerial erosion and landforms in mid-Wales. *Transactions of the Institute of British Geographers* 55, 47–68.

**Spence, M. T.** 1955: Wind erosion in the Fens. *Meteorology Magazine* 84, 304–307. 1957: Soil blowing in the Fens in 1956. *Meteorology Magazine* 86, 21–22.

**Spoor, G.** 1975: Fundamental aspects of cultivation. In *Soil physical conditions and crop production*, MAFF Technical Bulletin no. 29, 128–44.

**Spoor, G.** and **Muckle, T. B.** 1974: Influence of soil type and slope on tractor and implement performance. In Mackney, D., editor, *Soil type and land capability*, Harpenden: Soil Survey Technical Monograph no. 4, 125–34.

**Stallings, J. H.** 1957: *Soil conservation*. Englewood Cliffs, New Jersey: Prentice-Hall. (575 pp.)

**Stamp, L. D.** 1962: *The land of Britain: its use and misuse*. London: Longman.

**Streeter, D. T.** 1975: Preliminary observations on rates of erosion on Chalk downland footpaths. Paper presented to Symposium on environmental impact on recreational areas, Annual Meeting of the Institute of British Geographers, Oxford. 1977: Gully restoration on Box Hill. *Countryside Recreation Review* 2, 38–40.

**Thomas, T. M.** 1965: Sheet erosion induced by sheep in the Pumlumon (Plynlimon) area, mid-Wales. In *Rates of erosion and weathering in the British Isles*, Symposium of the British Geomorphological Research Group, 11–14.

**Tricart, J.** 1973: La géomorphologie dans les études intégrées d'aménagement du milieu naturel. *Annales de Géographie* 82, 421–53.

**Tricart, J.** and **Kilian, J.** 1979: *L'éco-géographie*. Paris: François Maspero. (326 pp.)

**Tuckfield, C. G.** 1965: Rate of erosion by gullying in the New Forest. In *Rates of erosion and weathering in the British Isles*, Symposium of the British Geomorphological Research Group.

**Verstappen, H. T.** and **van Zuidam, R. A.** 1968: *ITC system of geomorphological survey*. Delft: ITC.

**Walling, D. E.** 1974a: Suspended sediment and solute yields from a small catchment prior to urbanization. In Gregory, K. J. and Walling, D. E., editors, *Fluvial processes in instrumented watersheds*, Institute of British Geographers Special Publication 6, 169–92. 1974b: Prediction and modelling of suspended sediment production in small catchments. Paper presented to a Meeting of the Basin Sediment Systems Sub-group of the British Geomorphological Research Group, London School of Economics.

**Warren, A.** 1974: Managing the land. In Warren, A. and Goldsmith, F. B., editors, *Conservation in practice*, Chichester: Wiley, 37–56.

**White, G.** 1788: *The natural history of Selborne*. Edition edited by Mabey, R. (1977) Harmondsworth: Penguin.

**Wilkinson, B., Broughton, W.** and **Parker-Sutton, J.** 1969: Survey of wind erosion on sandy soils in the East Midlands. *Experimental Husbandry* 18, 53–9.

**Williams, A. M.** 1972: Wind protection for soil and crops. *Agriculture* 79, 148–54.

**Wischmeier, W. H.** and **Smith, D. D.** 1965: Predicting rainfall erosion losses from cropland east of the Rocky Mountains. *USDA Agricultural Research Service Handbook no. 282*.

**Young, A.** 1958: A record of the rate of erosion on Millstone Grit. *Proceedings of the Yorkshire Geological Society* 31, 149–56. 1960: Soil movement by denudational processes on slopes. *Nature* 188, 120–22.

Reprinted from *Geog. Annaler* **54-A:**203-220 (1972)

# MEASUREMENTS OF RUNOFF AND SOIL EROSION AT AN EROSION PLOT SCALE WITH PARTICULAR REFERENCE TO TANZANIA

BY PAUL H. TEMPLE

Department of Geography, University of Dar es Salaam

ABSTRACT. The principal results of a series of experiments designed to measure runoff and soil losses at an erosion plot scale in tropical and sub-tropical African environments are tabulated and discussed. Most of these results are not widely known, even in Tanzania.

Methodology as well as data are commented upon. Erosion plot data are particularly significant as a guide to local land use practice at the farm plot or field scale and as indicators of the magnitude and processes involved in runoff and soil erosion. The problems of extrapolating the results to larger scales and different areas are discussed.

In zones such as semi-arid central Tanzania where soil moisture availability is the chief limitng factor on crop yields and where soils are readily eroded under careless cultivation or through overgrazing, quantitative data on runoff and soil erosion are of particular value. In zones like the mountain footslopes of Kilimanjaro and Meru, where rainfall is higher, moisture deficiency is still a problem at the end of the dry season and soil erosion can be serious when steep land is cleared for cultivation.

## Introduction

The basic data discussed in this paper are drawn from reports of agricultural and veterinary research stations in Tanzania and neighbouring areas. These data were generally obscurely published and many of the results are now lost. The available data merit wider circulation and further analysis as they provide quantitative information of considerable scientific interest on the magnitude and process of soil erosion and water losses by runoff at the small farm plot (*shamba*) or field scale. The results are thus also of practical value to local farmers. This review of data collected from erosion plots is aimed at complementing data collected at different scales (eg. small catchments and large river basins) reported elsewhere in this volume.

Measurements on the erosion plot scale aim to provide quantitative data on water and soil loss from hillslope segments under differing but carefully monitored conditions. Erosion plot data have been widely employed as a means of determining the relative merits of:

a) different crop covers,

b) different conservation measures under the same crop

They have also been employed as a means of comparing such losses from otherwise uniform plots when one parameter (eg. slope, soil etc) is varied. Erosion plot data have provided very significant practical information on cultivation and conservation methods as well as on slope and soil evaluation.

The design and scale of such plots renders them most suitable for studies of raindrop impact and sheet wash effects on limited slope segments. They are unsuited to the measurement of other erosional processes eg. gullying or soil creep for which other techniques are used (cf. Seginer 1966; Young 1972).

Furthermore considerable caution must be exercised in the extrapolation of results obtained from such studies to different localities or to different scales. Processes operating on limited slope segments are not the same either in nature or magnitude to those operating over whole slopes or in river basins. (Carson & Kirkby 1972).

In order to facilitate comparison with other results set out in this volume, all original data were transformed into metric equivalents. Conversion of tons to $m^3$ was made by division by 1.5. Presentation of data on soil loss in $m^3$/ha should not be assumed to apply uniformly to larger areas: the scale of the plots where the data were obtained should be noted. Thus quantities tabulated have a greater relative than absolute value, as is reflected by comments on them below.

## Methodology

Accurate measurements of runoff and soil loss through splash effects and sheetwash on natural hill slopes are difficult to obtain and frequently difficult to interpret because of the complex controls operating. In order to isolate particular controls for detailed examination, erosion plots have been set up to monitor both processes and rates of water and soil loss during different types or runoff. Water and soil are most commonly collected in tanks, or pans: this sediment and water loss can then be measured and results obtained for both parameters and related to the size of the plot. The detailed techniques of such experiments are described by Gerlach (1967), Leopold and Emmett (1967) and Schick (1967). Data obtained by comparable methods are available for a number of localities in Tanzania.

Other methods have been employed in an attempt to obtain a more comprehensive picture of changes over whole slopes. Slopes may be periodically resurveyed between fixed points and the resultant profiles compared (Menne 1959, King & Hadley 1967). To be meaningful, surveys must be done with very considerable accuracy. In addition, changes of ground surface level caused by erosion and deposition can be measured at points identified by fixed and graduated rods placed vertically in the ground (Evans 1967, Schumm 1967). This method is the easiest of the three to operate but measurements may often be difficult to interpret even if they are carefully made and the rod placing well-designed.

## Some results obtained from plot measurements of runoff and soil erosion

### Rhodesia

Experiments designed to measure runoff and erosion, the mechanics of erosion and methods for its prevention in Rhodesia are of relevance to proper land use practices in the drier parts of Tanzania. The design, instrumentation and operation of these experiments are described by Hudson (1957a), and some of the principal results are set out by Hudson (1957b & 1959) and Hudson and Jackson (1959). Besides presenting unexpected data on the relation between maize yield and soil loss in terms of maize yield per unit soil loss, Hudson and Jackson present data on soil loss in relation

Table 1. Soil erosion from three clay soil plots with different slopes under identical treatment (continuous maize); soil loss in m³/ha (modified after Hudson & Jackson 1959).

| Season | Slope in degrees | | |
|---|---|---|---|
| | 3.5 | 2.5 | 1.5 |
| 1953/54 | 6.6 | 4.1 | 4.6 |
| 1954/55 | 2.8 | 1.0 | 1.8 |
| 1955/56 | 6.9 | 2.5 | 1.7 |
| 1956/57 | 11.0 | 7.1 | 3.6 |
| 1957/58 | 1.2 | 0.5 | 3.3 |
| 1958/59 | 11.5 | 6.9 | 3.5 |
| Average | 6.7 | 3.7 | 3.1 |

to slope (Table 1). These data document a clear relationship between slope angle and soil loss even over low gradients; a plot of 3.5° lost more than double the soil lost from a plot of 1.5°. The authors comment "that the effects of many factors controlling erosion are, in comparison with American results (eg. Zingg, 1940), exaggerated by the greater erosivity of tropical rainfall".

Data on processes are also presented, concerning the relative importance of raindrop splash and surface flow in a sub-tropical savanna environment (100 cm average annual rainfall). Two identical plots were kept free of weeds so that there should be no impedance of surface flow: over one plot a fine wire gauze was suspended which broke the force of the raindrops and allowed rain to fall through as fine spray: the other plot was left without such a cover. The results are presented in Table 2: the uncovered plot lost over 100 times more soil than the gauze-covered plot over a period of 6 years. Hudson and Jackson comment that "in conditions of subtropical thunderstorms with high intensities and large drop sizes very severe erosion occurs when there is no protective cover, and this can be completely controlled by complete cover". A further interesting result was that on a comparable plot with a dense grass sward, giving both full cover *and* impedance to surface flow, soil loss was not significantly less than that from the gauzed plot. They therefore concluded that "the well known ability of grass to reduce erosion is thus almost entirely due to its cover effect, since no (significant) additional benefit is obtained from the surface impedance effect ... erosion should be influenced more by cover

**185**

Table 2. Soil erosion from two plots, one bare, one with a gauze cover; soil loss in $m^3$ ha (modified after Hudson & Jackson 1959).

| Season | Bare | Gauzed |
|---|---|---|
| 1953 54 | 101.3 | 0.0 |
| 1954 55 | 372.0 | 1.5 |
| 1955 56 | 99.7 | 3.3 |
| 1956 57 | 90.6 | 0.2 |
| 1957 58 | 36.4 | 0.2 |
| 1958 59 | 147.8 | 1.8 |
| Average | 141.3 | 1.2 |

than soil factors, eg. slope and infiltration rate, which (thus only) affect quantity and rate of runoff" (Ibid, p. 580). The authors present other data substantiating this view, particularly significant being the data presented in Table 3, which tabulates soil loss and runoff from various crops, arranged horizontally in order of increasing cover and vertically in terms of land class. The table indicates that there is a decrease in soil loss and runoff on each soil type as cover increases and also that, over progressively poorer land classes, soil loss and runoff increase, except over the coarser sandy soils with high infiltration capacities, when runoff is reduced compared to that from finer sands. Even over coarse sands, runoff is inversely related to cover and soil loss is greater than from finer sands.

## Uganda
Experimental data are available from Uganda (Namulonge, 0°30′N, 32°37′E: 1300 m a.s.l.) on the high intensities of tropical rainstorms. The significance of rainfall intensity in determining runoff and the amount of soil lost by erosion was recognised and measured (Hut-

chinson, Manning, & Farbrother 1958) using special instrumentation over a period of 6 years.

The data demonstrate important characteristic relationships between intensity, duration and amount of rainfall in tropical storms. It was not however found possible to predict peak rate from amount for any specific storm due to the great variations. The collected data demonstrate a straight line log relationship between log duration and intensity for individual storms and between log amount and rate.

Small erosion plots of circular shape, smooth surface and low slope (1°) and $3m^2$ area were given different treatments (Table 4) and runoff and percolation rates recorded. The authors argue that while "it is usual in studies of the effects of cropping practices and soil treatments to measure susceptibility to erosion in terms of runoff ... in view of the dependence of runoff on rainfall intensity, the use of percolation rate is to be preferred as an index of a surface treatment" (Ibid, p. 258). Percolation rates showed greater consistency than runoff variations and may thus be a better and more significant index of moisture conservation (though not necessarily of the susceptibility of the soil to erosion).

Furthermore such rates are not a function of plot scale as is runoff.

Table 4 indicates that at Namulonge, ten times more runoff occurred from bare plots than from grass-covered plots, and that a grass-mulch cover was twice as effective as a stone mulch in terms of runoff control. Infiltration of water into the soil was most rapid under grass cover being three times as rapid as on a bare plot thus showing that a grass cover

Table 3. Runoff and soil erosion from various crops, soils and slopes; runoff (a) as percentage of seasonal total; soil loss (b) in $m^3/ha$ (modified after Hudson & Jackson, 1959).

| Land class and soil type | Late planted green manure crop | | Maize (ordinary) | | Maize (good) | | Grass row cropped | | Grass uniform cover | |
|---|---|---|---|---|---|---|---|---|---|---|
| | a | b | a | b | a | b | a | b | a | b |
| Class II Clay | 10 | 2.3 | 6 | 1.8 | 4 | 1.2 | 1 | 0.3 | — | — |
| Class III Clay: 1.5° | 19 | 4.6 | 19 | 2.6 | — | — | 1 | 0.3 | — | — |
| Class III Clay: 2.5° | 19 | 6.8 | 20 | 3.6 | 3 | 1.3 | 2 | 0.7 | — | — |
| Class III Clay: 3.5° | 21 | 17.5 | 23 | 6.8 | 7 | 2.0 | 3 | 0.8 | 3 | 0 |
| Class III Sand | — | — | 23 | 8.2 | 17 | 6.1 | — | — | 6 | 0.5 |
| Class IV Sand | — | — | 15 | 21.9 | 12 | 10.6 | — | — | 3 | 1.0 |

186

Table 4: Runoff in mm (a) and overall percolation rates in mm/hour (c) for small plots under various treatments at Namulonge (modified after Hutchinson, Manning & Farbrother 1958).

| Year | Rain mm (1) | Rain-storms > 6 mm | Bare soil | | Stone mulch | | Grass mulch | | Grass (Cynodon) | |
|---|---|---|---|---|---|---|---|---|---|---|
| | | | a | c | a | c | a | c | a | c |
| 1951 | 1840 | 1148 | 666 | 23 | 158 | 53 | 94 | 64 | 165[1] | 53[1] |
| 1952 | 1180 | 633 | 264 | 33 | 51 | 58 | 41 | 66 | 20 | 76 |
| 1953 | 1220 | 805 | 493 | 18 | 201 | 43 | 61 | 79 | 69 | 74 |
| Average | 1410 | 862 | 474 | 25 | 137 | 51 | 65 | 70 | 45[2] | 75[2] |

[1] Before full turf cover established:
[2] Average of two years.
(1) annual rainfall totals after E.A. Met. Dept., 1966.

not only retains water at the surface but also allows it to penetrate more rapidly into the soil than all the other treatments considered. A grass mulch was however almost as effective in encouraging infiltration and percolation.

*Tanzania*
A considerable volume of valuable data exists for certain sites in Tanzania, data which have not been previously assembled or properly discussed. A full analysis of the early Mpwapwa data was never published (Staples 1939). Data exist in various detail for the following sites:

(a) Mpwapwa; (b) Lyamungu; (c) Tengeru; (d) Shingyanga; (e) Mfumbwe (Temple & Murray-Rust 1972).

(a) *Mpwapwa* (6°20'S, 36°30'E: c. 1128 m a.s.l.)
Experiments designed to obtain quantitative data on runoff and erosion in a semi-arid area of Tanzania were initiated at Mpwapwa in 1933 (Staples 1934).

Comparative data for erosion plots differing with regard to plant cover and cultivation treatment were obtained under controlled conditions for at least 7 years (up to at least 1939): unfortunately only 2 years results are now available. Data collection from the erosion plots was resumed in 1946 under different treatments and a further 8 years results published in summary form (van Rensburg 1955).

Six plots were established on a uniform pediment or fan slope of 3.5° at the foot of the Kiboriani mountains. The slope was underlain by a coarse, red sandy, friable loam (clay percentage 33, silica—sesquioxide ratio 1 : 92).

The soil dried to at least 2 m by the end of the dry season and had a highly porous structure but showed no cracking or fissuring (Milne 1932). Staples observed that this soil appeared to be the least erodible of the three principal soil types of the area. Two further plots under deciduous thicket cover on the same site but having a slope of 4.5° were added in February 1935. Plots 1 to 5 measured 27.7 m downslope by 1.8 m wide (area 50 m²). Plot 6 had the same width as the others but only half their length. Plots 7 and 8 had the same length as plots 1—5 but had double the width of all other plots.

The plots were enclosed by low walls and runoff and soil collected in tanks built flush with the soil surface at the bottom of each plot. Runoff was measured after each storm by reference to the water level in the covered tanks and soil loss by weighing dry collected sediment after each rainy season.

Plot treatments were as follows:

Plot 1: Bare: no cultivation: all weeds carefully removed until 1938 when the plot was allowed to regenerate naturally to thicket in order to assess the speed of this process.
Plot 2: Bare: flat cultivation to between 7—15 cm depth: no cropping: kept weed free until 1938, then allowed to regenerate like plot 1.
Plot 3: Bulrush millet (*Pennisetum typhoides*): ridge cultivated down slope after initial flat cultivation; weeded.
Plot 4: Sorghum (*Sorghum vulgare*): after initial experimentation, ridge cultivated downslope until 1938: then flat cultivated.
Plot 5: Grass (*Cenchrus ciliaris*) after flat cultivation, grass seeded giving 50 % surface cover at 8 cm above surface by year two: cut subsequently for hay and burnt at the end of each dry season: the grass has an erect and tufted habit of growth and grows up to 1 m high.
Plot 6: Treatment identical to plot 2 but the plot was half the length of the rest.

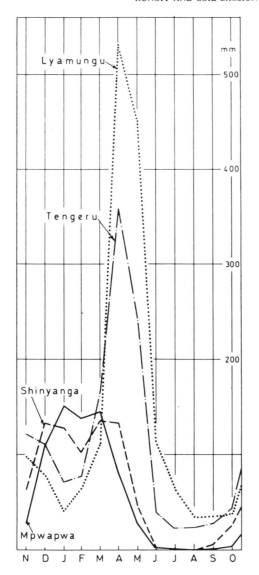

Figure 1. Mean monthly rainfalls at Mpwapwa, Lyamungu, Tengeru and Shinyanga (1931—60) (after E. Afr.Met.Dept., 1966).

Plot 7: Deciduous thicket: a closed formation of well-developed secondary growth: under African cultivation some 30 years before: dominant tree species *Grewia platyclada*, *Commiphora pilosa*, *Commiphora* sp. nr. *lindensis* with a few specimens of *Acacia tortilis* ssp. *spirocarpa*, *A. senegal*, *Albizzia petersiana* and *Hippocratea buchananii*: trees from 6 to 7.5 m high, below which a dense shrubby stratum up to 3 m high of *Grewia bicolor*, *Vitex* sp., *Allophylus rubifolius* and *Gnidia* sp. (= *Lasiosiphon emini*). Plots 7 and 8 were twice the width of previous plots to obtain a representative area of vege-

tation: no grazing: added to the experiment in January 1935 as was Plot 8.
Plot 8: Deciduous thicket as above but browsed by goats.

Slight inconsistencies should be noted between the various reports on the treatment of some plots at particular times. These treatments were designed to simulate the major land use variations of the area. Plot 1 was designed to resemble the bare slopes and hard compacted soil common in the more densely populated areas of central Tanzania. Plot 2 represented a variant of the above with surface compaction broken by cultivation. Plots 3 and 4 were cropped in rough accordance with local Gogo practice. Plot 5 was designed to test grass cover effects in a region where stock are of great importance. Plots 7 and 8 were representative of the vegetation covering large areas of dry central Tanzania (Staples 1935).

Mpwapwa has an average annual rainfall of 690 mm, 91 % of which falls on average over 65 days in the period December through April (Fig. 1). Rainfalls are very erratic even during this period: downpours of over 75 mm in an hour have been recorded.

During the 1933/34 rains 31 storms generated runoff, the smallest totalling 3 mm, the largest 72 mm (Fig. 2). During the 1934/35 rains, 28 storms generated runoff, the smallest totalling 1 mm, the largest 57 mm (Fig. 2).

Data from plot 1 (bare ground) are presented in Fig. 2. This demonstrates a clear relationship between rainstorm amount and runoff, particularly significant being the very rapid multiplication of runoff amounts in association with the larger storms. The majority of these are convectional, presumably showing a very strong correlation between storm amount and rainfall intensity (Hutchinson & others, see above).

The scatter of points may be explained partly by differences in rainfall intensity and partly by soil water conditions. For example, comparing storms A1 and A31, the first and last runoff generating rains of the 1933/34 season, the last storm generated nearly 8 times more runoff from the plots than did the first storm though it was only 5 % heavier. This indicates that soil moisture storage was inhibiting infiltration at the end of the rains and favouring greater runoff, precipitation being normally just sufficient during the rains to

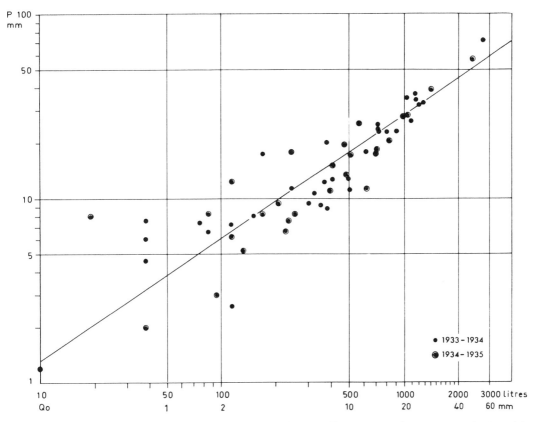

Figure 2. Runoff (Qo) in litres and mm for all storms causing runoff over two rainy seasons against precipitation (P) in mm on bare ground at Mpwapwa (data from Staples 1934, 1936). Size of plots are 50 m².

saturate the soil (Fig. 3).

Runoff from bare crusted soil occurred 59 times in 2 seasons compared to 7 times from a grass cover (mostly before the grass cover was fully developed) and an average of 13 times from the cultivated plots. There was no runoff from plot 4 while it was cultivated (for half a rainy season) with contour ridging; these ridges obstructed the wash flow downslope and afforded a larger soil surface for moisture infiltration than flat cultivation. The results show that grass cover assists infiltration (see above) while cultivation, by opening up the surface functions in the same way, but less effectively.

Extremely severe water losses from bare ground and cultivated plots were recorded. For example Storm B 20, the heaviest of the 1934/35 rains (57 mm) caused a percentage loss of 83.7 from plot 1, 70.9 from plot 2, 48.0 from plot 3 but only 9.2 from the grass plot. If

soil wash losses had been measured for each storm rather than for each season, there is no doubt that excessive soil loss would also have been demonstrated to be related to the heaviest storm (Temple & Rapp 1972). One such heavy storm of 51 mm in 30 minutes caused a loss of nearly 3500 m³ from a 10 hectare Kenya coffee field with a slope of 8°; this approximates to a soil loss of 35 mm (Brook 1955).

Table 5a shows the percentage of the annual total precipitation lost by each plot. Over two season's recordings, runoff from bare uncultivated soil was over half of the total season's precipitation and showed more than 53 times the loss recorded from an established grass cover, which lost less than 1 % of the seasonal rainfall supply to runoff (see Hudson & Jackson 1959, Table 3 above). Runoff loss from the sorghum plot was 29 times the loss from established grass. The deciduous ungrazed thicket plots lost only half the amount

**189**

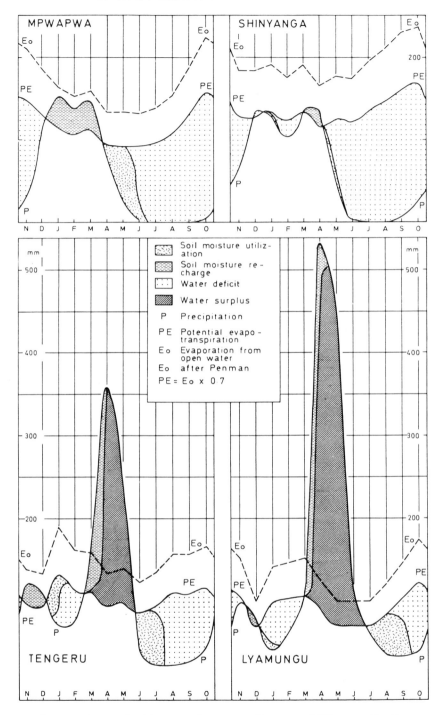

Figure 3. Water balance diagrams for Mpwapwa, Lyamungu, Tengeru and Shinyanga. For $E_0$ conversion to PE see Penman (1963).

Table 5. Runoff, soil erosion and sediment concentration, Mpwapwa erosion plots (modified after Staples 1936).
a) Percentage runoff of total precipitation: rainfall 1933/34, 675 mm; 1934/35, 564 mm.
b) Soil erosion in m³/ha. Quantity relates to less than half the 1933/34 rainy season, before which, ridges on contour and no soil loss.
c) Sediment concentration total in runoff in mg/l.

| Season | Plot 1 Bare: uncult. | Plot 2 Bare: flat cult. | Plot 3 Bare: ridge cult. | Plot 4 Bulrush millet or sorghum | Plot 5 Grass | Plot 6 Bare: flat cult. | Plot 7 Decid. ungr. | Plot 8 Thicket |
|---|---|---|---|---|---|---|---|---|
| a. | | | | | | | | |
| 1933/34 | 52.9 | 34.8 | 25.2 | 29.1 | 2.8 | 33.3 | ? | ? |
| 1934/35 | 47.8 | 28.2 | 20.8 | 22.9 | 0.9 | 27.1 | 0.5 | 0.4 |
| Average | 50.4 | 31.5 | 23.0 | 26.0 | 1.9 | 30.2 | 0.5 | 0.4 |
| b. | | | | | | | | |
| 1933/34 | 97.0 | 80.4 | 20.9* | 66.5 | 0 | 37.5 | ? | ? |
| 1934/35 | 98.7 | 90.0 | 44.7 | 37.5 | 0 | 25.0 | 0 | 0 |
| Average | 97.8 | 85.2 | 43.3+ | 52.0 | 0 | 31.3 | 0 | 0 |
| c. | | | | | | | | |
| 1933/34 | 42,205 | 53,180 | 53,280 | 37,515 | — | 25,745 | — | — |
| 1934/35 | 56,275 | 86,805 | 73,935 | 52,285 | — | 30,235 | — | — |
| Average | 49,240 | 69,993 | 63,608 | 44,900 | — | 27,990 | — | — |

Average determined by doubling 1933/34 value and probably an underestimation.

of water through runoff that was lost from the grass plot and less than 1 % of the water lost from bare uncultivated ground. This indicates that complete clearing of thicket by fire and felling on even small slope segments may be expected to increase runoff loss by two orders of magnitude.

Soil loss was, for technical reasons, determined only on a seasonal basis and not for individual storms. Table 5b presents the data. Despite a lower seasonal rainfall (15 % below average) and lower rainfall intensities in 1934/35 compared to 1933/34 and lower runoff, soil losses were greater in the second year of the trial than in the first. This anomaly may be explained by the progressive reduction in soil organic matter (primarily grass roots) causing a greater erodibility of the soil (decreased infiltration and moistureholding capacity and decreased binding effect) in 1934/35. Soil loss was negligible from the thicket and grass plots despite some runoff from both plots. (Cf. Rapp, Murray-Rust et al., 1972, Fig. 5).

The effect of slope length on soil loss is clearly demonstrated by reference to plots 2 and 6 which had identical treatments. Plot 6, half the length of plot 2, lost only 30 % of the soil lost from plot 2. The effect of length of slope on soil loss has been established by Zingg (1940).

Table 5c shows the relationship of total runoff and total soil loss for both seasons in the form of sediment concentration values for each plot. The wide contrast in values between the two seasons data is evident, the rise in concentration on all plots showing measurable loss being a function of the lower runoff and great soil loss experienced in the second year of the trials. Concentration values are greatest for plot 2, indicative of the fact that more water percolated into the soil on this plot than into the compact soil of plot 1, causing 60 % lower runoff (Table 5a) but that the lower runoff was able to transport almost as much soil, as a result of the surface soil being loose due to cultivation.

In late 1946 recording of soil and runoff losses were recommenced on Plots 1 to 6, under modified treatments. The results are reported by van Rensburg (1955).

Plot 1 was cultivated and planted with sorghum without any conservation method. Plot 2 was left under grass cover (mainly *Cynodon plectostachyus,* a robust creeping grass growing up to 1 m tall. Plot 3 was divided into two; the top half under sorghum as on plot 1, the bottom half under grass as on plot 2. Plot 4 was treated in the same way as plot 1. Plot 5 was under grass cover (mainly *Cenchrus ciliaris;* see above). The shorter plot 6 was

**191**

Table 6. Runoff and soil erosion from various crops on identical plots at Mpwapwa; runoff (a) as percentage of annual total; soil loss (b) in m³/ha (modified after van Rensburg 1955).

| Year | Rainfall mm | Cultivated plot | | Cultivated plot + narrow grass belts across slope | | Plot 50 % cultivated 50 % grass | | Grass plot | |
|------|------|------|------|------|------|------|------|------|------|
| | | a | b | a | b | a | b | a | b |
| 1946 47 | 780 | 9.0 | 18.5 | — | — | 3.9 | 3.8 | 1.2 | 0.6 |
| 1947 48 | 530 | 16.4 | 14.3 | — | — | 10.4 | 2.1 | 3.4 | 0.3 |
| 1948 49 | 650 | 21.8 | 44.1 | — | — | 19.2 | 1.8 | 7.3 | 0.5 |
| 1949 50 | 580 | 12.8 | 8.3 | — | — | 13.1 | 0.6 | 4.7 | 0.2 |
| 1950 51 | 670 | 26.7 | 65.3 | 19.1 | 7.7 | 15.4 | 2.2 | 5.3 | 0.9 |
| 1951 52 | 860 | 28.9 | 86.4 | 10.0 | 30.3 | 9.0 | 1.8 | 6.5 | 0.7 |
| 1952 53 | 410 | 13.3 | 6.2 | 11.8 | 3.0 | 6.9 | 0.9 | 3.9 | 0.3 |
| 1953 54 | 520 | 25.5 | 48.8 | 21.2 | 37.2 | 14.3 | 3.4 | 7.2 | 0.4 |
| Average | 690 | 19.3 | 36.5 | 15.5 | 19.6 | 11.6 | 2.1 | 4.9 | 0.5 |
| Average[1] | | — | 36.9 | — | 20.2 | — | 2.2 | — | 0.5 |

[1] Including soil washed to the bottom of the plot but prevented by the tank lip from entering the tank: for grass plot none.

divided into two; the top half treated as Plot 1, the bottom half as plot 5. At the beginning of the 1950/51 rainy season the treatment of plot 5 was modified; previously under grass for 4 years, it was cultivated but two 2 m wide grass belts were left a third and two-thirds of the distance down the plot.

By the time of this later series of measurements soil transfer *on the cultivated plots only* had caused vertical erosion at the plot tops of over 30 cm and accumulation at the base of the plots which affected the functioning of the experiments. This accumulation of debris had to be removed in order to prevent runoff flowing over the containing side walls at the lower ends of these plots and thus not entering the collecting tanks. At the end of the experiments this accumulation was measured and is added to the material collected in the tanks to give a more realistic average for soil loss over the 8 year period of data collection.

Percentage runoff for these plots over 8 seasons is presented in Table 6. Van Rensburg unfortunately does not tabulate the data for individual plots but aggregates them. Runoff losses from the cultivated sorghum plots without conservation averaged four times those from grass plots, while the plot half cultivated and half grass lost double the runoff that the grass plot showed. The extent to which runoff loss was reduced by introducing narrow grass strips into cultivated plots was very significant.

Soil erosion data over the same period are also presented in Table 6. Taking first the values measured in the collecting tanks, the cultivated plots lost 76 times the amount of soil that was lost from the grass plots. Narrow grass strips incorporated in the cultivated plots reduced soil loss by 53 % compared to unprotected cultivation. The plot with half cultivation, half grass showed only 4 times the soil loss of the grass plot and one twentieth of the loss from the wholly cultivated plots. If the soil which accumulated at the base of the slope is included in the calculations, these contrasts become even more striking eg. the cultivated plots then lost nearly 120 times the soil lost by the grass plots.

Van Rensburg emphasises that the data provided by these measurements may not be relevant for large slope segments. He also recorded the progressive decline in crop yields from the cultivated plots, indicative of the agricultural effects of the heavy water and soil loss recorded.

(b) *Lyamungu* (3°15′S, 37°15′E: 1300 m a.s.l.)

Experimental plots designed to measure soil erosion and runoff under young coffee using various conservation treatments were established at Lyamungu near Moshi in 1934. Results were reported in the Annual Reports and Quarterly Notes of the Coffee Research and Experimental station (1937, 1938 & 1939)

Table 7. Runoff soil erosion from coffee plots established in 1934 at Lyamungu; runoff (a) as percentage of annual total: soil loss (b) in m³/ha (modified after Mitchell 1965 & Annual Reports of Lyamungu 1937 1938)

| Plot No. | Plot treatment: Coffee[1] | 1935 a | 1935 b | 1936 a | 1936 b | 1937 a | 1937 b | 1938 b | 1939 b |
|---|---|---|---|---|---|---|---|---|---|
| 1 | Clean weeding | 2 9 | 18 0 | 8 1 | 53.4 | 26.2[1] | 55.4 | 4.8[2] | 1.2[2] |
| 2 | 3 contour ridges 10m apart on which were hedges (*Crotalaria* sp.) lowest ridge 0.6m from tank | 1.9 | 1.4 | 4.4 | 8.9 | 7.8[3] | 0[3] | 3.0[3] | 1.4[3] |
| 3 | Clean weeding | 3.0[4] | 13.8[4] | 5.3[4] | 35.6[4] | 26.0[4] | 33.6[4] | 2.6[5] | 0.3[5] |
| 4 | As plot 2 + procumbent cover of *Dolichos hosei* | 0.5 | 0.1 | 1.8 | 0.7 | 7.3 | 0 | 2.4 | 0.5 |
| 5 | Clean weeding | 2.7 | 23.2 | 11.1 | 47.5 | 7.0[6] | 0[6] | 1.7[6] | 0.4[6] |
| 6 | Procumbent cover of *Dolichos hosei* | 1.1 | 0.3 | 2.0 | 0.5 | 7.7 | 0 | 1.6 | 0.3 |
| 7 | Mixed erect cover of *Crostalaria* sp. & *Canavalia ensiformis* | 1.7 | 0.3 | 3.3 | 9.9[6] | 10.9 | 0 | 6.1 | 2.1 |
| 8 | Clean weeding | 2.8 | 30.6 | 7.3[4] | 26.7[4] | 25.6[4] | 33.6[4] | 2.6[4] | 0.3[4] |

[1] Estimate as tank overflowed at least once.
[2] Excessive weed growth (*Commelina* sp.) uncontrolled.
[3] Contour ridge spacing reduced by half & hedges dug up as ridges had silted and hedges were breached.
[4] Weeds placed in lines across slope.
[5] Procumbent cover crop established.
[6] Mulched with banana trash; only 60 % surface cover due to poor reseeding. In 1935 and 1936 all coffee trees and plots 4, 6 & 7 envelope forked: plots 1, 2, 3, 5 & 8 *jembe* forked: from 1937 onwards routine weeding only.

and are summarised by Mitchell (1965). Similar data from Kenya are available (Gethin-Jones 1936; Maher 1950).

Eight adjacent plots were located on a uniform slope of 9.5°, cleared from thick bush in March 1934. The soil was a deep, free-draining, partially laterised volcanic red earth of clay-loam texture and well-aggregated structure, possessing a fairly high moisture-retaining capacity. The subsoil was a deep, freely-draining, friable chocolate-red clay. The plots measured 30.5 m long downslope by 4.5 m wide (area 137.25 m²). The individual plots were walled round and tanks at their foot constructed to collect eroded soil and runoff. Coffee was planted on all plots at 3 m × 3 m spacing in April 1934 and treatments begun. Measurements began in February 1935 when the treatments were deemed effective. Unfortunately treatment of plots was not consistent over the full period of the trials. (Table 7 & notes).

Lyamungu has an average annual rainfall of 1660 mm, 73 % of which falls on average in March, April and May (Fig. 1). Heavy rainstorms are common during this period, many totalling well over 100 mm in 24 hours with intensities of up to 76 mm/hour (Table 8).

Runoff associated with each runoff—generating storm over the 3-year period is shown in Table 8 for each plot. During many of the heavier storms, the water-collecting tanks overflowed and runoff had to be estimated. In several of these instances, calculations show that these estimates are too high (i.e. in excess of the total rainfall on the plot according to the station raingauge some distance away).

The results require careful analysis due to changes in plot treatment (see Table 7). The clean weeded plots (1, 3, 5 & 8) lost most water, plots 3 and 8 despite the lines of weeds placed across the slope. This measure was thus shown to have little value in the prevention of runoff. The results confirm the value of heavy vegetative mulching indicated by the Namulonge data (above) and the Tengeru results (below).

Of the cover crops, the mixed erect cover gave poorest control, next came contour ridges and hedges, then procumbent (growing along the ground) cover. The best protection against runoff loss was a combination of procumbent

Table 8. Percentage runoff at Lyamungu associated with individual storms.

| No. | Date | Rain mm | Int. mm/hr | Plots 1 | 2 | 3 | 4 | 5 | 6 | 7 | 8 |
|-----|------|---------|------------|---------|---|---|---|---|---|---|---|
| 1 | 35.04.13 | 104 | 76 | 9.0 | 9.0 | 14.1 | 8.3 | 15.0 | 10.0 | 11.7 | 16.9 |
| 2 | 35.05.07 | 143 | ? | 26.1 | 14.3 | 23.1 | 0 | 19.1 | 5.5 | 10.5 | 19.3 |
| 3 | 36.04.10 | 50 | 20 | 11.5 | 11.0 | 15.8 | 8.6 | 38.7 | 10.0 | 12.9 | 17.2 |
| 4 | 36.04.18 | 125 | 50 | 75.7 | 54.5 | 53.3 | 20.1 | 78.0 | 21.8 | 38.7 | 75.7 |
| 5 | 36.05.19 | 43 | 20 | 80.0 | 20.0 | 63.4 | 9.2 | 100+ | 7.5 | 18.4 | 48.4 |
| 6 | 36.06.13 | 47 | 12 | 48.1 | 5.3 | 3.8 | 5.3 | 30.5 | 6.1 | 3.1 | 37.4 |
| 7 | 37.04.07 | 108 | 40 | 85.0 | 25.6 | 79.0 | 23.6 | 20.6 | 20.9 | 23.9 | 80.3 |
| 8 | 37.04.14 | 91 | 48 | 100+ | 34.7 | 100+ | 29.2 | 26.8 | 28.0 | 50.4 | 100+ |
| 9 | 37.04.20 | 120 | 50 | 75.0 | 31.0 | 68.4 | 27.8 | 36.4 | 47.5 | 55.3 | 58.6 |
| 10 | 37.04.28 | 82 | 60 | 100+ | 21.9 | 100+ | 24.5 | 19.2 | 21.9 | 41.5 | 100+ |
| 11 | 37.05.04 | 99 | 25 | 95.6 | 19.9 | 97.0 | 21.4 | 15.9 | 16.3 | 25.3 | 97.0 |
| 12 | 37.05.11 | 182 | ? | 52.0 | 21.7 | 52.8 | 18.9 | 17.7 | 17.3 | 24.0 | 52.8 |

cover and contour ridges (plot 4); this plot showed only 28 % of the losses experienced by the clean weeded plots.

The influence of soil moisture is demonstrated in Table 8. The storm of maximum intensity (No. 1) early in the 1935 rains caused only a limited surface runoff averaging less than 14 % on the clean weeded plots, yet a storm (No. 5) of only 43 % the size and 26 % of the intensity, coming the day after a very heavy downpour the previous day, generated an average of 73 % run off from the clean weeded plots. The three seasons show both a progressive increase in rainfall (161 cm in 1935; 195 cm in 1936 and 221 cm in 1937) and an increasing number of storms generating runoff (2 : 4 : 6). This in itself may explain the runoff increase over the period: it may also be due to soil deterioration after initial clearance in 1934 (see Mpwapwa results above).

As at Mpwapwa, soil loss was not measured in relation to individual storms but over the whole rainy season. Table 7 presents the results. The clean weeded plot (1) lost most soil, averaging 38 m³/ha and year. Even during the first year of the experiments, the clean weeded plots (1 & 8) showed rill development and the bare plot with ridges (2) showed a silting up of the banks and a consequent breaching of some of the hedges. Weed lines across such plots, although they did little to reduce runoff, apparently cut down soil loss to an average of 24 m³/ha (i.e. by 27 %). Banana mulch on clean weeded coffee reduced soil loss to negligible quantities (i.e. 1 m³/ha and yr).

Of the conservation measures, all proved valuable in reducing soil loss but to varying degrees. Widely spaced ridges & hedges (plot 2) were less successful than all other treatments emphasising the validity of the Rhodesian results on the value of cover. Nonetheless the plot protected in this way lost only 5 m³/ha and yr over 2 years—an 86 % reduction in soil loss over the clean weeded plot 1. Closer spacing of the ridges after 1937 cut soil loss to very small amounts (< 2 m³/ha and yr), emphasising that it is not the technique itself but its proper application that is important.

Erect cover was the least effective of the various cover treatments, approximating to ridges in its control over soil loss (Table 7). Soil loss from beneath an erect cover averaged 6 m³/ha and yr. But during 1936 poor reseeding gave only 60 per cent cover, probably causing excessive soil loss. A properly maintained erect cover is probably more effective a conservor of soil than widely-spaced ridges but less effective in this role than closely-spaced ridges. Most effective of all in terms of soil conservation was a procumbent cover crop: the addition of ridges showed no advantage. In fact plot 4 lost slightly more soil on average than plot 6.

But it was noted that in contrast to soil moisture conditions recorded at 38 cm depth at the end of the rains, when no difference was observed between the plots, 4 months later the clean weeded plots showed a significantly higher soil moisture retention than all other treatments, and coffee on these plots maintained more than double the leafage of all other plots. The coffee on the bare plots flowered earlier in response to rain (by 5 days) and was in better condition at the end of the

**194**

dry season. Thus, although heavy soil erosion and runoff losses occurred from the clean weeded plots indicating that conservation measures were necessary during the wet season, these had to be designed to conserve soil moisture during the long dry season. It was noted that soil moisture conditions in nearby terraced plots were no better after 4 months drought than those of clean weeded flat plots. Closely spaced contour ridges without hedges were consequently selected as the best and cheapest conservation practice, costing less than a tenth the price to establish on such slopes and the same price to maintain as bench terraces (Ann. Rept. for 1937, 1938, p. 30), and causing vastly less disturbance of the soil and no exposure of subsoil (Temple 1972).

(c) *Tengeru* (3°22'S, 36°48'E: 1463 m a.s.l.)

Experimental plots designed to measure runoff and soil erosion under different crops and different conservation methods were established at Tengeru near Arusha in 1954. Results were reported by Anderson (1962) and Mitchell (1965). Though full data on the experiments are not now available, a summary table (Table 9) lists the main findings for a three year period. Plots were of uniform size (not available) and slope (18°) and underlain by deep red volcanic soil.

Tengeru has an average annual rainfall of 1300 mm, 59 % of which falls on average over 47 days in March, April and May (Fig. 1). Rain days (> 0.25 mm) average 100 annually.

Table 9, which presents the plot treatments in inverse ranked order of control over soil and water loss, reveals striking contrasts between the different crops and treatments. Both crops (coffee, maize, bananas and grass) and conservation methods were representative of local land use practices.

Soil and water losses were least on the grass plots. The banana plot lost 15 times more soil than the grass plot and 30 % more water. But bananas showed a much better control over soil erosion and water loss than all other food crops tested, even over the short period reported. It is no accident that bananas are the principal subsistence crop on most steep high rainfall mountain slopes in East Africa (e.g. Mt. Meru, Kilimanjaro, Tukuyu, Mt. Kenya, Mt. Elgon and Kigezi) and support locally very

Table 9. Runoff and soil erosion from various crops on identical plots at Tengeru for the three years, 1958–1960; runoff (a) as percentage of annual total; soil loss (b) in m³/ha (modified after Mitchell, 1965). Average annual rainfall 1958–1960: 1070 mm.

| Plot treatment | a | b |
|---|---|---|
| Coffee: pruned; 3×3 m spacing; clean weeded continuously; no shade | 5.0 | 22.4 |
| Maize: stover removed; no conservation measures | 3.4 | 12.0 |
| Maize: stover & dead grass removed; grass bunds of *Pennisetum purpureum* at 3 m VI | 2.3 | 7.2 |
| Maize: stover & dead grass removed; grass bunds of *Pennisetum purpureum* at 2.1 m VI | 2.2 | 5.0 |
| Maize: stover and trash bunds at 3 m VI | 2.1 | 3.9 |
| Maize: stover & trash bunds at 2.1 m VI | 2.0 | 1.0 |
| Bananas: banana trash mulch | 1.8 | 0.5 |
| Grass (*Chloris gayana*): cut for hay: average of 8 plots | 1.4 | 0.0 |

high rural population densities. As bananas are a perennial crop, the soil is mantled by a permanent cover and, due to the build-up of soil fertility with the normal heavy mulching, may be expected to show progressively improving results over time relative to other treatments.

Annual food crops showed a poorer degree of control over soil erosion and runoff. A maize plot lost over twice the amount of soil and 10 % more water, even under the most effective conservation practice tested, compared to the banana plot. Table 9 indicates most strikingly the value of proper but simple and inexpensive conservation techniques. Trash bunds were clearly more effective than grass bunds in preventing losses. The maize plot protected by grass bunds at 2.1 m vertical interval with dead grass and stover removed, lost 5 times as much soil and 10 % more water than the maize plot protected by trash stover bunds of the same spacing. The contrast between maize with closely spaced trash bunds and maize grown without any conservation method is even more striking. The unprotected plot lost 12 times more soil and 70 % more water than the well-conserved plot. These data are of great importance for proper agricultural extension advice on comparable slopes and soils

Table 10. Loss of fertility under various crops at Tengeru 1954—1961 (after Anderson 1962).

| Plot treatment | pH (CaCl$_2$) | Organic carbon % | Available P (ppm)[1] | Total P (ppm) | Organic P (ppm) | % Organic in total |
|---|---|---|---|---|---|---|
| Coffee (clean weeded) | 5.86 | 2.76 | 428 | 3208 | 450 | 13.6 |
| Maize (no conservation) | 5.85 | 3.44 | 514 | 4688 | 782 | 17.0 |
| Maize + 3m VI grass bunds | 5.97 | 3.20 | 382 | 4584 | 524 | 11.4 |
| Maize + 2.1m VI grass bfnds | 5.95 | 3.22 | 451 | 4188 | 396 | 9.9 |
| Maize + 3m VI trash bunds | 5.99 | 3.00 | 366 | 4084 | 925 | 23.0 |
| Maize + 2.1m VI trash bunds | 6.07 | 3.22 | 412 | 4063 | 912 | 22.1 |
| Bananas | 6.06 | 2.90 | 402 | 5813 | 2111 | 31.9 |
| Grass | 5.99 | 3.54 | 454 | 5229 | 2454 | 46.5 |

[1] Extractable in 0.3 N HCl.

on the slopes of Mt. Meru and probably also on Kilimanjaro.

A further striking point shown by the table is that the cleanweeded coffee plot showed by far the greatest losses of soil and water. These losses were far above those recorded from the maize grown without conservation measures. Coffee lost nearly double the soil and almost 50 % more water than unprotected maize. These data call in question the commonlyheld assumption that a perennial bush crop is invariably better conservation practice than an annual crop grown using traditional farming methods. The coffee plot lost well over 700 times as much soil and nearly 4 times as much water as an equivalent plot under grass.

Anderson (1962) analysed the effect of these treatments on soil fertility by taking three composite samples of soil from 0—15 cm depth from each plot, one at the top, a second in the middle and a third at the bottom of each of the eight plots. He measured pH, the percentage of organic carbon, available, organic and total potassium and the percentage of organic potassium in the sample. The major results are shown in Table 10.

Anderson summarises his results as follows...

"Maize without soil conservation and clean-weeded coffee plots have the lowest mean pH which is in accord with their higher soil and water losses. Organic carbon is highest under Rhodes grass (*Chloris gayana*) with the control which has received the greatest return of maize stover coming second, despite the soil loss. The coffee soil on the other hand is the lowest in organic matter, showing that con-

tinuous clean-weeding under this crop can bring about a marked loss of organic matter in a few years. Total phosphorus in these soils is highest under the bananas and grass and it is surprising that over a third of the total phosphorus has been lost from the coffee plot in 7 years. The much greater content of organic phosphorus under the bananas and grass than on the other plots indicates how efficient these crops are in preventing losses. Taking the grass and bananas as a rough standard of the original status of the soil about three quarters of the organic phosphorus has been lost from the coffee and elephant grass (*Pennisetum*) bund plots and over half of it from the other maize plots. Phosphorus extractable in 0.3 N HCl bears little or no relationship to the total organic or inorganic phosporus in these soils and is highest in the control plot where the greatest opportunity for mineralisation of the phosphorus in the maize stover exists. The percentage organic phosphorus in the total phosphorus like the total and organic phosphorus is highest on the banana and grass plots and lowest on the elephant grass bund plots. (Thus) Rhodes grass and bananas are not only better in soil conservation than maize or coffee but much better in maintaining the fertility of the soil" (Anderson 1962, p. 2).

These data, in conjunction with economic analysis should be of value in planning more intensive land use in the area.

(d) *Shinyanga* (3°35'S, 33°25'E: 1200 m a.s.l.)

Erosion plots were established in Shinyanga (Rounce, King & Thornton 1942, p. 15), but the data obtained were apparently neither published nor analysed. Shinyanga has an average annual rainfall of 780 mm, 81 % of which falls on average in the period December through April (Fig. 1). The only known result is that erosion plot data had demonstrated a 60 % reduction of soil erosion as the result of contour hedge planting by *Euphorbia tirucalli*. Closely planted *Euphoribia* reduced wash from the time of its planting. As this plant has a caustic and semi-poisonous latex it is never grazed, hence part of its value. By contrast sisal hedges across the contour, favoured in this area, only become effective as a control on soil loss when the plants are large enough to protect the grass and weeds beneath them from grazing. In other words, their effect was indirect. It may be noted here that sisal hedges are employed to conserve the soil in the Kondoa area. This area has some of the most severely-eroded slopes in East Africa. On the evidence of the Shinyanga erosion plots, *Euphorbia* should have been employed.

## Discussion

### Runoff and soil erosion
The data from Mpwapwa, Lyamungu and Tengeru are not exactly comparable as plot sizes, plot slopes, soil types and treatments varied. Thus no quantitative comparisons are possible. However, some generalisations can be usefully made. Indications of the climatic differences between the various localities are given in Figs. 1 and 3.

The Mpwapwa and Lyamungu plots were approximately the same lengths (27.7 and 30.5 m); unfortunately the dimensions of the Tengeru plots are not mentioned in the published report. The Lyamungu plots were twice as wide as most of the Mpwapwa plots. This difference may well have affected the soil erosion measurements but not the percentage runoff. The Lyamungu and Tengeru plots had similar soils but were steeper (9.5 and 18°) than the Mpwapwa plots (3.5—4.5°).

Despite this gentler slope, a sandy soil and the low rainfall at Mpwapwa, runoff from grass was nearly 3 times as great and soil loss averaged 8 times as much as at Tengeru. The runoff percentage from cultivated plots of bulrush millet at Mpwapwa averaged 10 times that from maize plots at either Lyamungu or Tengeru. Differences in ground cover

were not measured and may have been contributory factors. More important causes were probably the differing infiltration capacities of the soils and the contrasts in the frequency of runoff-generating rainstorms. It would appear from the limited data presented above that in semi-arid central Tanzania runoff-generating storms are much more frequent than they are on the better-watered mountain footslopes in the north. This may be a function of different types and intensities of rainfall received in the two areas but must also be a function of soil differences. More comparative data are necessary to verify these suggestions.

Clean weeded coffee plots lost approximately double the water and soil at Lyamungu as at Tengeru. This difference is largely explained by the greater cover provided by the more mature Tengeru coffee (4 years old at the start of the measurements reported). A further influence may be the greater water balance surplus at Lyamungu as compared to Tengeru.

Other results of the data are treated below.

### Land use and conservation
The principal value of the experimental results set out above is in providing quantitative data at the *shamba* (small cultivated plot) scale of the effects of various cropping and conservation practices in particular localities. As both crops and conservation measures employed were selected to match or improve local agricultural practices, the results are of practical agricultural relevance. The localities chosen for the experments were representative of the two contrasted environments identified earlier as critical, namely semi-arid interior plains and deforested mountain slopes. But much more work on the geomorphology, soils and land use practices of both areas would be necessary to establish how representative of local conditions the plots actually were.

Nonetheless some further comments on cropping and conservation strategies for both environments seem justified. The assembled data indicate the seriousness of soil and water losses associated with improper land use in both central Tanzania and on the northern mountain footslopes.

It seems probable that the Mpwapwa data are relevant to cultivation practices over considerable areas of central Tanzania, particularly to those areas known as the "cultivation steppe", where the natural vegetation has been largely destroyed. Cultivation steppe was estimated to cover 30,000 km² in 1929 (Phillips 1929) but there are no later estimates. The Lyamungu and Tengeru results are likewise relevant to cultivation practices on the northern mountain footslopes.

The Mpwapwa data demonstrate extremely severe losses of soil and water associated with clearance, cultivation and overgrazing. Thicket and good grass, both ecologically well-adapted to the local environment, show low runoff and little soil loss even on steep gradients or when they are lightly grazed or burnt annually. But under thicket water is lost by other means. Staples (1934, p. 101) calculated that 30 % of the rainfall was lost by evaporation from interception storage and a further 65 % lost by transpiration leaving only 5 % to feed springs and water courses. Thus under such a cover the bulk of the water supplied by precipitation is used for the physiological requirements of the thicket. Thicket has only a limited value as a source of firewood, timber, wax, honey, game and poor grazing.

Presumably grass, being shallower rooted than thicket would loose less water by transpiration. As grass appears almost as effective as thicket in inducing infiltration, such a cover would conserve water much more effectively, releasing greater amounts to springs and water courses. Grazing by stock should be so managed as to maintain a good grass cover: this would ensure more and better fodder as well as retarding soil and water loss. Proper management would be assisted by the planting of sisal hedges around fields. But proper grazing management is difficult in such an area where people are sedentary and where cattle are kept for prestige and bride-price. Artificially high stock numbers together with rather limited mobility of herds encourages overgrazing of both grass and thicket. As Staples' data showed, this is rapidly reflected in increased soil and water losses, which become extreme if cover is completely destroyed (Table 5). Bare uncultivated land should be encouraged to regenerate naturally by control of fire and grazing. Staples' data showed that such regeneration may reduce runoff by half and soil loss to one thirtieth in a single year, at least in some sites.

The data indicate clearly the importance of some form of cover, as bare plots, whether cultivated or uncultivated lost extremely high amounts of soil and water (Table 5). Losses were much reduced under a grain crop (bulrush millet or sorghum). Bulrush millet is very drought resistant, gives reasonable yields on light, infertile, sandy soil unsuitable for other cereals, matures quickly (3—4 months) and is thus also drought-evading. Its initial growth is rapid and it tillers freely, thus causing a suppression of weeds and an increase in cover (Acland, 1971, 27—28). But it is lower yielding than maize, more prone to bird damage and less easy to thrash and winnow. Sorghum is also very drought resistant, yields reasonably well on infertile soils and has a very efficient well-branched root system. Sorghum is second only to bulrush millet in its ability to withstand drought and to give satisfactory yields on poor or exhausted soils. It matures in 3 to 6 months. Like bulrush millet, it is heavily attacked by birds and is more difficult to harvest, thresh and clean than maize (Acland, op.cit.).

On the Mpwapwa erosion plots these grains were planted as clear stands, presumably sown broadcast as is the local custom. But according to Rigby (1969, p. 27—28) clear cropping is not the local custom of the Gogo people. Sweet potatoes, pigeon peas, grams, marrows and gourds are often interplanted with grain on Gogo fields, while many uncultivated plants grow up with the crops and are left standing during cultivation. All of these crops except groundnuts are drought resistant, and all except sweet potatoes give good yields on infertile sandy soils. Many are procumbent eg. sweet potatoes and marrows and thus give excellent soil cover: others are leguminous eg. cowpeas and pigeon peas and enrich as well as protect the soil. Cassava is very drought resistant and can give good yields on poor soils. As it removes very few nutrients from the soil, it is equivalent to fallowing even if clear weeded (Acland, op.cit., pp. 33—38).

Thus the soil and water losses (Table 5) recorded from the bulrush millet/sorghum erosion plot are likely to be higher than experienced under traditional cultivation practice with interplanting and lesser weeding. Maize according to Rigby (op.cit., p. 26) is spreading in this area; this could have adverse effects as maize requires good soil and adequate rainfall and will have a greater tendency to failure in

Ugogo. Its cultivation encourages ox-ploughing, row cultivation and clean weeding, all of which would favour increased losses of soil and water except on the flattest slopes.

Keeping fields small conserves the soil but has limited effects on runoff. The introduction of grass strips even across small plots has beneficial effects for both soil and water conservation. Staples noted that such a measure may have only short-term advantages as farmers are tempted to dig up the strips once they become fertile through silt deposition. Also cattle damage them unless an unpalatable and densely-tufted grass like *khus-khus (Vetiveria zizanoides)* is employed. Hoe-built contour ridges at suitable spacing (1—2 m) and well-graded provide an excellent form of control, inadequately employed in this area.

Data from Sukumaland indicate that if such ridges are cross tied every 2.5 m "this is a complete answer to soil erosion and increases the effectiveness of rainfall and the yield of crops" (Peat & Brown 1960, p. 103). The authors quote mean yield increases resulting from tied-ridging as compared to normal contour ridging as follows: bulrush millet 128 %; sorghum 57—87 % according to soil type; groundnuts 59 %; cotton 39 % and maize 15 %. Tied ridging gives up to three times the yield from non-tied ridging in drought years as it reduces runoff and increases infiltration and soil moisture (le Mare 1954, Brown 1963). This method should be used in this drought-plagued area. Ridges retain their utility when the land is returned to fallow.

The Lyamungu and Tengeru results can be discussed together, as environmental conditions and cultivation practices are similar. Both areas have soils of high fertility, a strongly aggregated structure, high permeability and water-holding capacity developed from volcanic parent material. Both areas appear to have large seasonal water surpluses (Fig. 3) and soil erosion may be serious in the absence of proper conservation techniques. Soil water deficiency at the end of the marked dry season may be a problem for perennial croups such as coffee.

Erosion was high under clean weeded coffee and water lost by runoff during the rains is unavailable to sustain treees during the critical fruiting period. Both water and soil losses are greatest when the trees are young and fall off as they mature. Proper conservation measures are therefore essential during the first 3—4 years after planting. As no cover crop has been found in East Africa which does not compete severely with coffee (Acland 1971, p. 72), control of erosion, runoff and weeds is best achieved by mulching. Grass mulch induces almost as great infiltration as grass (Hutchinson & others, above) but requires both land and labour. One hectare of grass is needed to supply adequate mulch for one hectare of young coffee (Acland, op.cit.). Banana mulch is better and it does not rot so quickly. Contour strip mulching of young coffee was found adequate for slopes between 1.5 and 5.5°. Contour ridging at 0.6 m vertical interval was found necessary for slopes between 5.5 and 11°. On slopes steeper than 11° coffee growing was not recommended, but either grass or bananas provided good safeguards against water and soil loss. While the interplanting of coffee and bananas reduce coffee yields, on steep slopes this disadvantage is far outweighed by much reduced soil and water losses and the reduced dependence on a single crop (Mitchell, 1965), a clear justification of the normal Chagga and Meru farming methods.

The data from Tengeru demonstrate most clearly the beneficial effects of bananas compared to an annual crop such as maize as a control over runoff and erosion and in terms of soil fertility.

## Conclusions

Quantitative experimental data on runoff and soil losses relevant to Tanzania conditions are presented in Tables 1—4. The Rhodesian data show the critical influence of vegetative cover in controlling soil and water loss and indicate the progressive problems of cultivating land of increasing slope. The Uganda data establish the relationships between duration, amount and intensity of rainfall in tropical rainstorms and relate these to soil and water losses. Data from the Mpwapwa experiments are presented in Tables 5—6. They provide quantitative indications of the seriousness of water and soil losses in the semi-arid central area of Tanzania. This area is characterized by soils of low fertility and a low and erratic rainfall. As soil moisture is the major limiting factor on crop yields, there is a need to limit surface runoff

**199**

losses on cultivated land, to increase infiltration and reduce soil erosion. Losses of water and soil, which are quantified for selected periods and treatments, can be drastically reduced if simple conservation practices are employed. These practices are described in the text. The data also show that an undisturbed vegetative cover of thicket or grass induces rapid infiltration and reduces soil loss. If this cover is damaged by overgrazing, water and soil losses rise rapidly.

Runoff and soil erosion losses are apparently much less serious on the mountain footslopes of the northern volcanic areas (Tables 7—9), than they are in semi-arid central Tanzania. On these footslopes soils are deep and fertile and possess a good structure. Rainfall is higher and much less erratic. Even in these areas however soil moisture conditions may become critical for perennial crop yields at the end of the dry season. There is thus a need to conserve water and prevent heavy runoff. During the heavy rainfalls of the wet season, losses of soil are heavy under some existing types of cropping (eg. young coffee and maize). Again such losses of water and soil can be drastically reduced by sensible farming practices; these practices are examined.

As such losses will increase with increasing slope, and as a large proportion of the cultivated land of the area is steeper than that of the experimental sites (eg. the intermediate slopes of Kilimanjaro and Meru) the need for effective conservation measures is reinforced.

## Acknowledgements

The taxonomic descriptions employed above have been revised and updated from the original identifications with the assistance of Dr. B. Harris. Professor P. W. Porter provided computed data on evaporation ($E_0$) for Mpwapwa, Lyamungu, Tengeru and Shinyanga, and made useful comments on an earlier draft of this paper. Mr. G. D. Anderson kindly made available to the author a copy of his unpublished paper on cropping systems and soil fertility at Tengeru.

*Professor P. H. Temple, Department of Geography, University of Dar es Salaam, P. O. Box 35049, Dar es Salaam, Tanzania.*

## References

*Acland, J. D.*, 1971: *East African crops, An introduction to the production of field and plantation crops in Kenya, Tanzania and Uganda.* FAO/Longmans, London.

*Anderson, G. D.*, 1962: The effect of various cropping systems on soil fertility. E.A. Soil Fertility Spec. Comm., Tengeru, unpub. manuscript.

*Brook, T. R.*, 1955: Soil and water conservation. *Mon. Bull. Coffee Bd. Kenya*, 20, 231—265.

*Brown, K. J.*, 1963: Rainfall, tie-ridging and crop yields in Sukumaland, Tanganyika. *Empire Cotton Growing Rev.*, 40, 34—40.

*Carson, M. A. & Kirkby, M. J.*, 1972: *Hillslope form and process.* Cambridge, C.U.P.

*E. Afr. Met. Dept.*, 1966: Monthly and annual rainfall in Tanganyika and Zanzibar during the 30 years, 1931 to 1960, Nairobi, EACSO.

*Evans, R.*, 1967: On the use of welding rods for erosion and deposition pins. in Field methods for the study of slope and fluvial processes. *Rev. Geomorph. dyn.*, 42, 165.

*Gethin-Jones, G. H.*, 1936: Conservation of soil fertility in coffee estates, with special reference to antierosion measures. *E. A. agric. J.*, 1, 456—462.

*Gerlach, T.*, 1967: Hillslope troughs for measuring sediment movement. *Rev. Geomorph. dyn.*, 42, 197.

*Hudson, N. W.*, 1957a: The design of field experiments on soil erosion. *J. agr. Eng. Res.*, 2,

— 1957b: Erosion control research. *Rhod. agr. J.*, 54

— 1959: Results of erosion research in Southern Rhodesia. *Advisory Leaflet* 13, *Fed. Dept. of Conservation & Extension*, Salisbury.

*Hudson, N. W. & Jackson, D. C.*, 1959: Results achieved in the measurement of erosion and runoff in Southern Rhodesia. *Third Inter-Afri. Soils Conf., Dalaba CCTA*, 575—583.

*Hutchinson, Sir J., Manning, M. L. & Farbrother, H. G.*, 1958: On the characterization of tropical rainstorms in relation to runoff and percolation. *Empire Cotton Growing Corp., Res. Mem.*, 30, & *Quart. J. Roy. Met. Soc.*

*King, N. J. & Hadley, R. F.*, 1967: Measuring hillslope erosion. *Rev. Géomorph. dyn.*, 42, 165—167.

*Le Mare, P. H.*, 1954: Tie-ridging as a means of soil and water conservation and yield improvement. *Proc. 2nd Inter.Afr.Soils. Conf., Leopoldville*, 595.

*Leopold, L. B. & Emmett, W. W.*, 1967: On the design of a Gerlach trough. *Rev. Géomorph. dyn.*, 42, 170 —172.

*Maher, C.*, 1950: Soil conservation in coffee. *Mon. Bull., Coffee Bd. Kenya*, 15, 283.

*Menne, T. C.*, 1959: A review of work done in the Union of South Africa on the measurement of runoff and erosion. *Third.Inter.Afr.Soils Conf., Dalaba, CCTA*, 612—627.

*Milne, G.*, 1932: A note on three soil profiles at Mpwapwa, unpub. paper, Agric. Res. Stat., Amani.

*Mitchell, H. W.*, 1965: Soil erosion losses in coffee. *Tanganyika Coffee News*, April/June, 135—155.

*Peat, J. E. & Brown, K. J.*, 1960: Effect of management on increasing crop yields in the Lake Province of Tanganyika. *E.Afr.agric.J.*, 26, 103—109.

*Penman, H. L.*, 1963: Vegetation and hydrology. Commonwealth Agric. Bur., Farnham Royal.

Phillips, J. F. V., 1929: Some important vegetation communities in the Central Province of Tanganyika Territory: a preliminary account. S.Afr.J.Sci., 26, 332—372.

Rapp, A., Murray-Rust, D. H., Christiansson, C. & Berry, L., 1972: Soil erosion and sedimentation in four catchments near Dodoma, Tanzania. Geogr. Ann. A, 54.

Rensburg, H. J. van, 1955: Run-off and soil erosion tests, Mpwapwa, central Tanganyika. E.A. agric.J., 20, 228—231.

Rigby, P., 1969: Cattle and kinship among the Gogo. A semi-pastoral society of central Tanzania, Cornell U. P., Ithaca.

Rounce, N. V., King, J. G. M. & Thornton, D., 1942: A record of investigations and observations on the agriculture of the cultivation steppe of Sukuma and Nyamwezi with suggestions as to the lines of process. Pamphlet 30, Dar es Salaam, Govt. Printer.

Schick, P. A., 1967: On the construction of troughs. in Field methods for the study of slope and fluvial processes. Rev. Geomorph. dyn., 42.

Schumm, S. A., 1967: Erosion measured by stakes. Rev. Geomorph. dyn., 42, 161—162.

Seginer, I., 1966: Gully development and sediment yield. J. Hydrol., 4, 236—253.

Staples, R. R., 1934: A run-off and soil erosion experiment. Ann. Rep. Dept. Vet. Sci. & Animal Husb. for 1933, 95—99 (with notes on water conservation in sub-arid Tanganyika, 100—103)

— 1935: (photos of erosion plots, Plate VI, Fig. 1) Ibid for 1934.

— 1936: Run-off and soil erosion tests in semi-arid Tanganyika Territory. Second report. Ibid for 1935, 134—141.

— 1939: Run-off and soil erosion tests, Ibid for 1938, 50 (not final write up: this promised but never published: 1936 report most comprehensive).

Tanganyika Terr., Dept. of Agric., 1937: Third annual report of the Coffee Research Experimental Station, Lyamungu, Moshi, 1936, 58—62.

— 1938: Fourth Ibid, 1937, 25—30.

— 1939: Quarterly notes on Ibid., 11, 5—6.

Temple, P. H., 1972: Soil and water conservation policies in the Uluguru mountains, Tanzania. Geogr. Ann. 54A, 3—4.

Temple, P. H. & Murray-Rust, D. H., 1972: Sheetwash measurements on erosion plots at Mfumbwe, eastern Uluguru mountains, Tanzania. Geogr. Ann. 54A, 3—4.

Temple, P. H. & Rapp, A., 1972: Landslides in the Mgeta area, western Uluguru mountains, Tanzania. International Geography ed. Adams, W. P. & Helleiner, F. M., 1034—1035. Toronto U.P.

Young, A., 1972: Slopes. Edinburgh, Oliver & Boyd (part p. 48—54).

Zingg, A. W., 1940: Degree and length of land slope as it affects soil loss in runoff. Agric. Eng. 21, 2.

# 13

Reprinted from *Soil Science and Agricultural Development in Malaysia*
Conference Proc., Malaysian Society of Soil Science, Kuala Lumpur,
1980, pp. 307–324

## STATUS OF SOIL CONSERVATION RESEARCH IN PENINSULAR MALAYSIA AND ITS FUTURE DEVELOPMENT

L.M. MAENE AND WAN SULAIMAN WAN HARUN

*Universiti Pertanian Malaysia*

## Abstract

*The steady increase in national and world needs for agricultural products means that high quality cropland will be used more intensively and that other land, including marginal areas, will be converted to cropland. Soil conservation aims to obtain the maximum sustained level of production from a given area of land whilst maintaining soil loss below a threshold level. Problems of soil erosion are very much in evidence in Malaysia. Several compound factors have been recognised as being responsible for soil erosion. This paper reviews the status of soil conservation research in Peninsular Malaysia, with special emphasis on the impact on agricultural production. At the same time, future lines of research are projected and briefly discussed.*

## INTRODUCTION

Similar to most of the other developing countries, Malaysia is characterised by a rapid transformation of vast areas of rainforest into agricultural land. This major change in land use has been instigated by the desire to meet the food requirements of the population, to provide large quantities of raw materials for export and to support the agrobased industries. Being a country with vast natural resources, Malaysia has presently opted for the exploitation and export of natural resource products to meet the demands for better lifestyle and the challenges of exponential population growth. In order to accelerate the pace of land development and settlement in the various states of Peninsular Malaysia, particularly in Pahang, Johore, Kelantan and Trengganu, thirteen state and other statutory land development agencies have been established. Therefore, the rainforest ecosystem is always under pressure to make way initially for timber extraction and subsequently for land development and settlement.

At present, agriculture in Peninsular Malaysia covers about 4.2 million hectares, out of a total land area of 7.1 million hectares considered to be suitable for agricultural development. Among the land development agencies, the Federal Land Development Authority (FELDA) predominates in the extent of land clearing and development and should therefore be very aware of the importance of environmental quality. While massive deforestation is justified as a developmental necessity, concern must be directed towards the long-term environmental consequences of rapid changes in the natural ecosystem. Changes in land use have profound effects on the physical, chemical and biological properties of the soil and the environment.

The aim of soil conservation is to obtain the maximum sustained level of production from a given area of land whilst maintaining soil loss below a threshold level, which theoretically permits the natural rate of soil formation to keep pace with the rate of soil erosion.[1] The prevention of soil erosion relies on selecting appropriate strategies for soil conservation and this in turn, requires a thorough understanding of the processes of erosion. The factors which influence the rate of erosion are, rainfall, runoff, wind, soil, slope, plant cover and the presence or absence of conservation measures. Conversation design should be based on a thorough assessment of erosion risk.

According to Langbein and Schumm[2], erosion reaches a maximum in areas with a mean annual precipitation of 300 mm. Below this value, erosion increases as rainfall increases. However, the vegetative cover increases as precipitation increases, resulting in better protection of the soil surface. Thus when total precipitation is above 300 mm, soil loss decreases as precipitation increases because of the improved protection provided by the vegetation. Douglas[3,4] pointed out that if the vegetative cover is destroyed, erosion increases rapidly with precipitation and reaches a maximum in humid tropical areas.

In Peninsular Malaysia, numerous instances of soil erosion have been documented, mainly in association with timber extraction, mining, agricultural and urban activities. However, few quantitative measurements of soil erosion have been made and most data available are derived from studies of sediment concentration in rivers. Morgan[1] pointed out that much of the sediment removed from hillsides is deposited before it reaches the rivers and therefore, data on sediment concentration in rivers almost certainly underestimate the rates of soil loss.

SOIL EROSION PROBLEMS

Lake[5] and Hartley[6] described the high silt concentrations in rivers in Peninsular Malaysia. Quantification of the amount of sediment being eroded was attempted by Fermor.[7] Cultivation of land too steep for agriculture inevitably results in erosion as reported by Soper.[8] These references indicate the early awareness of the menace of soil erosion.

The soil erosion system is controlled by factors such as the erosivity of the rainfall, the erodibility of the soil, the slope of the land and the nature of the plant cover. These factors are briefly discussed below.

Rainfall

In Malaysia, the rainfall station network can be considered to be satisfactory for certain areas, but the coverage in general is somewhat uneven. Many stations have records for over thirty years, which is considered to be the minimum for most statistical studies. In spite of an apparent uniformity of climate over the country, important regional variations exist. The east coast has a strongly seasonal rainfall pattern. More than a quarter of the annual rain falls in December and January. The rest of the country has a more evenly distributed rainfall throughout the year, but significant variations in intensity are common. Morgan[9] reported daily totals with a ten-year return period of 125-150 mm for most of the country, but more than 250 mm can be expected along the east coast and less than 100 mm in the highlands. The role of rain intensity is not always obvious. Erosion appears to be related to two types of rain event, short intense storms and prolonged storms of low intensity. In the first case, the infiltration capacity of the soil may be rapidly exceeded, whereas during a prolonged storm, the soil may be saturated.

Previous meteorological conditions may affect the response of the soil in terms of erosion to rainfall. Ghulam and Erh[10] showed the effect of antecedent soil moisture status for Munchong series on soil erosion. Bols[11], in a multiple linear regression analysis, examined the effect of antecedent rainfall on runoff for six sites in Java and concluded that the soil moisture content at the onset of rain had a limited effect on surface runoff. The soils under consideration were classified as ultisol (4), entisol (1) and vertisol (1). Tajudin[12] found no significant relationship between soil loss from a Petroferric tropudult (Padang Besar) in Serdang and cumulative rainfall figures up to five days before the erosion sampling. Morgan[13] also reported that no relationship could be obtained between soil loss and antecedent precipitation in mid-Bedfordshire.

A number of erosivity indices have been forwarded to characterise erosion by overland flow and rills. Most indices are based on the kinetic energy of the rain. In a study under mature oil palm on an Orthoxic tropudult in North Johore, Maene et.al.[14] found that the daily rainfall was better correlated with surface wash along harvesting paths than erosivity indices such as total kinetic energy of the rain event, Wischmeier's $EI_{30}$ - index, Hudson's KE > 25-index or Lal's $\sum$ai-index. This has also been confirmed on bare control plots in the same area and in Serdang (Petroferric tropudult). Morgan[15] reported that Hudson's threshold value of 25 mm h$^{-1}$ as a critical rainfall intensity for erosion by overland flow and rills is appropriate in Malaysia. Morgan[16] has shown that such overland flow occurs in the Kuala Lumpur area with rainfall intensities of 60 to 75 mm h$^{-1}$. The return period of overland flow in Kuala Lumpur is 60 days. However, to initiate gullying, events with return periods in excess of 10 years are in order.

## Soil Erodibility

Some soils erode more readily than others. This difference in resistance to detachment and transport is determined by the properties of the soil such as texture, aggregate stability, shear strength, infiltration capacity, organic matter content and chemical status. Numerous indices of erodibility have been devised. They are either based on soil properties determined in the laboratory or the field, or on the response of the soil to rainfall. Every soil property which can be quantitatively measured has at one time or another, been considered for this purpose. Bryan[17] recommended a dynamic laboratory test whereby the water-stable aggregate content is determined and used as an indicator of erodibility. Very little information is available on the erodibility of Peninsular Malaysian soils.

Wong[18] based on field observations, classified soils with clay contents exceeding 27% and sand contents of less than 45% as 'less erodible.' Such soils have surface textures of clay loam, silty clay loam, silty clay and clay. Soils with more than 45% sand and less than 27% clay were classified as 'more erodible.' Such soils include the textural classes of sandy clay, sandy clay loam, sandy loam, loamy sand, loam, silt loam and silt.

Maene et.al.[19] used a rainfall simulating method in the laboratory to conclude that Rengam series soil (Oxi tropudult) is much less susceptible to erosion than Durian (Typic tropudult) and Serdang (Oxic tropudult) series. Rengam series soil had more water-stable aggregates after simulated rainfall than the other soils. Abdul Rashid[20] used a similar set-up to study the relative erodibility of five Malaysian soils (Telemong, Kedah, Serdang, Malacca and Rengam).

He compared five parameters which express the erodibility of the soil with actual soil losses by splash and runoff under simulated rainfall. The ratio of percentage silt and sand over percentage clay (clay ratio) was best correlated with splash losses. Based on splash and runoff losses, the order of erodibility among the five experimental soils was: Telemong, Kedah > Serdang > Malacca > Rengam. Chee[21] used a rainfall simulator with full-cone spray nozzles to determine the erodibility of four Malaysian soils. The simulated rainfall had fairly similar characteristics as natural rainfall. He found that aggregate stability as determined by wet-sieving, was the most reliable index of erodibility and therefore best correlated with soil losses under simulated rainfall. A commonly used index is the K-value which represents the soil loss per unit $EI_{30}$. Chee[21] found from his simulation studies a K-value (metric equivalent) of 0.33 for Serdang series, 0.28 for Munchong series, 0.26 for Malacca series and 0.24 for Kedah series soil.

After seven months of observation in the field on standard bare soil plots in North Johore, a K-factor (metric equivalent) of 0.20 was calculated for Durian series soil (Orthoxic tropudult). However, using the nomograph of Wischmeier et.al.[22], a K-value of 0.37 was found for the same soil.

Many attempts to devise a simple index of erodibility have been made and the procedures range from static laboratory measurements to dynamic field tests. However, very few of the methods have been used in Peninsular Malaysia, partly because simulation studies are still at an experimental stage and partly because of the magnitude of the work in field erosion experiments.

## Slope Effect

Both degree and slope length are important factors influencing soil erosion. The effect of length of slope, slope form and angle of slope on soil erosion have not been extensively studied in the humid tropics and most investigations relate to erosion by overland flow. Hudson[23] reported that erosion generally increases exponentially with increase in slope. For tropical soils, the exponent approaches 2 although values ranging from 1.3 to 2.1 have been reported. Maene et.al.[19] reported an increase in soil loss from 43.5 tonnes ha$^{-1}$ on a slope of 17% to 63.5 tonnes ha$^{-1}$ on a slope of 34% for Munchong series soil (Typic tropudult). These results were obtained from field plot studies whereby the plots were planted with cuttings of <u>Pennisetum purpureum</u> (Napier grass) and the losses were recorded during the first 60 days after planting.

## Plant Cover Effect

The major role of vegetation is in interception of raindrops to reduce the amount of kinetic energy upon impact with the soil surface.

The mosquito gauze experiment by Hudson and Jackson[24] illustrates this role. Over a six year period, the mean annual soil loss was 141.3 $m^3 ha^{-1}$ on bare soil and 1.2 $m^3 ha^{-1}$ on bare soil over which a fine wire gauze had been suspended.

The effectiveness of a plant cover for erosion protection depends on the type of canopy, the density of the ground cover and the root density. Shallow[25] observed soil losses from three hilly catchments in the Cameron Highlands, in the order of 24.5, 488 and 732 $m^3 km^{-2} yr^{-1}$, respectively under jungle, tea plantation and vegetable farming. Hunting Technical Services Ltd. (reported by Daniel and Kulasingam[26]) estimated the rate of soil removal from hilly catchments under jungle in the Jengka Triangle to be of the order of 50 $m^3 km^{-2} yr^{-1}$. In a study in Johore Tenggara, soil losses from catchments were reported to be 31 $m^3 km^{-2} yr^{-1}$ in an undisturbed jungle catchment, 46 $m^3 km^{-2} yr^{-1}$ in a small catchment with mature rubber and oil palm and 87 $m^3 km^{-2} yr^{-1}$ in a large catchment comprising of the same crops. In these studies, only suspended load in the streams was measured so that the soil loss values are under-estimated. Moreover, the physiography and the soil types for the various catchments need not have been the same which makes it difficult to assess the exact influence of the type of landuse.

Generally, forests are the most effective in reducing erosion because of their canopy, but a dense undergrowth may be almost as efficient. Fournier[27] and Elwell and Stocking[28] reported that for adequate erosion protection, at least 70 percent of the ground surface must be covered. The presence of several tree storeys as well as a dense undergrowth in the rainforest results in a larger interception of precipitation than in a cultivated plantation. Low[29] reported a mean interception value of 36% for a primary forest at Lawin in South Selangor. Teoh[30] mentioned a mean interception value of 25% for 23 years old Hevea trees, whereas in North Johore, a value of 17.2% for 13 year old oil palms (Elaeis guineensis) was found. In a study of drop size distribution under the mature oil palm canopy, Maene and Chong[31] found no significant difference between kinetic energy of the rain in the open field and under the oil palm canopy, saturated with intercepted rain water. Raindrops intercepted by the canopy coalesce on the leaves to form larger drops. Despite the fact that a certain percentage of the rainfall is used for stemflow and that bigger drops fall at velocities less than terminal velocity, the larger proportion of bigger drops under the canopy is able to compensate for the loss of energy load. The canopy by itself does not reduce the striking force of the rain. The height of the canopy is important because water drops falling from a height of 7 m may attain more than 90% of their terminal velocity.

Apart from the influence of the canopy on the erosion process, ground surface cover is also an important factor. Maene et.al.[14]

reported soil loss values under mature oil palm on a standard slope of 9%, varying from 1.1 tonnes ha$^{-1}$ yr$^{-1}$ under pruned fronds to 14.9 tonnes ha$^{-1}$ yr$^{-1}$ on the harvesting paths. Runoff volume under these conditions was respectively 2.8% and 30.6% of the rainfall, whereas a steady state infiltration rate of 0.9 cm h$^{-1}$ was recorded on the harvesting paths against 43.0 cm h$^{-1}$ under the pruned fronds.

Establishment of cover crops to reduce erosion has been widely practised. There has been a great deal of work done on the effects of different cover crops on rubber growth and yield under Malaysian conditions.[32 to 35] Creeping leguminous cover, such as <u>Pueraria</u> <u>phaseoloides</u> and <u>Calopogonium</u> <u>mucunoides</u> have better residual effect than grasses or natural covers. Soong and Yap[35] found that leguminous creeping covers improved soil physical characteristics. Selection of the cover should depend on its easy establishment and adaptation to local soil and climatic conditions. Pushparajah et.al.[36] reported the following effect of covers on soil erosion under rubber (<u>Table 1</u>).

TABLE 1.   EFFECT OF COVERS UNDER MATURE RUBBER ON
SOIL EROSION[1]

| Soil Series | Slope | Rainfall (15 months) (cm) | Soil loss (kg ha$^{-1}$) | | |
|---|---|---|---|---|---|
| | | | Bare | Grass | Nephrolepis |
| Rengam | 4°-5° | 292 | 103 | 44 | negligible |
| Serdang | 3°-4° | 325 | 132 | 117 | 59 |

After Pushparajah et.al.[36]

Ling et.al.[37] studied the effect of the stage of cover establishment on runoff and soil loss (<u>Table 2</u>). However, the effect of different stages of cover were observed at different points. Nevertheless, the general conclusion was that a greater spread of cover was more effective in reducing erosion.

### ASSESSMENT OF EROSION RISK

There is a general awareness of soil erosion as a problem in Peninsular Malaysia, as illustrated by the widespread use of bench terracing and ground covers as conservation measures. Several laws deal with soil conservation and erosion, such as the National Land Code of 1965, the Land Conservation Act of 1960, the Forest Enactment 1934 and the Mining Enactment 1935. Erosion risk is closely related

TABLE 2.    MEAN RUNOFF AND SOIL LOSS AT DIFFERENT
STAGES OF COVER ESTABLISHMENT[1]

| Percent Ground Cover | Rainfall (mm) Bare or Legumes | Runoff (mm) Bare | Runoff (mm) Legumes | Soil Loss(kg ha$^{-1}$) Bare | Soil Loss(kg ha$^{-1}$) Legumes |
|---|---|---|---|---|---|
| 0 - 30% | 269 | 57 | 47 | 13503 | 9043 |
| 30 - 90% | 311 | 71 | 19 | 30201 | 1763 |
| 90 - 100% | 287 | 64 | 3 | 11237 | 9 |

1. After Ling et.al.[37]

to soil type and plant cover.  Therefore, erosion surveys often come
under land resources studies and may be carried out at different scales.
A general assessment of erosion hazard is largely based on the analysis
of climatic data.

Erosivity indices may be used to study regional variations
in erosion potential.  Fournier[38] showed that the ratio $p^2/P$, where
p is the highest mean monthly precipitation and P is the mean annual
precipitation, is significantly correlated with sediment yield in rivers.
Lal[39] found that the index $AI_m$, where A is the total rainfall (cm) and
$I_m$ is the maximum intensity (cm h$^{-1}$), is better correlated with runoff
and soil loss from field plots at Ibadan than $EI_{30}$ and KE > 25.  Bols[11]
carried out a similar study in Java and found no significant difference
between $AI_m$, $EI_{30}$ and KE > 25 in their correlation with soil losses.
However, the correlation between total rainfall R and soil loss was
significantly different from the other indices.  In a preliminary study
in North Johore and Serdang, the total rainfall R (mm) was found to be
better correlated with soil loss than the three above-mentioned erosivity
indices.

In order to plan soil conservation work, a land area is divided
into smaller regions, similar in degree and type of erosion hazard.
Morgan[1] described two approaches for erosion assessment at the
reconnaissance level in Peninsular Malaysia.  First, regional variations
in various indices of erosion intensity are examined.  Two measures of
erosion intensity were tested: drainage density, defined as the length
of streams per unit area and drainage texture, defined as the number of
first order streams per unit area.  The second approach employs rainfall
data to express the rainfall aggressiveness.  He produced a map of $p^2/P$
and used the following equations to predict daily erosivity (KE > 25)
from daily totals (Table 3).

**209**

TABLE 3.   REGRESSION EQUATIONS FOR PREDICTING DAILY
EROSIVITY FROM DAILY RAINFALL TOTALS IN
PENINSULAR MALAYSIA[*]

| Area | Equation |
|------|----------|
| Peninsular Malaysia | Evd = 16.64 Rd - 173.82 |
| West coast climate | Evd = 34.42 Rd - 1121.97 |
| East coast climate | Evd = 16.16 Rd - 357.17 |
| Port Dickson climate | Evd = 26.06 Rd - 553.85 |

[*] After Morgan[1]

Evd = daily erosivity (J.m$^{-2}$)       Rd = daily rainfall (mm)

By summing the daily values, the annual erosivity can then be calculated,
which was found to be related to the mean annual rainfall, following the
equation:

EVa = 9.28P - 8838.15       EVa = annual erosivity (J.m$^{-2}$)

P = annual rainfall (mm)

This equation can then be used to estimate mean annual erosivity values
from mean annual rainfall totals and an iso-erodent map can be drawn.

Morgan[1] stated that both $p^2/P$ and drainage texture may be
regarded as indicators of gully erosion risk, whereas the mean annual
erosivity values reflect the risk of erosion by rainsplash, overland
flow and rills. He located areas of high gully erosion risk in Perlis,
Kedah and North Perak and scattered areas around Klang and Kajang.
Areas of high overland flow risk occur in Middle Perak and the Kinta
Valley, north of the Pahang river and in south-western Johore.

A reconnaissance erosion survey only indicates the potential
of an area for erosion. An analysis of soils, slopes and land use is
required to study the real erosion pattern. Land capability assessment
should be geared towards the identification of the risks attached to
the cultivation of the land and the required soil conservation measures.
However, it is difficult to incorporate both agricultural and non-
agricultural activities into a single land capability classification.
Wong[18] pointed out that a soil-crop suitability classification is an
essential contribution to the more comprehensive land capability

classification, whereby the concept of land evaluation includes the exploitability of minerals and timber besides agricultural development.

Based on rainfall-runoff ratios of 0.25 for rainforest, 0.35 for rubber and 0.75 for urban areas and an acceptable sediment yield of 0.2 kg $m^{-2}$ $yr^{-1}$, for central Pahang, the maximum permissible slope angles at which acceptable sediment yields can be maintained were estimated[1,3] to be as follows:  $3^O$ for urban landuse, $13^O$ for rubber and $24^O$ for rainforest. The analysis of erosion risk enables strategies for landuse planning and soil conservation. The recommendations require detailed design work before they can be implemented.

In a small catchment on the University of Malaya campus, Morgan[16] used rainfall-runoff ratios and a maximum acceptable sediment yield to determine maximum permissible slope angles. From the results, he recommended that tree crops can be grown with the use of ground cover on slopes up to $7^O$, with bench terracing on land over $7^O$. Above $14^O$ slope, the catchment should remain or be allowed to revert to a forest cover.

## EROSION CONTROL STRATEGIES

Conservation techniques can be divided into agronomic measures, soil management and mechanical methods. Agronomic measures are based on the protective effect of plant cover. Crops which help produce an early ground cover are certainly more useful in controlling runoff and erosion than those which take longer for full canopy cover to develop. Aina et.al.[40] found that the number of days required for 50 percent canopy cover to form were approximately 38 for soyabean (Glycine max), 46 for pigeon peas (Cajanus cajan) and 63 for cassava (Manihot esculenta) grown at Ibadan, Nigeria. Maene et.al.[41] found Pueraria phaseoloides to be much faster in cover establishment on slopes than cuttings of Pennisetum purpureum.

Plant population, time of planting and fertility level are cultural practices which affect erosion control.[42] Lal[43] stated that a soil-depleting crop grown with proper soil-conserving techniques such as mulching, could give rise to less runoff and soil loss than a soil conserving crop grown with erosion-promoting practices. Mokhtaruddin and Maene[44] found, over two cropping seasons of maize, a soil loss of 7.5 tonnes $ha^{-1}$ under conventional tillage and 0.5 tonnes $ha^{-1}$ under surface mulch of lalang applied at 3 tonnes $ha^{-1}$.

Ling et.al.[37] studied the effect of land clearing, burning and subsequent plant establishment on certain soil properties and erosion losses. On a soil with a high infiltration capacity as well as infiltration rate, they found soil losses of 79 tonnes $ha^{-1}yr^{-1}$ under bare conditions compared to 10 tonnes $ha^{-1}$ $yr^{-1}$ under plant cover (Table 4).

TABLE 4.   MEAN RUNOFF AND SOIL LOSSES UNDER DIFFERENT
COVERS*

| Treatment | Rainfall (mm) | Runoff (% rain) | Soil loss (tonnes ha$^{-1}$ yr$^{-1}$) |
|---|---|---|---|
| Bare | 1854 | 15 | 79 |
| Legumes | 1854 | 5 | 11 |
| Natural cover | 1854 | 3 | 10 |

*After Ling et.al.[37]

They concluded that current land clearing and preparation
methods by FELDA, destroy much of the inherent nutrient status of the
forest ecosystem.

The aim of sound soil management is to maintain the fertility
and the structure of the soil. Highly fertile soils result in high
crop yields, good plant cover. The latter creates conditions which
minimise the erosive effects of raindrops and runoff. Hence, soil
fertility can be seen as the key to soil conservation.[1] This has much
relevance in Peninsular Malaysia where highly weathered soils are
prevalent. Removal of the fertile topsoil leads to the exposure of
the subsoil which, more often than not, has a poor physical status
apart from being low on nutrients. Chin and Tan[45] carried out a
simulation study whereby maize and groundnut were grown on soil from
which topsoil had been removed. They reported a decrease in the yield
of maize by respectively 18%, 68% and 89% for a corresponding loss of
7.5 cm, 15 cm and 30 cm of topsoil. The mortality rate and the
percentage of plants without cobs increased with increasing loss in
volume of topsoil. Similarly, the yield of groundnut decreased
significantly by 3%, 34% and 45% for a corresponding loss of 7.5 cm,
15 cm and 30 cm of topsoil.

Temporary stability of the soil can be obtained on most soils
by using soil stabilisers or soil conditioners. They are helpful on
special sites where the cost is warranted. Soong and Yeoh[46] recommended
the use of latex-oil emulsions for controlling soil erosion on exposed
surfaces in small areas such as road cuttings, land embankments in
housing estates, embankments of irrigation canals and natural waterways,
and steep slopes in rubber or oil palm plantations. They found that
application of a formulation containing 10% rubber, 4% aromatic oil and
0.5% emulsifier was able to reduce soil loss by 70 - 87% on Serdang and
Sungei Buloh series soil during the monsoon season. They further
reported that incorporation of grass seeds with the formulation increased
their effectiveness. Depending on the erodibility of the soil, different
formulations can be used. Maene et.al.[41] found that application of a

212

bitumen emulsion during the establishment stage of legume cover
or grass on slopes was effective in controlling soil erosion.  They
further observed that bitumen application had a beneficial effect on
the establishment of Napier grass on Munchong series soil.  Soil
conditioner application reduced the soil losses to approximately one-
third of the value obtained on plots without soil conditioner for all
slopes considered.

Soil conservation strategies are aimed at reducing erosion to
an acceptable level.  Hudson[23] recommended that for sensitive areas
where soils are thin or highly erodible, a mean annual soil loss of 2
to 5 tonnes ha$^{-1}$ was the permissible maximum.  This target may be
unrealistic for areas where erosion rates are naturally high as in
mountainous terrain with high rainfall.  Under these conditions, Morgan[1]
proposed a rate of 25 tonnes ha$^{-1}$ yr$^{-1}$ as reasonable.

Agronomic treatments are preferred as conservation measures
because they are less expensive and have a direct effect on raindrop
impact, infiltration, runoff volume and water velocity.  Mechanical
measures should follow a thorough assessment of erosion risk.


EROSION MODELS


When planning conservation work, it would be very useful if
a method of predicting soil loss under a wide range of conditions were
available.  Morgan[1] described some of the models used in soil erosion
studies.  Most of the models are parametric and based on defining
important factors which can be related to measurements of soil loss.
One of the best known and widely used parametric models is the Universal
Soil-Loss Equation of Wischmeier and Smith.[47]  The equation is normally
used to predict soil loss.  It has practical limitations and there is
considerable interdependence between the variables.  Basic mathematical
models are being developed that combine fundamental principles, concepts
and relationships of erosion mechanics, hydrology, hydraulics, soil
science and meteorology to simulate the erosion and sedimentation
processes.  The models have not become operational in the field and
additional research is needed.  However, once perfected, such models
will produce a more comprehensive and widely applicable technique than
parametric models.

RESEARCH NEEDS

In the humid tropics, soil erosion is a serious problem likely to increase in importance as the need to use soils for intensified food production expands. At present, knowledge of the many factors determining the magnitude of erosion is still inadequate to design readily applicable and secure control systems. One of the basic principles which can be used is the maintenance of a vegetative or mulch cover over the soil at all times.

Every agricultural development scheme involving large scale clearing operations should be assessed in terms of erosion risk, while control methods should be included as part of the cost. In order to design appropriate soil management systems and erosion control methods, quantitative information on the factors involved is needed. It is very important to study the critical period of soil erosion from the time of jungle clearing, through the seedling and establishment stage of the crop until maturity. Joint projects, such as the "Sungei Tekam Experimental Basin Study," whereby environmental monitoring takes place from the time catchments are still under rainforest until the establishment of a perennial crop, are of extreme value to the understanding of the erosion processes in this country. Tang et.al.[48] reported the start of hydrological studies of landuse change to quantify the consequences of logging activities in the hill forests, in order to make suitable recommendations so that these consequences remain minimal.

More information has to be acquired on climatic erosivity in the humid tropics so that iso-erodent maps can be prepared to guide agricultural practice. Further, detailed studies of rainfall-runoff relationships under different landuse and of geomorphological processes operative in small drainage basins and on hillslopes should be undertaken.

There is also a serious lack of information on the erodibility of Malaysian soils. The most satisfactory data is obtained from standard runoff plots for which a small number have been established in Peninsular Malaysia. Greenland and Lal[49] pointed out that the number of soils where such plots can be sited and data collected over several years, are very limited. Mobile rainfall simulators which can be taken to different sites may accelerate the collection of data. At the same time, more research should be carried out on intrinsic soil properties in relation to erodibility. A study of the effect of soil genesis and its chemistry on erodibility should be very useful. The lack of well-defined erodibility values for soils of the humid tropics presents difficulties in developing valid land capability classifications. Land-form factors, such as length of slope, slope

**214**

form and angle of slope should be studied to evaluate their effect on erosion.

Greenland and Lal[49] pointed out that in the humid tropics, relevant data on control methods in relation to engineering design are seriously lacking and more research is urgently needed.

Finally, the various agronomic prediction systems which are compatible with increased crop production and erosion control should be investigated.

ACKNOWLEDGEMENTS

The authors would like to thank the Vice-Chancellor of Universiti Pertanian Malaysia and the Belgian Development Co-operation Administration, for their permission to present this paper.

REFERENCES

1.  MORGAN, R.P.C. (1979)    Soil erosion.  Longman Group Ltd., London.

2.  LANGBEIN, W.B. and SCHUMM, S.A. (1958)    Yield of sediment in relation to mean annual precipitation. Trans. Am. Geophys. Un. 39, 1076-1084.

3.  DOUGLAS, I. (1967)    Natural and man-made erosion in the humid tropics of Australia, Malaysia and Singapore.  Int. Assoc. for Sci. Hydrol. 75, 17-30.

4.  DOUGLAS, I. (1967)    Man, vegetation and sediment yield of rivers. Nature 215, 925-928.

5.  LAKE, H.M.(1894)    Johore. Geogr. J. 3, 281-302.

6.  HARTLEY, C.W.S. (1949)    Soil erosion in Malaya.  Corona 1, 25-27.

7.  FERMOR, L.L. (1939)    Report upon the mining industry of Malaya. Govt. Printer, Kuala Lumpur.

8.  SOPER, J.R.P. (1938)    Soil erosion on Penang Hill.  Malay. Agr. J. 26, 407-413.

9.  MORGAN, R.P.C. (1971)    Rainfall of West Malaysia: A preliminary regionalisation using principal components analysis. Area 3, 222-227.

10. GHULAM, M.H. and ERH, K.T. (1979)    Rainfall intensity and its significance to soil erosion. Proc. Symp. on Water in Malaysian Agriculture, Kuala Lumpur, pp.31-40.

11. BOLS, P. (1979)    Bijdrage tot de studie van de bovengrondse afstroming en erosie op Java. Ph.D. thesis, University of Ghent.

12. TAJUDIN, A. (1980)    Kesan pengurusan tanaman ke atas hakisan tanah. Project paper (B. Agr. Sc.), Universiti Pertanian Malaysia, Serdang.

13. ·MORGAN, R.P.C. (1977)    Soil erosion in the United Kingdom: field studies in the Silsoe area, 1973-1975. Occasional Paper 5, Nat. Coll. Agr. Engng., Silsoe.

14. MAENE, L.M., THONG, K.C., ONG, T.S. and MOKHTARUDDIN, A.M. (1979) Surface wash under mature oil palm. Proc. Symp. on Water in Agriculture in Malaysia, Kuala Lumpur, pp.203-216.

15. MORGAN, R.P.C. (1974)    Estimating regional variations in soil erosion hazard in Peninsular Malaysia. Malay. Nat. J. 28, 94-106.

16. MORGAN, R.P.C. (1972)    Observations on factors affecting the behaviour of a first-order stream. Trans. Inst. Br. Geogr. 56, 171-185.

17. BRYAN, R.B. (1968)    The development, use and efficiency of indices of soil erodibility. Geoderma 2, 5-26.

18. WONG, I.F.T. (1974)    Soil-crop suitability classification for Peninsular Malaysia. Soils and Analytical Services Bulletin No.1, Min. of Agric. and Fisheries Malaysia, Kuala Lumpur.

19. MAENE, L.M., MOK, C.K. and CHEAH, K.F. (1975)    The application of a rainfall simulating method in erosion studies on three Peninsular Malaysian soils. Proc. Third ASEAN Soil Conf., Kuala Lumpur, pp.331-340.

20. ABDUL RASHID, F.D. (1975)    Studies on the correlation between the erodibility of standard sand and five Malaysian soils. Project paper (B.Agr.Sc.), University of Malaya, Kuala Lumpur.

21.  CHEE, B.W. (1977)    The design and use of a rainfall simulator
     in soil erosion studies. M. Agr. Sc. thesis, University of
     Malaya.

22.  WISCHMEIER, W.H., JOHNSON, C.B. and CROSS, B.V. (1971)    A soil
     erodibility nomograph for farmland and construction sites.
     J. Soil and Water Conserv. 26, 189-193.

23.  HUDSON, N.W. (1971)    Soil conservation. Bt  Batsford Ltd., London.

24.  HUDSON, N.W. and JACKSON, D.C. (1959)    Results achieved in the
     measurement of erosion and runoff in Southern Rhodesia.
     Proceedings, Third Inter-African Soils Conference, Dalaba,
     pp.575-583.

25.  SHALLOW, P.G.D. (1956)    River flow in the Cameron Highlands.
     Hydro-electrical Technical Memorandum 3, Central Electricity
     Board, Kuala Lumpur.

26.  DANIEL, J.B. and KULASINGAM, A. (1974)    Problems arising from
     large scale forest clearing for agricultural use - The
     Malaysian Experience. Malay. Forester 37, 152-160.

27.  FOURNIER, F. (1972)    Soil conservation. Nature and Environment
     Series, Council of Europe.

28.  ELWELL, H.A. and STOCKING, M.A. (1976)    Vegetal cover to estimate
     soil erosion hazard in Rhodesia. Geoderma 15, 61-70.

29.  LOW, K.S. (1972)    Interception loss in the humid forested areas
     (with special reference to Sungei Lui Catchment, West Malaysia).
     Malay. Nat. J. 25(2), 104-111.

30.  TEOH, T.S. (1973)    Some effects of Hevea plantations on
     rainfall redistribution. Proc. Symp. on Biological Resources
     and National Development, Kuala Lumpur, pp.73-82.

31.  MAENE, L.M. and CHONG, C.P. (1979)    Drop size distribution and
     erosivity of tropical rainstorms under the oil palm canopy.
     Lapuran Penyelidikan Jabatan Sains Tanah 1977-78, Universiti
     Pertanian Malaysia, Serdang, pp.81-93.

32.  WATSON, G.A. (1963)    Cover plants and tree growth. Plrs' Bull.
     Rubb. Res. Inst. Malaya No.68, 123.

33.  WATSON, G.A., WONG, P.W. and NARAYANAN, R. (1964)    Effect of
     cover plants on soil nutrient status and on growth of Hevea.
     III.  A comparison of leguminous creepers with grasses and
     Mikania cordata.  J. Rubb. Res. Inst. Malaya 18(2), 80.

**217**

34. PUSHPARAJAH, E. and CHELLAPAH, K. (1969)    Manuring of rubber in relation to covers.  J. Rubb. Res. Inst. Malaya 21(2), 126.

35. SOONG, N.K. and YAP, W.C. (1976)    Effect of cover management on physical properties of rubber growing soils.  J. Rubb. Res. Inst. Malaysia 24(3), 145-159.

36. PUSHPARAJAH, E., TAN, K.H. and SOONG, N.K. (1977)    Influence of covers and fertilisers and management on soil.  In "Soils under Hevea in Peninsular Malaysia and their Management," Rubber Research Institute of Malaysia, Kuala Lumpur, pp.75-93.

37. LING, A.H., TAN, K.Y., TAN, P.Y. and SYED SOFI (1979)    Preliminary observations on some possible post-clearing changes in soil properties.  Seminar on Fertility and Management of Deforested Land, Kota Kinabalu (preprint).

38. FOURNIER, F. (1960)    Climat et erosion: la relation entre l'erosion du sol par l'eau et les precipitations atmospheriques. Presses Universitaires de France, Paris.

39. LAL, R. (1976)    Soil erosion problems on an alfisol in western Nigeria and their control.  IITA Mongraph I, Ibadan.

40. AINA, P.O., LAL, R. and TAYLOR, G.S. (1976)    Soil and crop management in relation to soil erosion in the rainforest of western Nigeria.  In 'Soil erosion: prediction and control,' Soil Conservation Society of America, Special publication No.21, pp.75-82.

41. MAENE, L.M., MOK, C.K., LAI, A.L. and CHEE, S.K. (1975) Erodibility and erosion control of two Peninsular Malaysian soils.  Proc. Third ASEAN Soil Conf., Kuala Lumpur, pp.323-330.

42. HUSDON, N.W. (1957)    The design of field experiments on soil erosion.  J. Agric. Engng. Res. 2, 56-65.

43. LAL, R. (1977)    Soil-conserving versus soil-degrading crops and soil management for erosion control.  In 'Soil Conservation and Management in the Humid Tropics,'  John Wiley & Sons Ltd., pp.81-86.

44. MOKHTARUDDIN, A.M. and MAENE, L.M. (1979)    Soil erosion under different crops and management practices.  International Conf. on Agricultural Engineering in National Development, Serdang (preprint).

45.  CHIN FATT and TAN, E.H. (1974)      Effects of simulated erosion
         on the performance of maize (Zea mays) and groundnut (Arachis
         hypogea):  a preliminary assessment.  Symp. on Classification
         and Management of Malaysian Soils, Kota Kinabalu (preprint).

46.  SOONG, N.K. and YEOH, C.S. (1975)      Latex-oil emulsions for
         controlling soil erosion on exposed soil surfaces.
         International Rubber Conference 1975, Kuala Lumpur (preprint).

47.  WISCHMEIER, W.H. and SMITH, D.D. (1962)      Soil loss estimation
         as a tool in soil and water management planning.  Int. Assoc.
         for scient. Hydrol. 59, 148-159.

48.  TANG, H.T., MANOKARAN, N. and BLAKE, G.J. (1979)      The status
         of hydrological studies at the Forest Research Institute,
         Kepong.  Proc. Symp. on Water in Agriculture in Malaysia,
         Kuala Lumpur, pp.21-29.

49.  GREENLAND, D.J. and LAL, R. (1977)      Soil erosion in the humid
         tropics:  The need for action and the need for research.
         In 'Soil Conservation and Management in the Humid Tropics,'
         John Wiley & Sons Ltd., pp.261-265.

Part IV

# MODELING

# Editor's Comments
# on Papers 14 Through 17

**14  MEYER and WISCHMEIER**
*Mathematical Simulation of the Process of Soil Erosion by Water*

**15  BENNETT**
*Concepts of Mathematical Modeling of Sediment Yield*

**16  FOSTER et al.**
*Estimating Erosion and Sediment Yield on Field-Sized Areas*

**17  MORGAN, MORGAN, and FINNEY**
*A Predictive Model for the Assessment of Soil Erosion Risk*

The Universal Soil Loss Equation (Wischmeier and Smith, 1978) has dominated erosion modeling for the last 20 years. The equation was developed as an empirical predictive model for use as a management tool in soil conservation planning. Its success is a result of its simplicity which makes it easy for nontechnical staff in advisory and extension services to understand and use.

The model has also been taken up by researchers to predict hillslope erosion rates, but as a research tool it has many limitations. This is because the model views erosion as a multiplicative function of rainfall erosivity (R), soil erodibility (K), slope steepness (S), slope length (L), crop management (C), and erosion control practices (P) such as contouring and strip-cropping. Unfortunately, erosion cannot be properly represented by taking values of six factors and multiplying them together. At is simplest this would assume that each factor has equal weight, that there are no interactions between them, and that erosion is linearly related to each. In fact, the Universal Soil Loss Equation is more complex than this because differences in the magnitudes and ranges of the factor values prevent equal weighting. Also, the equation used to determine values for the slope steepness factor is nonlinear in form. However, increasing recognition of the limitations of the Universal Soil Loss Equation has led to a move to replace it.

Meyer and Wischmeier (Paper 14) laid the foundation for a new generation of erosion models by calling for a return to the fundamental approach proposed by Ellison (1947) more than 20 years previously. This approach considers erosion to result from the detachment of soil particles from the soil mass and their transport from the place of detachment. The Meyer-Wischmeier scheme allows for raindrop impact and surface runoff as agents of detachment and transport. As presented, the scheme cannot be used for predicting erosion rates because input data are lacking on the soil parameters, and the method can be applied only to rather simple conditions of bare soil, no conservation practice, a slope that is uniform in plan, and a slope from which no material is removed from the base by river action. However, the Meyer-Wischmeier scheme can predict the location of erosion and deposition on a hillslope and, therefore, the main sources and sinks for sediment. It can also indicate whether erosion on any given slope segment is controlled by the rate of detachment or by the transport capacity. This information should help in directing erosion control practices to the dominant process and to the critical parts of a hillside. The most important achievement of the scheme, however, is that it simulates erosion in a reasonably realistic manner and therefore shows that the approach is a feasible basis for future model development. Such development has resulted in the model presented by Foster et al. (Paper 16) in which physically based equations for detachment and transport by interrill and rill erosion are combined with those factors in the Universal Soil Loss Equation representing soil erodibility, crop management, and erosion control practices to predict soil loss from hillslopes on a daily basis.

The advantages of physically based or partially physically based models over empirical ones have been much debated. Hudson (1981) argues that the two types should be clearly distinguished. Models such as the Universal Soil Loss Equation should not be used as a research tool while physically based models should not be used to plan soil conservation schemes that can be more easily designed using empirical techniques. In Paper 15 Bennett states that physically based models have an advantage in their ability to relate to an individual event or even part of an event. This arises from the use of modeling equations that can be applied only to virtually instantaneous conditions. There is no way in which physically based models can be used for conditions averaged over long time periods such as a year. Bennett suggests that one hour is about the maximum time period for which averaging is acceptable although Foster et al. apply their equations to conditions averaged over a single storm. However, since they use daily rainfall totals as surrogates for storm rainfall amounts their model operates as a daily simulation model. This means that predicting mean annual soil loss over a period of 20 or 25 years requires a vast quantity of daily rainfall data.

Since the Universal Soil Loss Equation has served its purpose well by predicting mean annual soil loss, the need for the greater resolution provided by a daily simulation model is questionable. Knowing how much sediment is lost in each storm is not necessary if all that matters is whether the annual soil loss exceeds a particular threshold or tolerance level. The main value of this extra information is in relation to pollution where the relevant effects of an erosive event last only a few hours to a few days. Thus the scheme of Foster et al. (Paper 16) should be seen as a major contribution to pollution modeling. It has been adopted as the erosion component of the CREAMS model where it is linked with hydrology and chemistry components to predict non-point-source pollution arising from different agricultural management systems (Knisel, 1980). It therefore forms part of a management tool for pollution control. Although the model has the apparent advantage of being able to predict the particle size of the eroded material and its enrichment with fines, the approach used to effect this is empirical and rather simplistic. This is because, although it is known that erosion by rainsplash, overland flow, and rill flow is particle-size selective, very limited data exist for developing or verifying operating equations to describe this aspect.

With the emphasis of research on developing sophisticated storm models or daily simulation models, the simplicity of the Universal Soil Loss Equation has been lost. With models like that of Foster et al. it is often difficult to determine what factor or factors have contributed most to a particular value of predicted soil loss. Also, the models are not suitable for field use by conservation workers because of the need for a large data input and computer facilities. Since they are also best operated by skilled researchers, their use requires the services of some form of modeling or predictive consultancy unit to support the field staff. Little attention has been given to an alternative approach of improving simple annual models in line with our better understanding of erosion processes. The scheme presented by Morgan, Morgan, and Finney in Paper 17 is an attempt to do this. The operating functions are of necessity empirical and were taken from previously published work. Thus the model is not new but is an amalgam of existing research brought within the already accepted framework of Meyer and Wischmeier. Morgan, Morgan, and Finney claim that, for a model of its type, the results of a comparison of predicted and measured values using erosion plot data from 12 countries are encouragingly good. Also, the model has the advantages of simulating the erosion process and the effects of conservation practice in a reasonably realistic manner. Its simplicity enables the most important factors contributing to a particular predicted value of erosion to be interpreted. It provides more information than the Universal Soil Loss Equation by giving an estimate of mean annual runoff and by indicating whether, on average, erosion is detachment-limited or transport-limited.

The basis for making a judgment on the success or nonsuccess of a model is still somewhat slender because, as pointed out by Bennett, much more attention needs to be devoted to calibration and verification. Too many models are never tested at all. Where tests are carried out, they most often take the form of calculating some measure of goodness of fit such as mean square error or coefficient of determination. As indicated by Morgan, Morgan, and Finney, this may have the disadvantages of assuming that errors are of equal importance over the whole range of the data and that there are no errors in the observed values. As seen in Part II, this last assumption is almost certainly invalid, particularly where the measured data are for a single erosion plot that may or may not be a representative sample of the surrounding area. Virtually no attention has been paid by users of predictive models to the level of accuracy required for specific purposes. Is it necessary to predict soil loss absolutely to plus or minus so many tons per hectare with a given confidence, i.e., the prediction will be correct 95% of the time? Must the effects of, say, an increase in rainfall be predicted absolutely, or is it sufficient to predict relative or percentage change correctly even if the absolute values of soil loss are an order of magnitude out? Are there occasions when it is enough to predict only the nature of the change, i.e., an increase or decrease in soil loss? Answers to these questions should be a prerequisite to model development. The required level of accuracy should be a design requirement of the model.

The selection of a particular model is likely to be a compromise between the accuracy required and that which the model can produce. Bennett, citing work by Dawdy, Lichty, and Bergman (1972), shows that predictions of a typical rainfall-runoff model have about a 30% standard error of estimate. Sediment yield models based on rainfall-runoff models are therefore unlikely to perform better than about a 40% standard error of estimate. Thus relatively sophisticated models give worse results than sediment yield predictions from a simple black-box model, the sediment-discharge rating curve, for which a 20% standard error of estimate is typical.

In evaluating how well a particular model performs, Bennett emphasizes the importance of a sensitivity analysis to determine how the predictions are affected by changes in the values of the input parameters. As shown by Morgan, Morgan, and Finney, this indicates which parameters need to be assessed most accurately. Again, because of errors in measurement, there is often a compromise between what is required by the model and what can be achieved in practice.

Proper analysis of what accuracy is required and what can be achieved within the constraints imposed by the operating functions and the accuracy of measurement of the input parameters used in the model brings out priorities for further research. These may rest with developing alternative mathematical relationships or devising better techniques of measurement. Priorities identified by Bennett involve the development of mathematical procedures for linking hillslope and channel processes, describing gully erosion, characterizing roughness effects, and simulating the role of vegetation.

The papers reviewed here pose an additional challenge. Following the success of the Universal Soil Loss Equation as a design tool, it is widely accepted that soil conservation strategies should be aimed at reducing mean annual soil loss to an acceptable level. The arguments presented earlier in favor of separating empirical or design tools from more physically based research tools and the rationale behind the annual model of Morgan, Morgan, and Finney give further support to this view. Yet certain aspects of soil conservation design are based on the individual event. Waterways are designed to convey the peak runoff with a given return period. Would soil conservation be better served if all its design work was related to such an event? Are there sound reasons for adopting the criterion of mean annual soil loss or is its use merely a reflection that suitable event models were not available when the Universal Soil Loss Equation was first published? If the latter is the case, it may be time to rethink the basis on which conservation systems are designed and allow models operating on an individual storm simulation to assume greater significance.

## REFERENCES

Ellison, W. D., 1947, Soil erosion studies. Part I., *Argric. Eng.* **28:**145–146.

Dawdy, D. R., R. W. Lichty, and J. M. Bergman, 1972, A rainfall-runoff simulation model for estimation of flood peaks for small drainage basins, *U. S. Geological Survey Professional Paper 506-B,* 28p.

Hudson, N. W., 1981, *Soil Conservation,* Batsford, London, 324p.

Knisel, W. G., 1980, CREAMS: a field scale model for chemicals, runoff and erosion from agricultural management systems, *USDA Conservation Research Report 26,* 643p.

Wischmeier, W. H. and D. D. Smith, 1978, Predicting rainfall erosion losses. A guide to conservation planning, *USDA Agricultural Handbook 537,* 58p.

# 14

# Mathematical Simulation of the Process of Soil Erosion by Water

L. D. Meyer and W. H. Wischmeier
MEMBER ASAE        EXEC. AFFILIATE ASAE

MORE than 20 years ago, W. D. Ellison defined soil erosion as "a process of detachment and transportation of soil materials by erosive agents" (4)[*]. For erosion by water, these agents are rainfall and runoff. He pointed out that each has both a detaching and a transporting capacity, and that these must be studied separately. Similarly, a soil's erodibility may be separated into its detachability and transportability components. Although this approach appears as sound today as it did when it was first presented, these subprocesses have not been distinguished in the soil-erosion relationships commonly used to date.

This report suggests an approach to erosion simulation that considers (a) soil detachment by rainfall, (b) transport by rainfall, (c) detachment by runoff, and (d) transport by runoff as separate but interrelated phases of the process of soil erosion by water. The dynamics of each phase then may be described by fundamental hydraulic, hydrologic, meteorologic and other physical relationships plus parameters describing the soil properties that influence erosion. The general model based on this approach can be expanded to a more detailed model by introducing other components.

## Background

Mathematical expressions useful for evaluating erosion systems and control practices originated more than a quarter century ago. In 1940, Zingg (14) published an empirical equation expressing soil loss as a function of slope length and steepness. Other researchers used this as a basis for developing erosion prediction equations applicable to specific climatic regions. Derivation of the erosion index, EI, (10) and a technique for bringing effects of local rainfall distribution into practice evaluations (11) overcame regional boundaries and led to development of a universal soil-loss equation (13). Nearly a decade of widespread use has established the value of this equation and has emphasized the importance of such mathematical expressions for planning soil and water conservation and management systems.

Since the purpose of these past equations was to provide a planning tool for conservation technicians, brevity and simplicity were more essential than great detail. Empirical relations were determined and were combined in equations designed to predict long-term soil losses from particular tracts of land under various combinations of land use and management. They were not designed to meet the present need for a detailed mathematical model to simulate soil erosion as a dynamic process and to describe soil movement at all locations along a slope at any time.

The approach suggested by Ellison (4) has not been followed, so data directly applicable to it are scarce. However, some research information is available that appears applicable for preliminary trials of this approach to modeling the erosion process. For instance, soil splash by raindrop impact may be considered a measure of particle detachment by rainfall. Free (6) found that sand-splash loss correlated best with rainfall energy but that splash of aggregated soil was more closely correlated with the erosion index, EI. For 30-minute rain applications, Ellison (5) determined the relationship: splash loss = $kV^{4.33}D^{1.07}I^{0.65}$, where k, V, D and I are the soil factor, drop velocity, drop diameter and intensity, respectively.

As a parameter to represent detachment by runoff, the average shearing stress along the wetted perimeter, known as the tractive force, $\tau$, (2) seemed appropriate.

For soil transport by rainfall, Ekern (3) reported that the percentage of total splashed soil that moved downslope equaled the percent slope plus 50. Ellison (5) found that 75 percent of the splashed soil moved downhill and 25 percent uphill on a 10 percent slope.

The capacity of flowing water to transport soil was shown by Laursen (7) to be approximately proportional to the fifth power of velocity.

The susceptibilities of different soils to detachment and to transport by rainfall and by runoff have not been evaluated separately. Although soils have been grouped according to their estimated relative erodibilities, the criteria for such classification were not well defined. Two empirical equations for estimating a soil's erodibility on the basis of interrelations of its physical and chemical properties were recently presented (1, 12). However, erodibility as defined combines effects of differences in infiltration rates, soil permeability, detachability and transportability in one factor. Therefore, the results are not adaptable to study of individual phases of the erosion process.

## Procedure

This study was concerned primarily with investigating the feasibility of separating the soil-erosion process into several component processes rather than with obtaining quantitative results. Therefore, only the major components of the erosion process were considered. Such effects as wind, varying rainfall and infiltration rates, and rill patterns will not be discussed so that the general approach can be emphasized.

Slope lengths were divided into short, finite increments, and steady-state conditions were assumed for initial tests of the model. All terms were expressed per unit of slope width, so area was proportional to slope length. Four values were computed for each increment: detachment by rainfall, detachment by runoff, transport capacity of rainfall, and transport capacity of runoff. Mathematical submodels of these four parts of the erosion process were developed, and pertinent parameter values were approximated from available information on basic relationships. The four submodels were then combined in a computer program to route the soil downslope. A brief discussion of the submodels follows.

Paper No. 68-732 was presented at the Winter Meeting of the American Society of Agricultural Engineers at Chicago, Ill., December 1968, on a program arranged by the Soil and Water Division.

The authors—L. D. MEYER and W. H. WISCHMEIER—are, respectively, agricultural engineer and analytical statistician, Soil and Water Conservation Research Division, Agricultural Research Service, U. S. Department of Agriculture.

*Author's Note:* This paper is approved as a contribution from the Corn Belt Branch, Soil and Water Conservation Research Division, Agricultural Research Service, U. S. Department of Agriculture, in cooperation with the Purdue Agricultural Experiment Station (Purdue journal series paper No. 3579).

[*] Numbers in parentheses refer to the appended references.

## SOIL EROSION PROCESS

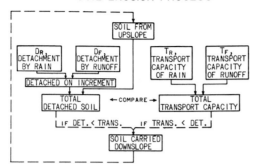

**Fig. 1 Approach used to simulate the process of soil erosion by water. Four subprocesses, $D_R$, $D_F$, $T_R$, and $T_F$, were evaluated for each successive slope-length increment, and soil movement was routed downslope as illustrated**

*Detachment by Rainfall, $D_R$.* Previously cited research (6) indicated that soil splash loss is approximately proportional to the parameter EI, the product of rainfall kinetic energy and maximum 30-minute intensity. For moderate to intense rainstorms, kinetic energy per unit of rain per unit of area varies approximately as $I^{0.14}$, where $I$ = intensity. For steady-state conditions, rain amount per unit area is proportional to $I$, so total energy, $E$, is proportional to $I^{1.14}$. For a uniform rainfall intensity, the EI parameter is then approximately proportional to $I^{1.14} \times I$ or $I^{2.14}$. A similar value for the exponent of $I$ was obtained by using the relations of $V$ and $D$ to $I$ in Ellison's equation (5), splash per unit time = $kV^{4.33} D^{1.07} I^{0.65}$. Using these as guides, but recognizing that a more rigorous relationship is yet to be defined, soil detachment by rainfall was assumed proportional to $I^2$.

The amount of soil detached is also a function of the area of the increment, $A_I$, and of the soil type. Since no mathematical expression was available for the latter function, soil effect was represented by a constant, $S_{DR}$. The complete relationship for soil detachment by rainfall then became, $D_R = S_{DR} A_I I^2$.

*Detachment by Runoff, $D_F$.* Tractive force was chosen to represent detachment by runoff. Since the tractive force, $\tau$, is proportional to flow-velocity squared, $D_F$ was assumed proportional to $V^2$. Using the continuity and Manning's equations and assuming turbulent flow, Meyer (8) showed that $V = C_t S^a Q^b n^{-c}$, where $V$, $S$, $Q$, and $n$ are velocity of flow, slope steepness, flow rate, and hydraulic roughness, respectively, and $C_t$, $a$, $b$, and $c$ are constants. For this equation, $a$ varied from 0.30 for shallow, infinitely wide channels to 0.35 for wide, shallow, parabolic channels to 0.38 for wide, shallow, tri-

angular channels. For the same channel conditions, $b$ varied from 0.40 to 0.31 to 0.25 and $c$ from 0.60 to 0.69 to 0.75, respectively. Thus, $V$ is approximately proportional to $S^{1/3} Q^{1/3}$, where the hydraulic roughness is constant. Therefore, $D_F$ was assumed proportional to $S^{2/3} Q^{2/3}$. Since information was not available for expressing a soil's susceptibility to detachment by runoff as a function of its properties, soil effect was expressed as a constant, $S_{DF}$. Detachment by runoff on a slope increment was computed as the average of the detachment capacities at the start, s, and the end, $E$, of the increment; that is, $D_F = S_{DF} A_I \tfrac{1}{2}(S_s^{2/3} Q_s^{2/3} + S_E^{2/3} Q_E^{2/3})$.

*Transportation Capacity of Rainfall, $T_R$.* The capacity of rainfall to transport soil by splashing is a function of slope steepness, amount of rain, soil properties, microtopography, and wind velocity. Slope effect was based on Ekern's (3) finding that net downslope movement is proportional to the slope steepness, $S$. For steady-state conditions, amount of rainfall is proportional to its intensity, $I$. Soil effect was included as a constant, $S_{TR}$, and the effects of wind and microtopography were omitted for lack of pertinent relationships. Thus, transportation by rainfall was expressed as $T_R = S_{TR} S I$.

*Transportation Capacity of Runoff, $T_F$.* Based on Laursen's (7) findings that the sediment-carrying capacity of flowing water is approximately proportional to the fifth power of flow velocity, $T_F$ was assumed proportional to $V^5$. When combined with the approximation, $V \propto S^{1/3} Q^{1/3}$, this gave $T_F \propto S^{5/3} Q^{5/3}$. A soil term, to account for the effect of particle size and density on the soil's transportability, was again included as a constant, $S_{TF}$. Transportation by runoff was then expressed as $T_F = S_{TF} S^{5/3} Q^{5/3}$.

*Computational Procedure* A com-

puter program was developed to route the soil and water from the top to the bottom of a slope, based on conservation of mass. The four preceding subprocesses of the erosion process were evaluated for each slope-length increment and used as illustrated in Fig. 1. On each increment, the soil available for erosion was the material detached on that increment by rainfall and by runoff $(D_R + D_F)$ plus the material carried to it from the increment upslope. This sum was compared with the sum of the transport capacities at the end of the increment $(T_R + T_F)$. If the total detached soil available for transport was less than the total transport capacity, available soil was the limiting factor on the slope increment, and the sediment load carried to the next increment equaled the amount of available material. However, if the total transport capacity was less than the soil available for erosion, transportation was the limiting factor, and the sediment load equaled the transport capacity.

To test this form of an erosion model, the terms representing the soil's detachability and transportability, $S_{DR}$, $S_{DF}$, $S_{TR}$, $S_{TF}$, were assigned values that gave results in seemingly proper relation to each other. The assumed values showed rainfall effects dominant near the top of slopes but runoff effects dominant beyond some point downslope, and showed detachment capacities greater than transportation capacities on the upper portions of moderate slopes. The four soil terms were held constant in all computer investigations reported hereafter except Figs. 2h and 2i, where they were varied to study effects of differences in soil detachability.

### Results and Interpretations

The accompanying table and graphs show how various physical parameters affect the results obtained by using this approach. They illustrate the individual and collective effects of detachment and transport of soil by rainfall and by runoff, based on the assumed relationships. The results are for steady-state conditions with respect to time.

The following abbreviations are used in Table 1, each of which refers only to the specific slope-length increment under consideration, unless otherwise specified:

LNGH, distance from top of slope to bottom of increment

STEEP, slope steepness computed as average steepness between top of increment and bottom of next increment

$D_R$, soil detachment by rainfall

$D_F$, soil detachment by runoff

DINC, total detachment on increment $(D_R + D_F)$

TABLE 1. EXAMPLE OF EROSION MODEL RESULTS FOR A 250-FT, COMPLEX-SHAPED SLOPE WITH AN AVERAGE STEEPNESS OF 8 PERCENT*

| LGTH | STEEP | $D_R$ | $D_F$ | DINC | $T_R$ | $T_F$ | TCAP | TOTD | SEDLD | EROS |
|------|-------|-------|-------|------|-------|-------|------|------|-------|------|
| 0 | | | | | | | | | | |
| 10 | 1.57 | 0.40 | 0.02 | 0.42 | 0.03 | 0.01 | 0.04 | 0.42 | 0.04 | 0.04 |
| 20 | 3.12 | 0.40 | 0.06 | 0.46 | 0.06 | 0.10 | 0.16 | 0.50 | 0.16 | 0.12 |
| 30 | 4.61 | 0.40 | 0.11 | 0.51 | 0.09 | 0.37 | 0.46 | 0.67 | 0.46 | 0.30 |
| 40 | 6.04 | 0.40 | 0.16 | 0.56 | 0.12 | 0.94 | 1.06 | 1.03 | 1.03 | 0.56 |
| 50 | 7.37 | 0.40 | 0.23 | 0.63 | 0.15 | 1.90 | 2.04 | 1.65 | 1.65 | 0.63 |
| 60 | 8.58 | 0.40 | 0.29 | 0.69 | 0.17 | 3.31 | 3.49 | 2.34 | 2.34 | 0.69 |
| 70 | 9.66 | 0.40 | 0.35 | 0.75 | 0.19 | 5.22 | 5.41 | 3.10 | 3.10 | 0.75 |
| 80 | 10.58 | 0.40 | 0.42 | 0.82 | 0.21 | 7.59 | 7.80 | 3.91 | 3.91 | 0.82 |
| 90 | 11.34 | 0.40 | 0.48 | 0.88 | 0.23 | 10.37 | 10.60 | 4.79 | 4.79 | 0.88 |
| 100 | 11.92 | 0.40 | 0.54 | 0.94 | 0.24 | 13.43 | 13.67 | 5.73 | 5.73 | 0.94 |
| 110 | 12.31 | 0.40 | 0.59 | 0.99 | 0.25 | 16.62 | 16.86 | 6.72 | 6.72 | 0.99 |
| 120 | 12.51 | 0.40 | 0.64 | 1.04 | 0.25 | 19.73 | 19.98 | 7.75 | 7.75 | 1.04 |
| 130 | 12.51 | 0.40 | 0.68 | 1.08 | 0.25 | 22.54 | 22.79 | 8.83 | 8.83 | 1.08 |
| 140 | 12.31 | 0.40 | 0.71 | 1.11 | 0.25 | 24.84 | 25.09 | 9.93 | 9.93 | 1.11 |
| 150 | 11.92 | 0.40 | 0.73 | 1.13 | 0.24 | 26.41 | 26.64 | 11.06 | 11.06 | 1.13 |
| 160 | 11.34 | 0.40 | 0.74 | 1.14 | 0.23 | 27.06 | 27.29 | 12.21 | 12.21 | 1.14 |
| 170 | 10.58 | 0.40 | 0.74 | 1.14 | 0.21 | 26.68 | 26.89 | 13.35 | 13.35 | 1.14 |
| 180 | 9.66 | 0.40 | 0.73 | 1.13 | 0.19 | 25.19 | 25.39 | 14.48 | 14.48 | 1.13 |
| 190 | 8.58 | 0.40 | 0.71 | 1.11 | 0.17 | 22.63 | 22.81 | 15.59 | 15.59 | 1.11 |
| 200 | 7.37 | 0.40 | 0.67 | 1.07 | 0.15 | 19.12 | 19.27 | 16.66 | 16.66 | 1.07 |
| 210 | 6.04 | 0.40 | 0.62 | 1.02 | 0.12 | 14.89 | 15.01 | 17.68 | 15.01 | —1.65 |
| 220 | 4.61 | 0.40 | 0.55 | 0.95 | 0.09 | 10.28 | 10.37 | 15.96 | 10.37 | —4.64 |
| 230 | 3.12 | 0.40 | 0.45 | 0.85 | 0.06 | 5.75 | 5.82 | 11.22 | 5.82 | —4.55 |
| 240 | 1.57 | 0.40 | 0.33 | 0.73 | 0.03 | 1.97 | 2.00 | 6.55 | 2.00 | —3.81 |
| 250 | 0.39 | 0.40 | 0.18 | 0.58 | 0.01 | 0.21 | 0.22 | 2.59 | 0.22 | —1.78 |
| 260 | 0. | 0.40 | 0.05 | 0.45 | 0. | 0. | 0. | 0.67 | 0. | —0.22 |
| 270 | 0. | 0.40 | 0. | 0.40 | 0. | 0. | 0. | 0.40 | 0. | 0. |
| 280 | 0. | 0.40 | 0. | 0.40 | 0. | 0. | 0. | 0.40 | 0. | 0. |
| 290 | 0. | 0.40 | 0. | 0.40 | 0. | 0. | 0. | 0.40 | 0. | 0. |
| 300 | 0. | 0.40 | 0. | 0.40 | 0. | 0. | 0. | 0.40 | 0. | 0. |

*Rainfall intensity = 2.0 in. per hr; infiltration rate = 1.0 in. per hr; $S_{DR} = 0.01$, $S_{DF} = 0.0005$, $S_{TR} = 0.01$, and $S_{TF} = 0.0001$. All values shown in this table are relative. Explanations of column headings may be found under subtitle "Results and interpretations." These results are shown graphically in Fig. 2a.

$T_R$, transportation capacity of rainfall
$T_F$, transportation capacity of runoff
TCAP, total transportation capacity at end of increment ($T_R + T_F$)
TOTD, total detached soil available for transport (DINC + previous SEDLD)
SEDLD, sediment load carried from increment (either TCAP or TOTD, whichever is smaller)
EROS, net soil loss from increment (SEDLD — previous SEDLD) Negative values indicate deposition.

In Fig. 2, the upper graphs show available detached soil (TOTD), transportation capacity (TCAP), and sediment load (SEDLD) plotted against slope length. The sediment-load curves follow the lower of the other two at all points. The middle graphs (omitted on d, e, h and i since they are identical to a) show the slope shape studied, with the vertical scale expanded to ten times the horizontal. The lower graphs show the net erosion or deposition (EROS) for each increment along the slope.

The key condition chosen for comparison with others is represented by the data in Table 1 and Fig. 2a. This is a complex-shaped slope, 250 feet long, with an elevation change of 20 ft. This slope was generated by solving the relationship:

Depth = 0.5 x total elevation change

x [1 — cos ($\pi$ x distance downslope / total slope length)].

Rainfall intensity and infiltration rate were constant at 2.0 and 1.0 in. per hr, respectively. Slope-length increments of 10 ft were studied. Assumed soil constants were: $S_{DR} = 0.01$, $S_{DF} = 0.0005$, $S_{TR} = 0.01$, and $S_{TF} = 0.0001$. In each succeeding figure, only one of these key conditions was changed.

On the complex slope averaging 8 percent steepness represented in Table 1 and Fig. 2a, 0.42 unit of soil was detached on the upper 10 ft of the slope, but the transport capacity was only 0.04. Thus, transportation was limiting, and the sediment load at the end of 10 ft was 0.04 unit. For the 10 to 20-ft increment, 0.04 unit was carried from the previous increment, and 0.46 was detached, so available detached soil totaled 0.50 unit. Again, the transport capacity of 0.16 unit was limiting, so this was the sediment load at the end of 20 ft. At 30 ft, the transport capacity also was limiting, but by 40 ft the available detached soil became limiting, so the sediment load there was 1.03 units.

In the upper graph of Fig. 2a, the available-soil and transport-capacity curves cross between 30 and 40 ft. The sediment-load curve, which followed the transport curve up to 30 ft, follows the detached-soil curve after 40 ft. Transport capacity continued to exceed available detached soil until the

slope flattened appreciably beyond 200 ft. Thereafter, transport capacity again became limiting, and the sediment-load curve follows it along the remainder of the slope. Net erosion increased for the first 200 ft to a maximum sediment load of 17 units, and then deposition occurred due to the slope flattening along its lower portion. The rate of net erosion was proportional to the slope of the sediment-load curve.

A uniform slope of the same length and average steepness as the complex slope of Fig. 2a is illustrated in Fig. 2b. Transport capacity was limiting for the first few increments, but available detached soil limited erosion for the remainder of the sloping portion. The maximum sediment load reached 20 units at the end of the slope. (The abrupt discontinuity at 250 ft would predict an absurd rate of deposition there if infinitesimal increments of slope length and a more rigorous definition of slope steepness were used.)

The sediment-load curves of Fig. 2a and 2b are very similar to those obtained by applying empirical soil-loss equations on successive increments of complex and uniform slopes (9). The major difference is the abrupt change in the sediment-load curve at about 200 ft in Fig. 2a, but such discontinuity could be expected to be smoothed by equations based on the empirical curve-fitting techniques used for the other study.

The erosion that this procedure predicted for an irregular slope is illustrated in Fig. 2c. The transport capacity and detached soil changed erratically as the steepness varied. Deposition occurred where the slope decreased considerably, as indicated at 60 and at 100 ft. Such deposition often occurs on irregular topography where the slope flattens.

Fig. 2d shows the effects of a more intense rainfall on the complex slope of Fig. 2a. When the intensity was increased to 3 in. per hr, the maximum sediment load increased from 17 to 33 units, but the maximum transport capacity increased from 27 to 86 units as compared to the conditions of Fig. 2a. Thus, the sediment load was much less than the runoff's transport capacity along most of the sloping portion.

For a rainfall intensity of 1¼ in. per hr, only slightly greater than the infiltration rate, the results are illustrated in Fig. 2e. Detachment and transport capacity were both low, but erosion was limited by transport capacity throughout the slope length. The maximum sediment load was relatively very low, less than three units. Since excess detached soil was available all along the slope, a succeeding period of high-

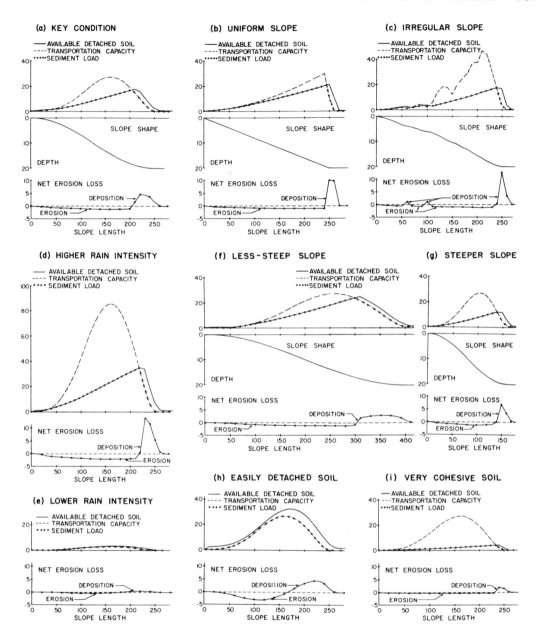

Fig. 2 Erosion-model results for various conditions. Graph ordinates are relative, but slope units may be assumed to be in feet and ordinates of the upper and lower graphs may be considered as pounds per foot of slope width: (a) Key condition for comparison with succeeding graphs: complex slope averaging 8 percent steepness; rainfall intensity, 2.0 in. per hr; infiltration rate, 1.0 in. per hr; soil of moderate detachability. (b) Same as (a) except uniform slope. (c) Same as (a) except irregular slope. (d) Same as (a) except rainfall intensity of 3.0 in. per hr. (e) Same as (a) except rainfall intensity of 1¼ in. per hr. (f) Same as (a) except a 400-ft slope averaging 5 percent steepness. (g) Same as (a) except a 160-ft slope averaging 12½ percent steepness. (h) Same as (a) except easily detachable soil with detachment values five times as great. (i) Same as (a) except very cohesive soil with detachment values only one-fifth as great

intensity rainfall would find considerable detached soil already available for transport. Therefore, the resulting sediment load would likely be greater than that predicted by the detachment capacity of the succeeding rainfall alone.

For the complex slope averaging 5 percent illustrated in Fig. 2f, transport capacity was limiting for the first 75 ft and became limiting again after about 300 ft of the 400-ft slope. It was not so much greater than the sediment load along the midportion of this slope as it was on the steeper slope of Fig. 2a. Maximum sediment load was 24 units at about 300 ft. For slopes averaging 3 percent or less, transport capacity was limiting for the entire length.

For the complex slope averaging 12½ percent illustrated in Fig. 2g, available detached soil was limiting for all but the first 20 ft and the last 25 ft. Maximum sediment load was 18 units at about 140 ft. The sediment load along most of the slope was much less than the transport capacity. Of course the transport capacity increases more rapidly on steeper slopes if other conditions are the same.

The next two figures (Fig. 2h and Fig. 2i) illustrate the effects of differences in soil detachability. For a relatively noncohesive soil, such as one with a very high silt or sand content, erosion would seldom be limited by the available detached soil; thus, the sediment load would depend on the transport capacity. Fig. 2h illustrates the erosion predicted by use of soil-detachability constants five times as great as those used for earlier figures. Since transport capacity was limiting all along the slope, the sediment load equalled the transport capacity and reached a maximum of 27 units at 160 ft. The sediment load along most of the slope was greater than for the soil of moderate detachability in Fig. 2a. Deposition occurred on a much longer portion of the lower slope than where available detached soil was also limiting·

Comparable results for a highly cohesive soil of one-fifth the detachability used for Figs. 2a through 2g are given in Fig. 2i. Detachment was limiting for all but the first and last increments, so erosion was relatively very low. Midway downslope, the sediment load was less than 10 percent of the transport capacity and reached a maximum of only four units at about 230 ft. The soil changes illustrated in Figs. 2h and 2i probably would also affect soil transportability and infiltration rates, but such changes were not studied.

The significance of knowing whether transport capacity or available detached soil is limiting when developing erosion

control practices is indicated by Fig. 2h and 2i. Where transportation capacity is limiting as in Fig. 2h, efforts to reduce soil detachment would be less effective than practices that reduce transport capacity. However, for conditions where detachment is limiting, as in Fig. 2i, decreasing detachment would be more effective. The fact that some erosion-control practices are much more effective for certain conditions than for others may be explained by such considerations.

### Discussion

Soil erosion by rainfall and runoff is a complex process that has not been descriptively simulated by empirical equations. The proposed mathematical model of the erosion-sedimentation process considers the relative contributions of four individual subprocesses and describes soil movement at all locations along a slope. However, each of these subprocesses is a function of complex parameter interrelations for which information is inadequate. Some information may be available from past erosion studies or research in related areas, but the applicability to an erosion model must be tested. Parameters for which mathematical relationships are needed in terms that can be quantitatively measured or calculated, include the detachment potentials and transport capacities of rainfall and of runoff ($D_R$, $D_F$, $T_R$, and $T_F$) and expressions for soil susceptibility to detachment and transport by rainfall and by runoff ($S_{DR}$, $S_{DF}$, $S_{TR}$, and $S_{TF}$).

The basic model, primarily conceptual at present, also needs to be broadened to include pertinent parameters in their time-dependent forms to proceed beyond the steady-state conditions considered herein. These include rainfall intensity, infiltration rate, and rill geometry as it affects the hydraulics of runoff. The time required for soil and water to move downslope and the manner in which some of the soil properties that affect detachment and transport change with time, must also be taken into account.

Other important items that will need to be introduced into an expanded model include:

1 Seepage and other subsurface flow phenomena as they affect soil erosion.

2 Vegetation and crop residues as they affect the erosive potential of rainfall and runoff.

3 Tillage, freezing and thawing, and other natural or man-caused actions as they influence soil detachment and transport.

4 Land topography and microtopography as they affect storage, overland flow, and exposure to rainfall.

5 Surface-water depth as it affects detachment and transport.

6 Accumulation of excess detached soil available for removal during subsequent periods of greater transportation capacity.

7 Additional interrelationships among the subprocesses.

Development of a detailed erosion model will involve the individual and collective efforts of specialists in many disciplines. Several phases are already under study by a team of researchers with its nucleus in the Corn Belt Branch of the Soil and Water Conservation Research Division.

### Application of Results

A mathematical erosion model would provide a comprehensive expression of the underlying physical, mechanical, and chemical processes involved in erosion and sedimentation. With it, data needed for thorough study of effects of important interactions among soil, rainfall, topographic, cover, and management variables could be provided rapidly and economically. Dependable predictions of locations of excessive soil erosion or deposition within a watershed should become feasible. Long-term economic analyses of proposed erosion and sediment-control systems might be made by basing solutions on synthesized weather events. The relative importance of various components of the soil erosion process could be evaluated as a basis for selecting the most effective method of erosion reduction. When adequately perfected, such a model will enable detailed analysis of total erosion systems and thereby provide basic information for designing erosion-control techniques commensurate with changing trends in agricultural production, suburban developments and recreational needs.

### Summary

The framework for a mathematical model to describe the process of soil erosion by water was developed. The approach treated (a) soil detachment by rainfall, (b) transport by rainfall, (c) detachment by runoff, and (d) transport by runoff as separate but interrelated parts of the soil erosion process. Mathematical relationships to describe the dynamics of these subprocesses were introduced into the basic model, and the resulting masses of soil (erosion) and water (runoff) were routed along successive increments of slope. The sediment load carried from each increment was the lesser of: (a) the sediment load from the previous increment plus the detachment on that

*(Continued on page 762)*

## MATHEMATICAL SIMULATION OF THE PROCESS OF SOIL EROSION

*(Continued from page 758)*

increment or (b) the transportation capacity from that increment. Net erosion (or sedimentation) for an increment was the difference between the sediment loads entering and leaving it.

Soil erosion was simulated for several combinations of slope shape, slope steepness and length, rainfall intensity, infiltration rate, and soil detachability, and results were presented graphically. The model described the resulting soil movement and the relative contributions of the four subprocesses all along

the slope. Present empirical methods cannot provide such descriptive simulation of this complex process. However, the described erosion-sedimentation model will require determination of improved mathematical relationships and introduction of additional components. The proposed framework lends itself to expansion into a more comprehensive model.

## References

1 Barnett, A. P. and Rogers, J. S. Soil physical properties related to runoff and erosion from artificial rainfall. *Transactions of the ASAE* 9:(1) 123-125, 1966.

2 Chow, V. T. Open channel hydraulics. McGraw-Hill, New York, p. 168, 1959.

3 Ekern, P. C. Problems of raindrop impact erosion. *Agricultural Engineering.* 34:(1) 23-25, 28, January 1953.

4 Ellison, W. D. Soil erosion studies—Part I. *Agricultural Engineering* 28:(4) 145-146, April 1947.

5 Ellison, W. D. Studies of raindrop erosion. *Agricultural Engineering* 25:(4-5) 131-136, 181-182, April-May 1944.

6 Free, G. R. Erosion characteristics of rainfall. *Agricultural Engineering* 41:(7) 447-449, 455, July 1960.

7 Laursen, E. M. Sediment transport mechanics in stable channel designs. Trans. ASCE 123:195-206, 1958.

8 Meyer, L. D. Mathematical relationships governing soil erosion by water (Resume). J. Soil and Water Conserv. 20:149-150, 1965.

9 Meyer, L. D. and Kramer, L. A. Erosion equations predict land slope development. *Agricultural Engineering* 50:(9) 522-523, September 1969.

10 Wischmeier, W. H. Rainfall erosion potential. *Agricultural Engineering* 43:(4) 212-215, 225, April 1962.

11 Wischmeier, W. H. Cropping-management factor evaluations for a universal soil-loss equation. Soil Sci. Soc. Amer. Proc. 24:322-326, 1960.

12 Wischmeier, W. H. and Mannering, J. V. Relation of soil properties to its erodibility. Soil Sci. Soc. Amer. Proc. 33:131-137, 1969.

13 Wischmeier, W. H. and Smith, D. D. A universal soil-loss equation to guide conservation farm planning. Trans. 7th Interntl. Cong. Soil Sci. 1:418-425, 1960.

14 Zingg, A. W. Degree and length of land slope as it affects soil loss in runoff. *Agricultural Engineering* 21:(2) 59-64, February 1940.

# 15

Reprinted from *Water Resources Research* **10**:485-492 (1974)

# Concepts of Mathematical Modeling of Sediment Yield

JAMES P. BENNETT

*U.S. Geological Survey, Bay St. Louis, Mississippi  39520*

A deterministic structure imitating a sediment yield model should mathematically approximate the behavior of two distinct phases of the phenomenon, the upland phase and the lowland channel phase. For upland erosion, research is most needed in quantifying gully erosion, whereas for lowland streams it is most needed to explain the influence of meanders on bed material transport and to develop a floodplain accounting component. In both phases, research is needed to explain the effects of unsteadiness and flow nonuniformities on transport. Sensitivity analysis studies of sediment yield models are needed to illustrate how well the models can be calibrated and what output precision can be expected from input data with known statistical characteristics. Present estimates indicate that although a regression of annual sediment discharge on annual water discharge might be expected to give a prediction within 20% of the observed value, even a well-calibrated digital model might give an error greater than 40% for a single storm on a small watershed.

A deterministic structure imitating mathematical model of sediment yield utilizes mathematical relationships to approximate the pertinent hydrologic, physiochemical, and biologic processes in some conceptually isolated system, such as an upland slope, a lowland stream, or an entire watershed, for predicting the amount and composition of sediment output from the system in a given time period. Before such a model can be used in simulation, it must be calibrated, and its predictions verified by comparison with observed outputs. Construction of a viable sediment yield model requires that as much or more attention be devoted to calibration and verification as to formulation and solution of the mathematical relationships. This paper is presented to discuss the first phase in the construction of a general deterministic structure imitating a sediment yield model, the mathematical formulation of the pertinent processes. The mathematics are such that solution of the resulting equations will almost certainly have to be implemented by digital computer; an attempt is made to outline the most feasible method of solving the equations. Several structure-imitating sediment yield models have appeared in technical publications; a brief discussion of them is included.

Discussions of other types of sediment yield models have appeared widely dispersed in the technical literature. Gross soil erosion equations such as the universal soil loss equation [*Agricultural Research Service*, 1961; *Wischmeier and Smith*, 1958; *Wischmeier*, 1960] used in combination with sediment delivery ratios [*Vanoni*, 1970; *Williams and Berndt*, 1972], predictive equations based on watershed parameters such as *Flaxman*'s [1972], or sediment transport versus water discharge curves [*Bennett and Sabol*, 1973; *Miller*, 1951] have been used extensively for some time by a number of agencies for predicting long-term sediment yield; they can be classed as deterministic black box models. Stochastic sediment yield models have appeared more recently [*Murota and Hashino*, 1969; *Woolhiser and Todorovic*, 1974]. These models provide invaluable information concerning the time variability of the sediment yield phenomenon. Such models can presently be formulated analytically only for small areas of land on which most of the variables pertinent to sediment yield are uniform. For larger areas, use must be made of Monte Carlo simulation procedures. These procedures employ artificially generated random inputs into some deterministic model of the physical processes to generate a number of equally likely time series of outputs, the statistics of which can then be analyzed. For those interested in constructing such models the formulation of the deterministic model is discussed here, and techniques for generating the simulated inputs are presented by, among others, *Hillier and Lieberman* [1967].

## SEDIMENT YIELD PROCESSES

The sediment yield phenomenon is generally divided into two broad categories, the upland phase and the lowland stream or in-channel phase. The upland phase is intimately related to discernible individual precipitation events. The mechanics of the precipitation event are of major importance in determining the amount of sediment yield in the upland phase. Major variables influencing the yield in this phase are soil type, soil condition, and soil moisture content at the start of the event; slope and slope length; vegetative and litter cover; and rainfall amount, intensity, and duration. In this phase the soil particle sizes of major interest are in the silt and clay ranges.

In the in-channel phase, aside from determining the gross amount of water available and the time distribution of its arrival, individual storms have little influence on the sediment yield. Pertinent variables in this phase are the velocity and depth of flow, channel slope, wash load and water temperature, and median size and grain size distribution of bed material. In this phase, calculation of the bed material transport capacity need be the only major concern, since channel flow can usually transport all the fine material supplied.

Sediment yield modeling is more complicated than other types of hydrologic modeling because before the sediment movement can be modeled, there must be detailed information concerning the amount and type of precipitation and the movement of water. Because empirical observations of such information are seldom available in the detail and variety required, a rainfall-runoff model for small watersheds and an in-channel flow routing model must be operated in conjunction with a sediment yield model for any basin of appreciable size. Because sediment transport and flow are so interrelated, it is difficult to discuss sediment yield modeling without also mentioning the modeling of the flow; flow modeling is discussed in what follows, but the discussion has been minimized

**233**

because the subject has been treated exhaustively elsewhere [*Amein and Fang,* 1970; *Liggett and Woolhiser,* 1967; *Strelkoff,* 1969, 1970].

Three differential equations form the mathematical basis for modeling both phases of sediment yield. They are the equations that describe the movement of suspended sediment particles in a one-dimensional, infinitely wide, free surface flow:

$$\frac{\partial h}{\partial t} + u \frac{\partial h}{\partial x} + h \frac{\partial u}{\partial x} = q \tag{1}$$

$$\frac{\partial u}{\partial t} + u \frac{\partial u}{\partial x} + g \frac{\partial h}{\partial x} = g(S_0 - S_f) - \frac{qu}{h} \tag{2}$$

and

$$\frac{\partial(hc)}{\partial t} + (1 - \lambda) \frac{\partial y}{\partial t} + \frac{\partial(hu_p c)}{\partial x} = \frac{\partial}{\partial x} h\epsilon_p \frac{\partial c}{\partial x} \tag{3}$$

Equations (1) and (2) are the standard conservation of mass and momentum equations for the flow. Equation (3) is the conservation of mass equation for the sediment. The quantities appearing in these equations are illustrated in Figure 1 and are defined in the notation list.

Equations (1)–(3) are completely general inasmuch as this statement can be made about an infinitely wide one-dimensional flow. Two- and three-dimensional equations could have been presented, but solution procedures for them are so complicated that in light of present uncertainties concerning the physics of two-phase flows the effort required for their solution is not warranted. Equations (1)–(3) are equally applicable to overland flow and in-channel flow sediment transport, although in certain channel flow situations the equations need to be modified to consider variations in cross-sectional shape and area with distance. Equation (3) differs from the conservation of mass equation for a dissolved substance in that the second term on the left accounts for deposition or erosion from the bed. Also for suspended sediment it cannot necessarily be assumed that the transport velocity of the suspended substance is the same as the flow velocity. As is customary, however, this assumption has been made herein. The term on the right-hand side of (3) accounts for dispersion of the material while it is suspended in the flow. In comparison with the effects of other processes embodied in (3), dispersion is normally negligible, and as a result it will not be considered here.

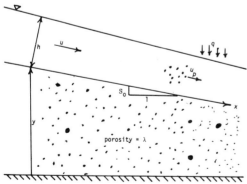

Fig. 1. Definition sketch of flow transport system.

## UPLAND PHASE

Upland erosion by water on an exposed or partly exposed soil surface such as a recently worked field occurs in three forms or stages. These stages tend to follow one another in time of appearance at a point and to a certain extent, in progression downslope at a particular time. The first of these stages is sheet erosion, which may be idealized as the removal by any process of a sheet of sediment of uniform thickness over an entire area. It is the first stage of erosion of material from a newly worked field. The second stage is rill erosion, the development of small channels in which runoff concentrates. The small channels form because of natural areal variations in the erosion resistance of the soil and because of small variations in elevation and slope. The flows from a large number of rills concentrate in gullies, larger upland channels. Gullies, the final stage of upland erosion, are fairly permanent topographic features; as opposed to rills they are not completely obliterated by normal agricultural tillage practices. Gullies may be further distinguished from rills in that in gully flow the influence of precipitation on material detachment and transport is negligible. On an upland area, sediment removal and transport are accomplished by four different processes during a storm. These are detachment by raindrop impact, detachment by runoff, transport by raindrop splash, and transport by runoff. In each of the stages of upland erosion a different mode of sediment removal and transport predominates, and ideally, each stage should be considered separately in a sediment yield model.

Variables that must be considered in predicting upland erosion are soil properties, climate, vegetation, topography, and human influences. The major soil properties are texture, structure, permeability, compactness, and infiltration capacity. Permeability and infiltration capacity primarily influence the amount of runoff; the other listed factors influence the behavior of the soil when it is subjected to raindrop impact and to the forces of running water. Rainfall, temperature, and snow are the primary climatic factors influencing upland sediment yield. The amount, area, and time distribution of rainfall are probably the most important factors in most areas, but temperature may also be significant, especially when a large number of freeze and thaw cycles occur. Sediment yield from a snowfall can be predicted in a manner analogous to that from a rainfall except that raindrop influences should be deleted and the time distribution of the melt should be considered. Vegetation influences sediment yield by dissipating some of the energy of the drops before they reach the soil, by retarding runoff water, by influencing permeability through root effects, and by influencing soil moisture through transpiration. The major topographic variables influencing sediment yield are slope and slope length. Human influences occur primarily through agriculture and construction practices. As a whole, it might be said with considerable understatement that these factors are difficult to quantify. Despite this difficulty a large body of research concerning this problem does exist. Some of it will be mentioned in what follows, but a more complete reference list may be found in the work by *Vanoni* [1970].

Equation (3) has been presented by *Foster and Meyer* [1972] as

$$D_F + R_{DT} = \partial G_F / \partial x \tag{4}$$

wherein $D_F$ is net flow detachment rate, $R_{DT}$ is net raindrop detachment rate, in terms of both weight per unit of time and unit area, and $G_F$ is weight transport rate per unit of width. In terms of (3),

$$G_F = \gamma \rho u c h \tag{5}$$

and

$$D_F + R_{DT} = (1 - \lambda) \gamma \rho \frac{\partial y}{\partial t} \tag{6}$$

where $\gamma$ is unit weight of water and $\rho$ is density of the suspended solids. In (4), *Foster and Meyer* [1972] have implicitly assumed that the first term on the left of (3) is negligible or alternatively, that the flow is quasi steady state. This is a common assumption made in calculations concerning predictions of bed form motion [e.g., *Simons et al.*, 1965]. In (4), $G_F$ is the dependent variable of interest; to calculate it, the independent variables $D_F$ and $R_{DT}$ will have to be obtained from other considerations. If variations in elevation of the surface over which flow occurs are of interest, a time integral of $D_F + R_{DT}$ will have to be maintained at each point of concern. In practice, (4) can probably be applied to combined sheet and rill erosion without too much loss of generality. Gullies, however, should be considered separately.

From physical considerations, derivation of (4) is relatively straightforward. Because present knowledge of the micromechanics of erosion of sediment particles is rudimentary, expression of the terms $D_F$ and $R_{DT}$ of (4) is not so simple, and considerable empiricism and uncertainty arise in their formulation. This usually results in the use of relatively simple equations with one or more coefficients derived from regression. For example, *Foster and Meyer* [1972] take an original approach to determining $D_F$. They state that

$$D_F/D_c + G_F/T_c = 1 \tag{7}$$

wherein $D_c$ is the ultimate detachment capacity of the overland flow and $T_c$ is its ultimate transport capacity. They point out that this approach is equivalent to assuming that a particular flow carrying less than its capacity to transport will fill this capacity according to a first-order reaction; that is,

$$D_F = G(T_c - G_F) \tag{8}$$

where $G$ is a reaction rate coefficient. This is a reasonable assumption; many processes in nature, such as satisfaction of biochemical oxygen demand and instream reaeration, follow the first-order reaction law. Furthermore, Foster and Meyer are able to cite experimental evidence of *Meyer and Monke* [1965] and *Willis* [1971] that illustrates a tendency for flows over noncohesive beds to entrain sediment according to (7). *Foster and Meyer* [1972] further assume on the basis of empirical evidence that $T_c \propto \tau^{3/2}$, where $\tau$ is the bottom shear stress, and that $D_c \propto T_c$. The term $R_{DT}$ also on the basis of empirical evidence [*Meyer and Wischmeier*, 1969] has been shown to be proportional to the square of the rainfall intensity. These expressions leave a number of coefficients to be evaluated, but although little is known concerning the generality of the coefficients, analytical or numerical solution of (4) is possible given knowledge of the overland flow and the rainfall intensity. Other similar approaches could be taken, and more of the factors influencing sediment yield such as vegetative cover could be accounted for in the empirical equations.

Equation (7) implies that the overland flow–sediment transport system requires a finite amount of travel time to respond to changes in flow characteristics. In light of present uncertainty concerning transient sediment transport

phenomena, it might appear to be just as reasonable to assume that at every point in space and time the transport capacity of the flow is instantaneously filled. If this assumption is made, the need for considering (7) is eliminated, and (4) may be determined exclusively on the basis of hydraulic considerations, along with an assumed sediment transport relationship. The main drawback to this approach is the possibility of introducing instability into the solution when we must model interactions between the flow and the deformable bed over which the flow moves.

Regardless of how the sediment transport and erosion by the flow are characterized, computation of sediment transport capacity must be based on detailed knowledge of the overland flow. This knowledge can be obtained from solution of (1) and (2). In their general form these equations have to be solved by numerical methods, even for steady flow. For unsteady flow, as was pointed out by *Liggett and Woolhiser* [1967], the numerical solution of (1) and (2) is extremely difficult for overland flows, but solutions can be obtained by the method of characteristics or an implicit scheme for a wide range of flow parameters. *Liggett and Woolhiser* [1967] provide a brief review of the considerable body of literature concerning the solution of the general forms of (1) and (2). In most modeling situations, however, the mathematical rigor provided by the numerical solution of (1) and (2) is not merited in light of its high cost, stability and convergence problems, and uncertainty concerning friction losses and the other phenomena being modeled. For example, *Foster and Meyer* [1972] assume that the local velocity and depth can be approximated by the normal velocity and depth corresponding to the local value of the discharge. This assumption is slightly more restrictive than their use of the quasi steady mass transport equation (4), but it allows for explicit description of the overland flow without resorting to a finite difference solution.

Completely aside from the mathematical problems associated with solving (1) and (2) are two major problems associated with the determination of the terms $S_f$ and $q$. The friction slope may be written

$$S_f = u^2/C^2 h \tag{9}$$

wherein $C$ is the Chezy discharge coefficient. For undisturbed shallow flows it is difficult to estimate $C$ on the basis of known physical measures of surface roughness, and $C$ is even more difficult to estimate when the flow is disturbed by raindrops. The parameter $q$, the lateral inflow, is the rainfall excess or rainfall minus infiltration, interception, and evaporation during the storm. Interception and evaporation are usually minor, but infiltration can be appreciable, and to model it is difficult. Infiltration rate is highly dependent on soil type, vegetative cover, and antecedent moisture conditions. Operational rainfall-runoff models such as those presented by *Dawdy et al.* [1972], *Holtan and Lopez* [1971], or *Crawford and Linsley* [1966] contain antecedent moisture accounting components. Such components should readily be adoptable for computation of infiltration in overland flow sediment yield models.

Unless appreciable changes in land surface elevations take place during the time of simulation for an overland flow sediment yield model, it is possible to operate the rainfall-runoff component independently of the sediment erosion and transport component. This observation is true because for small elevation changes the erosion does not appreciably influence the slope of the surface on which flow occurs and because the sediment concentrations entrained are usually small enough that they do not appreciably influence the mechanics of flow. Once flow characteristics have been computed, transport capacity can be obtained (and if flow erosion rate is desired, it can be calculated by using (1) and (3)). For instance, if flow transport capacity is proportional to the 3/2 power of bottom shear stress, as was assumed by *Foster and Meyer* [1972], then

$$T_c \propto (\gamma h S_0)^{3/2} \qquad (10)$$

Or if the flow transport capacity is proportional to stream power, the product of bottom shear and flow velocity, as was assumed by *Bagnold* [1966] and *Yang* [1972], then

$$T_c \propto (\gamma h S_0)u \qquad (11)$$

If (7) is assumed to represent the interaction of the transport capacity, the load, the detachment capacity, and the detachment rate of the flow, additional information concerning rainfall energy must be available for computing rainfall detachment rate in solving (4). If transport is always assumed to be at capacity, such information is not required. The transport capacity of rainfall by raindrop splash is usually negligible.

For purposes of simulation, computer time and storage requirements for solving (1), (2), and (4) will decrease with the number of simplifying assumptions made. Solutions can range in complexity from the closed-form analytical solutions given by *Foster and Meyer* [1972] to the complicated numerical solutions of the complete forms of (1)–(3) requiring large parts of computer memory capacity and considerable execution time. In light of the idealizations of the physical processes implied in the governing equations and of the fact that the upland flow produces only one component of the sediment yield of a basin, it might be better to use the simplest formulation of the governing equations justifiable and to account for basin complexity by segmenting it into a number of small uniform subbasins. In fact, this approach is commonly taken in hydrologic modeling [e.g., *Crawford and Linsley*, 1966].

The mechanics of gully formation and the prediction of sediment to be supplied by gully erosion are, if possible, even more difficult to describe and quantify than the two phases of upland erosion discussed above because the predominant supply and transport mechanisms are extremely variable in space and time and because the amount of material available for transport by a particular storm runoff may not even be related to the intensity or duration of the storm but to such factors as time available for bank sluffing and soil mositure and to weather conditions since the last runoff event. In fact, as was pointed out by *Vanoni* [1970], 'the gully erosion process has been admirably described for several regions of the U.S., but the cause-effect interrelationships of gully formation have never been put into proper perspective.' For the purpose of simulation it is probably not possible at the present time to outline a procedure that will provide a workable model in all cases, but it should be possible to outline several alternative models that can be used to solve most of the problems encountered.

Gully flows in general are similar to sheet and rill flows because the bed material of the gullies is such that transport capacities for this material are very high and flow distances are probably small in relation to the distance required for the load to adjust to changes in transport capacity. They differ from sheet and rill flows in that raindrop impact is probably not an important factor either in flow resistance or in sediment particle detachment. It is suspected that it is more important to model the phenomena associated with flow unsteadiness for gully erosion than for sheet and rill erosion.

Gully flows are generally ephemeral; but when there is flow, its mechanics are the same as they are in lowland channels except that the channel slopes may be quite a bit steeper in the gullies. Bed sediments in gullies are often finer than those found in lowland streams, and they are often cohesive. Sediment supply mechanisms such as bank sluffing, head cutting, and piping are different from those encountered in lowland channel flows. Gully sediment yield modeling appears to require a mixture of the techniques used in sheet and rill modeling and in lowland channel modeling. Selection of the proper proportions of the mixture depends on the particular combination of situations occurring on the watershed.

In many cases the contribution of gully erosion to watershed sediment yield is small in comparison with the contribution of sheet and rill erosion [*Vanoni*, 1970]. In such cases its simulation can probably be lumped in with the sheet and rill simulation. For cases in which gully erosion is appreciable, its simulation should be accomplished separately from sheet and rill erosion. When the gully sediment yield is eroded from the gully floor by the flow in the course of a runoff event, the flow from the upland areas should be routed through the gully by using (1) and (2), and the sediment yield should be computed by using (4) and (7). In the latter two equations the coefficients will probably not be the same as those for sheet and rill erosion, even on the same watershed. When the gully flow essentially only transports sediment supplied by sheet and rill erosion and carries off material accumulated in the gully from bank sluffing and wind deposits since the previous storm, an antecedent condition accounting component can be constructed to simulate the accumulation of deposits, and it can be hypothesized that the gully flow will transport the deposits at capacity until they are exhausted or until flow stops. If flow continues following exhaustion of the deposits, only the material supplied by sheet and rill erosion will be transported by the gully. Of course, the simulation of a combination of the above situations would require only an appropriate combination of the simulation algorithms of the individual cases. Finally, it might be possible in certain geographical areas to modify existing equations for prediction of new annual gully area or annual rate of advancement to predict sediment yield on a per storm basis. Samples of annual curves are given by *Vanoni* [1970].

## LOWLAND STREAM PHASE

A lowland stream is herein defined as a well-established water course flowing through a valley in which the material composing the valley floor has been deposited by the stream in the recent geologic past. In an undisturbed state the bed profiles of such streams will be relatively stable in terms of engineering time, in contrast to the longitudinal profiles of gullies, which can change drastically in short time periods. The bed material in lowland streams is predominantly coarser than that of the silt sizes (0.062 mm). Flows in lowland streams are usually perennial in comparison with those of gullies, which are dry most of the time. Lowland stream channels are characterized by a rapidly varying cross-sectional shape in the direction of flow and are usually sinuous in plan. Procedures for calculation of flow velocity and depth in such channels are given by *Henderson* [1966], *Chow* [1959], *Amein and Fang* [1970], and *Strelkoff* [1970]. Sediment movement and the flow of water in such channels are characterized by a number of complex interactions; for example, between the sediment transported by a flow and the shape or bed form assumed by the deformable channel bottom or between bed form and resistance to flow. Such interactions are discussed in the work edited by *Shen* [1971]. Inchannel sediment transport phenomena are reviewed by *Graf* [1971].

Water flow and sediment transport in lowland streams are characterized by a large number of complexly interrelated variables. Variables influencing flow velocity and depth are discharge, cross-sectional shape, viscosity of the water-sediment mixture, and frictional resistance of the flow boundary. For flows confined to the main channel, resistance to flow is a function of flow depth and the form assumed by the alluvial boundary material, which is in turn a function of the flow parameters. For overbank flows, in addition to the bed form resistance of the main channel the resistance is a function of type and degree of submergence of floodplain vegetation. Primary variables known to influence sediment transport in lowland streams are flow velocity, flow depth, slope of energy gradient, density and viscosity of water-sediment mixture, mean fall diameter and graduation of bed sediment, and seepage force on the stream bed. Again these variables are interdependent, and the details of the dependence are not completely understood, but they have been investigated empirically for wide ranges of most of the important variables.

Equations (1)–(3) written for a finite width stream channel are

$$\frac{\partial A}{\partial t} + u \frac{\partial A}{\partial x} + A \frac{\partial u}{\partial x} = q_A \qquad (12)$$

$$\frac{\partial u}{\partial t} + u \frac{\partial u}{\partial x} + g \frac{\partial h}{\partial x} = g(S_0 - S_f) - \frac{q_A u}{A} \qquad (13)$$

and

$$\frac{\partial (Ac)}{\partial t} + B(1 - \lambda) \frac{\partial y}{\partial t} + \frac{\partial (Au_p c)}{\partial x} = \frac{\partial}{\partial x} A\epsilon_v \frac{\partial c}{\partial x} \qquad (14)$$

where all symbols are the same as those for (1)–(3), $A$ is cross-sectional area, $q_A$ is lateral inflow per unit length of channel, and $B$ is width of channel bottom. The assumptions inherent in (12) and (13) are discussed by *Strelkoff* [1969]. With the exception of the fact that the flow resistance in an alluvial boundary channel is extremely sensitive to flow conditions, the solutions of (12) and (13) should be no different for an alluvial boundary channel than for one with a rigid boundary. The one-dimensional formulation, which is necessary because of cost and computer memory limitations, unfortunately, precludes consideration of cross-sectional velocity variations that might be important in determining sediment supply. In (14), only the sediment sizes found in appreciable quantity in the bed material are considered in computing the concentration $c$. It is commonly assumed that sediments finer than these are transported by the flow in whatever quantities they are supplied by upland erosion. The finer material is called wash load; the coarser is called bed material load. In the usual case the dividing line between wash load and bed material load falls at the upper end of the silt size range (0.062 mm). It is usually further assumed that the bed material transport velocity is the same as the mean stream velocity and that the longitudinal dispersion term on the right side of (14) is negligible in comparison with other terms in the equation. Because the transport equation is one-dimensional, it must necessarily ignore or more precisely treat 'in the average'

such important sediment production processes as bank sluffing due to meander advancement. For overbank flows or for wide streams with nonuniform bed conditions it is, however, possible to obtain a quasi two-dimensional representation, as is now done in flood routing, by treating the flow as though it were moving in two or more parallel stream tubes, each with its own resistance and transport characteristics.

Sediment load measurements have shown that the major part of the annual sediment discharge of many streams may be transported in a relatively short period of time by a few storm runoffs during which the discharge of the stream is continuously changing. It is considered necessary that at least the flow computation part of the sediment yield algorithm for a lowland stream solve the unsteady state versions of (12) and (13). How the concentration component, (14), should be treated is not so clearly defined. For example, this author knows of no theory relating the instantaneous values of flow variables to the time rate of change of bottom elevation at a point. This information is necessary for computing $c$ by means of (14). In the past, considerable attention has been devoted not to predicting local rates of scour or deposition under nonequilibrium flow and sediment transport conditions but to developing equations for predicting equilibrium sediment transport with steady uniform conditions. Probably the best known and the most firmly based on physical principles is *Einstein*'s [1950] total bed load transport equation. By using Einstein's or any other of a number of available total load transport equations applicable to steady uniform flow [*Graf*, 1971], it should be possible to avoid the problem of lack of information about local erosion rates by formulating (14) in a manner analogous to *Foster and Meyer*'s [1972] approach as

$$\frac{\partial c}{\partial t} + u \frac{\partial c}{\partial x} = K \frac{(1 - \lambda)B}{A} \left( \frac{Q_s}{Q} - c \right) - \frac{c q_A}{A} \qquad (15)$$

In (15), $Q_s$ is the equilibrium total load transport rate corresponding to the local instantaneous flow conditions, and $Q$ is the corresponding local water discharge. The coefficient $K$ in (15) is a first-order reaction rate coefficient; it is probably a strong function of bed sediment fall velocity, mean stream velocity, and depth of flow. If the sediment transport is computed by using (15), the bed surface elevation at a point $x = x_A$ at time $t$ may be computed by using

$$y(x_A, t) = y(x_A, 0) - \int_0^t \left\{ K \left( \frac{Q_s}{Q} - C \right) \right\}_{x_A} dt \qquad (16)$$

The subscript $x_A$ inside the integral of (16) indicates that the quantities inside the braces are to be evaluated at $x = x_A$. For (15) and (16) to be used in a simulation algorithm, $K$ would have to be determined by calibration from observed data; the author knows of no discussions concerning such a coefficient. Equations (15) and (16) could be used individually for each of a number of different size fractions of the bed material. By keeping track of the size fractions independently the algorithm could be used to predict the occurrence of bed armoring. Given an initial condition of known concentration for all $x$ and an upstream boundary condition for all $t$, (15) can be solved by a fully centered explicit finite difference scheme. If the ratio of the distance increment to the time increment is approximately equal to the mean velocity in the channel, the solution should be well behaved, even for steep concentration gradients. Solution of (16) is by simple numerical integration, for example, by using Simpson's one-third rule.

**237**

As an alternative to (15), one may assume the first term on the left of (14) to be negligible and write the equation as

$$B(1 - \lambda)\frac{\partial y}{\partial t} + \frac{\partial Q_s}{\partial x} = 0 \qquad (17)$$

This version of (14) has been used by *Simons et al.* [1965] to predict bed load transport from observed bed form movement and by *Kennedy* [1963] for computations concerning theoretical bed form mechanics. In (17) it is assumed that the sediment load of the stream is everywhere at transport capacity as given by some equilibrium total load transport equation. The sediment discharge for any cross section at any time is therefore simply calculated from the chosen equilibrium transport relation, and the local bed elevation is computed by numerical integration of (17).

It is important in lowland stream sediment yield modeling to consider the temporal variation of the stream discharge because, as was mentioned previously, it is common for a large part of the sediment discharge of a lowland stream to be transported in a relatively short time period when the discharge is large and varying rapidly. This observation means that to adequately describe the system, (12) and (13) should be solved in their general forms, that is, for unsteady nonuniform flow. Such solutions are possible, as was pointed out by *Strelkoff* [1970], *Amein and Fang* [1970], and many others, but the suggested procedures are too complicated to be discussed further here. A possible alternative to complete solution is the technique discussed by *Thomas* [1972] wherein the input hydrograph is idealized as a series of constant discharges of different magnitudes and durations with instantaneous changes in discharge from one level to another. Flow characteristics for each level of discharge can be generated by using a standard backwater curve computation technique. It is felt that to obtain reliable simulations of stream sediment discharge, scour or fill amount, and armoring, no matter what flow computation technique is used, there should be available considerably more detailed channel cross-sectional information than is presently used for flood-routing computations. The author knows of no discussions in the literature of detailed solutions of the nonuniform unsteady flow equations aimed at predicting sediment transport capacity; hopefully, in the near future this shortage will be alleviated.

In this section we have discussed the processes governing the movement of sediment from the watershed divide down through the flow system and out the mainstream draining the watershed. In so far as possible the equations that relate these processes have been presented, and we have attempted to outline the difficulties involved in their solution. With a knowledge of the options available for constructing a general sediment yield computation algorithm for a watershed the sediment yield modeling problem becomes one of allocation of resources to obtain the required detail in the solution of a particular problem.

## ALGORITHMS FOR PREDICTING SEDIMENT YIELD

Designing a sediment yield algorithm for execution by a digital computer is a complicated and time-consuming process, which increases rapidly in difficulty with increasing generality of the algorithm, that is, with increasing diversity. In addition to solving the equations discussed in the previous section the algorithm must provide a number of bookkeeping services. The algorithm must accept and manipulate input information to make it useful in the solution procedure. The inputs might describe time and space distribution of precipitation, channel segmenting and geometry, and characteristics

and subdivisions of upland areas. The algorithm should contain rules for sequencing computations and combining flows based on these inputs and for converting input geometries into a useful form for the solution routines. The algorithm should contain components for auxiliary computations, such as an interpolation routine, a curve-fitting routine, and an optimization routine. Finally, it must contain rules for providing the results of the calculations in the desired form.

Once an algorithm for sediment yield prediction has been constructed, it must be calibrated for use in a particular basin. Calibration consists of determining from observed inputs and outputs the best values of certain model parameters for the particular basin. Selection is usually accomplished on the basis of the minimization of a mean square error criterion between observed and calculated outputs. An optimization routine should be included in the algorithm for accomplishing the calibration. Usually, a sophisticated one will be required, such as the Rosenbrock procedure discussed by *Dawdy et al.*, [1972]. Before the calibrated algorithm is used as a predictive tool, it should be verified by using observed input and output data not used in calibration. The verification criterion might be some minimum acceptable level of mean square error of prediction.

The earliest digital computer sediment yield model was reported by *Negev* [1967]. This model uses the Stanford watershed model [*Crawford and Linsley,* 1966] for the water phase of both its upland flow component and its lowland stream component. In the upland phase, flow is assumed to be steady state for each hour based on individual values of hourly rainfall, and sediment yield is assumed to be a power function of hourly overland flow and of amount of material eroded in the previous hour. Channel flow is computed by using a modification of the Clark time-area-histogram method, and the contribution of sediment from channels (including rills and gullies) is computed by using power functions based on daily water discharge. The parameters of the model must be adjusted by trial and error. The performance of the model in reproducing observed sediment records for two California watersheds (the same records that were used for calibration) was quite good.

*Thomas* [1972] reported a lowland stream algorithm designed to predict scour and fill within a particular study reach. Inputs are cross-sectional geometry, bed material properties, a downstream stage-discharge relation, the input flow hydrograph, and an upstream flow–sediment concentration rating. Flow computations are made by using the standard step-backwater method and by assuming that the input hydrograph can be approximated by a series of steady state discharges. Cross-sectional bottom elevations and bed material gradations are adjusted for scour and fill at the end of each discharge increment. One of the sediment transport relations that can be used in the algorithm is *Toffaleti*'s [1969] modification of *Einstein*'s [1950] bed load function. Outputs from the program are channel bed elevations and sediment load and composition.

*Meyer and Wischmeier* [1969] describe a steady state model that segments a slope into a finite number of increments and for each increment computes detachment capacity and transport capacity of rainfall and detachment capacity and transport capacity of runoff. Depending on whether local transport capacity is greater than or less than the sediment available for transport, the model either routes all the sediment available in one increment to the next downslope increment or routes it to the next increment at transport capacity.

The model shows considerable promise as a basis for development into a general upland erosion algorithm.

## RESEARCH NEEDS

In upland erosion, development of a rationale for quantifying gully erosion would be most helpful. Attention is most necessary in the investigation of the mechanics of gully head cutting and the definition of amount of material supplied by bank caving as a function of antecedent weather conditions, plant cover, and soil characteristics.

For lowland streams, there is a need to investigate the influence of plan nonuniformities, that is, of meanders, on bed material transport characteristics. Another quite useful tool for use in lowland stream modeling would be the development of a floodplain deposition component.

In both the upland and the stream sediment yield modeling, there is a need for investigation of the influence of unsteadiness and of flow nonuniformities on sediment transport characteristics. This would clarify whether or not it is proper to define the response of the sediment load to changes in transport and detachment capacity by a first-order reaction law (8). If such an expression is proper, the investigation should define the coefficient $C$ of (8) or $K$ of (15). This investigation should also define the relationships between the time scales of flow and sediment transport for the various types of sediment yield phenomena. This should point out which shortcuts can and cannot be taken in solving, say, the flow equations commensurate with known allowable simplifications in the concentration equation.

Finally, an investigation of the solutions of various forms of the pertinent equations for a wide range of conditions is necessary to determine their sensitivity to the various parameters. This should aid in delimiting our ability to calibrate sediment yield models by using observed data with known statistical characteristics. It should also aid in defining the amount of precision that can be obtained from input data with known characteristics. As an example, *Dawdy et al.* [1972] report that with presently available rainfall information as input their rainfall-runoff model gives about a 30% standard error of estimate. This might be typical of any rainfall-runoff model. Because a sediment yield model must be based on some rainfall-runoff model, is it reasonable to expect that the standard error of estimate of a sediment yield model can ever be better than 40%? Of course, the standard error of estimate of annual discharge obtained from individual simulations by using a sediment yield model should be considerably less. It would be very interesting to see how this compares to the 20% standard error of estimate [*Bennett and Sabol,* 1973] obtainable for annual sediment yields from the sediment transport rating curve technique.

Sediment yield modeling is a valuable tool for expanding knowledge of our surroundings. By using realistic mathematical models, segmenting a river basin until reasonably uniform conditions exist in each segment, and properly combining outputs, acceptably accurate predictions of basin behavior can be obtained. These predictions may be used for filling in missing information, for predicting response to hypothetical statistically possible inputs, or for defining the consequences of various alternative plans of action that would alter conditions in the basin.

## NOTATION

$A$    cross-sectional area, $L^2$.
$B$    channel bottom width, $L$.

$c$    sediment concentration.
$C$    Chezy $C$, $L^{1/2}/T$.
$D_c$    flow detachment capacity, $FL^{-2}T^{-1}$.
$D_F$    flow detachment rate, $FL^{-2}T^{-1}$.
$g$    gravitational constant, $LT^{-2}$.
$G$    reaction rate coefficient, $L^{-1}$.
$G_F$    weight transport rate of sediment, $FL^{-1}T^{-1}$.
$h$    depth of flow, $L$.
$K$    entrainment rate coefficient, $LT^{-1}$.
$q$    distributed water inflow, $LT^{-1}$.
$q_A$    distributed water inflow, $L^2T^{-1}$.
$Q$    water discharge, $L^3T^{-1}$.
$Q_s$    volume sediment discharge, velocity in $x$ direction assumed to be equal to zero, $L^3T^{-1}$.
$R_{DT}$    net raindrop detachment rate, $FL^{-2}T^{-1}$.
$S_f$    friction slope.
$S_0$    local bed slope.
$t$    time, $T$.
$u$    flow velocity, $LT^{-1}$.
$u_p$    average velocity of sediment, $LT^{-1}$.
$x$    distance positive in the direction of flow, $L$.
$y$    local bed elevation.
$\gamma$    unit weight of water, $FL^{-3}$.
$\epsilon_p$    sediment particle mass transfer coefficient, $L^2T^{-1}$.
$\lambda$    porosity of deposited sediment.
$\rho$    sediment particle density.

*Acknowledgment.* This document derives from an invited paper presented at the Sediment Yield and Sources Symposium, AGU Spring Annual Meeting, 1973. It is intended to form a part of the proceedings of that symposium.

### REFERENCES

Agricultural Research Service, A universal equation for predicting rainfall-erosion losses, *Rep. ARS 22-66,* 11pp., U.S. Dep. of Agr., Washington, D. C., 1961.

Amein, M., and C. S. Fang, Implicit flood routing in natural channels, *J. Hydraul. Div. Amer. Soc. Civil Eng.,* 96(HY12), 2481–2501, 1970.

Bagnold, R. A., An approach to the sediment transport problem from general physics, *U.S. Geol. Surv. Prof. Pap. 422-I,* 37 pp., 1966.

Bennett, J. P., and G. V. Sabol, Investigation of sediment transport curves constructed using periodic and aperiodic samples, in *Proceedings of the International Association for Hydraulic Research International Symposium on River Mechanics,* vol. 2, pp. 49–61, Asian Institute of Technology, Bangkok, Thailand, 1973.

Chow, V. T., *Open Channel Hydraulics,* pp. 525–609, McGraw-Hill, New York, 1959.

Crawford, N. H., and R. K. Linsley, Digital simulation in hydrology, Stanford watershed model 4, *Tech. Rep. 39,* 186 pp., Dep. of Civil Eng., Stanford Univ., Stanford, Calif., July 1966.

Dawdy, D. R., R. W. Lichty, and J. M. Bergman, A rainfall-runoff simulation model for estimation of flood peaks for small drainage basins, *U.S. Geol. Surv. Prof. Pap. 506-B,* 28 pp., 1972.

Einstein, H. A., The bed load function for sediment transportation in open channel flows, *Tech. Bull. 1026,* 71 pp., Soil Conserv. Serv., U.S. Dep. of Agr., Washington, D. C., Sept. 1950.

Flaxman, E. M., Predicting sediment yield in western United States, *J. Hydraul. Div. Amer. Soc. Civil Eng.,* 98(HY12), 2073–2087, 1972.

Foster, G. R., and L. D. Meyer, A closed-form soil erosion equation for upland areas, in *Sedimentation Symposium to Honor Professor Hans Albert Einstein,* edited by H. W. Shen, pp. 12-1–12-19, Colorado State University, Fort Collins, 1972.

Graf, W. H., *Hydraulics of Sediment Transport,* McGraw-Hill, New York, 1971.

Henderson, F. M., *Open Channel Flow,* pp. 285–394, Macmillan, New York, 1966.

Hillier, F. S., and G. J. Lieberman, *Introduction to Operations Research,* pp. 439–479, Holden-Day, San Francisco, Calif., 1967.

Holtan, H. N., and N. C. Lopez, USDAHL-70 model of watershed hydrology, *U.S. Dep. Agr. Tech. Bull. 1435,* 84 pp., 1971.

**239**

Kennedy, J. F., The mechanics of dunes and antidunes in erodible bed channels, *J. Fluid Mech., 16,* 521–524, 1963.

Liggett, J. A., and D. A. Woolhiser, Difference solutions of the shallow water equation, *J. Eng. Mech. Div. Amer. Soc. Civil Eng., 93*(EM2), 39–71, 1967.

Meyer, L. D., and E. J. Monke, Mechanics of soil erosion by rainfall and overland flow, *Trans. ASAE, 8*(4), 572–577, 1965.

Meyer, L. D., and W. H. Wischmeier, Mathematical simulation of the process of soil erosion by water, *Trans. ASAE, 12*(6), 754–762, 1969.

Miller, C. R., Analysis of flow-duration sediment-rating curve method of computing sediment yield, technical report, 55 pp., U.S. Bur. of Reclam., Washington, D. C., 1951.

Murota, A., and M. Hashino, Studies of a stochastic rainfall model and its application to sediment transportation, *Technol. Rep. Osaka Univ., 19,* 231–247, 1969.

Negev, M., A sediment model on a digital computer, *Tech. Rep. 76,* 109 pp., Dep. of Civil Eng., Stanford Univ., Stanford, Calif., March 1967.

Shen, H. W. (Ed.), *River Mechanics,* vol. 1, 2, Colorado State University, Fort Collins, 1971.

Simons, D. B., E. V. Richardson, and C. F. Nordin, Bedload equation for ripples and dunes, *U.S. Geol. Surv. Prof. Pap. 462-H,* 9 pp., 1965.

Strelkoff, T., One-dimensional equations of open-channel flow, *J. Hydraul. Div. Amer. Soc. Civil Eng., 95*(HY3), 861–877, 1969.

Strelkoff, T., Numerical solution of the Saint-Venant equations, *J. Hydraul. Div. Amer. Soc. Civil. Eng., 96*(HY1), 223–253, 1970.

Thomas, W. A., Using a scour on deposition model to determine sediment yield, paper presented at Sediment Yield Workshop, U.S. Dep. of Agr. Lab., Oxford, Miss., Nov. 1972.

Toffaleti, F. B., Definitive computations of sand discharge in rivers, *J. Hydraul. Div. Amer. Soc. Civil Eng., 95*(HY1), 225–249, 1969.

Vanoni, V. A., Chairman, Task committee on preparation of manual on sedimentation, sediment engineering, *J. Hydraul. Div. Amer. Soc. Civil Eng., 96*(HY6), 1283–1331, 1970.

Williams, J. R., and H. D. Berndt, Sediment yield computed with universal equation, *J. Hydraul. Div. Amer. Soc. Civil Eng., 98*(HY12), 2087–2099, 1972.

Willis, J. C., Erosion by concentrated flow, *Rep. ARS 41–179,* 38 pp., Agr. Res. Serv., U.S. Dep. of Agr., Washington, D. C., 1971.

Wischmeier, W. H., Cropping-management factor evaluations for a universal soil loss equation, *Soil Sci. Soc. Amer. Proc., 24*(4), 176–193, 1960.

Wischmeier, W. H., and D. D. Smith, Rainfall energy and its relationship to soil loss, *Eos Trans. AGU, 39*(2), 285–291, 1958.

Woolhiser, D. A., and P. Todorovic, A stochastic model of sediment yield for ephemeral streams, Proceedings USDA-IASPS Symposium on Statistical Hydrology, *U.S. Dep. Agr. Misc. Publ.,* in press, 1974.

Yang, C. T., Unit stream power and sediment transport, *J. Hydraul. Div. Amer. Soc. Civil Eng., 98*(HY10), 1805–1827, 1972.

(Received January 15, 1974;
revised February 19, 1974.)

# 16

# Estimating Erosion and Sediment Yield on Field-Sized Areas

G. R. Foster, L. J. Lane, J. D. Nowlin, J. M. Laflen, R. A. Young

MEMBER
ASAE

MEMBER
ASAE

MEMBER
ASAE

## ABSTRACT

A model for field-sized areas was developed to evaluate sediment yield under various management practices. The model provides a tool for evaluating sediment yield on a storm-by-storm basis for control of erosion and sediment yield from farm fields. The model incorporates fundamental principles of erosion, deposition, and sediment transport. The procedures allow parameter values to change along complex overland flow profiles and along waterways to represent both spatial variability and variations that occur from storm to storm. Many of the model parameter values are obtained from topographic maps or directly from the Universal Soil-Loss Equation (USLE). Thus we feel that the model has immediate applications without extensive calibration.

Individual components of the model were tested using experimental data from studies of overland flow, erodible channels, and small impoundments. These results suggest that the model produces reasonable estimates of erosion, sediment transport, and deposition under a variety of conditions common to field-sized areas. The procedures developed here can be used to evaluate alternative management practices such as conservation tillage, terracing, and contouring.

## INTRODUCTION

Estimates of erosion and sediment yield on field-sized areas are needed so that best management practices (BMPs) can be selected to control erosion for maintaining soil productivity and to control sediment yield for preventing excessive degradation of water quality. The field is typically the management unit used by most farmers to select management practices. For several years, soil conservationists have used the Universal Soil-

Article was submitted for publication in September 1980; reviewed and approved for publication by the Soil and Water Division of ASAE in March 1981.

Contribution from USDA-SEA-AR in cooperation with the Purdue Agricultural Experiment Station, Purdue Journal No. 8111.

The authors are: G. R. FOSTER, Hydraulic Engineer, USDA, and Associate Professor, Agricultural Engineering Dept., Purdue University, Lafayette, IN; L. J. LANE, Hydrologist, USDA, Tucson, AZ; J. D. NOWLIN, Computer Programmer, USDA, Agricultural Engineering Dept., Purdue University, Lafayette, IN; J. M. LAFLEN, Agricultural Engineer, USDA, Ames, IA; and R. A. YOUNG, Agricultural Engineer, USDA, Morris, MN.

**Acknowledgements:** The erosion-sediment yield model described herein is a component of CREAMS, a comprehensive field-scale model including hydrologic, erosion, pesiticide, and nutrient components developed by USDA-SEA-AR scientists under the leadership of W. G. Knisel, Jr. Complete documentation on the comprehensive model is given by USDA (1980). Additional support for this project, including computer programming, was provided by K. G. Renard, Research Leader, Southwest Rangeland Watershed Research Center, USDA-SEA-AR, Tucson, AZ; W. C. Moldenhauer, Research Leader, Erosion Research Unit, USDA-SEA-AR, Lafayette, IN; and G. W. Isaacs, Head, Agricultural Engineering Dept., Purdue University, W. Lafayette, IN.

Loss Equation (USLE) (Wischmeier and Smith, 1978) to select erosion control practices tailored for a given farmer and his fields. If sediment yield tolerances for maintenance of water quality are established for local areas, a model is needed to select BMPs based on a farmer's site specific conditions, his needs, and tolerable loading rates for streams in his area.

On a given field, sediment yield is controlled by either sediment detachment or sediment transport capacity (Ellison, 1947; Meyer and Wischmeier, 1969), depending on factors such as topography, soil, cover, and rainfall/runoff characteristics. The effects of these factors change from season to season and from storm to storm. The need to consider detachment and transport processes on a storm-by-storm basis limits the accuracy of lumped equations such as the USLE (an erosion equation), or Williams' (1975) modified USLE (a flow transport sediment yield equation) on field-sized areas.

Several detailed models (Beasley et al., 1980; Donigian and Crawford, 1976; Li, 1977) compute erosion and sediment transport at various times over a runoff event. Although these models are powerful, their considerable use of computer time prohibits the practical simulation of long periods of record on many fields to select a BMP for specific fields.

The purpose of this paper is to describe a model that, while simply constructed and usable over a broad range of situations at reasonable cost, embodies the latest knowledge on the fundamentals of erosion mechanics. The model may be used without calibration or collection of data to determine parameter values. It can be linked to hydrologic and chemical transport models, and was developed for that specific purpose as a component of CREAMS, a field scale model for Chemicals, Runoff, and Erosion from Agricultural Management Systems (USDA, 1980).

## BASIC CONCEPTS

The basis of this model is that USLE storm erosivity, EI, and the peak runoff rate at the watershed outlet can be used to characterize a storm's rainfall, runoff, and sediment yield. Quasi-steady state is assumed. Thus, sediment movement downslope obeys continuity expressed by:

$$dq_s/dx = D_L + D_F \dots \dots \dots \dots \dots \dots \dots \dots \dots [1]$$

where $q_s$ = sediment load (mass/unit width/unit time), $x$ = distance, $D_L$ = lateral inflow of sediment (mass/unit area/ unit time), and $D_F$ = detachment or deposition by flow (mass/unit area/unit time). Mathematically, detachment and deposition differ only in sign; detachment is positive, deposition is negative.

Hydrologically and hydraulically, a typical watershed may be divided into areas or elements of overland flow, channel flow, or impounded runoff. Each type of flow is

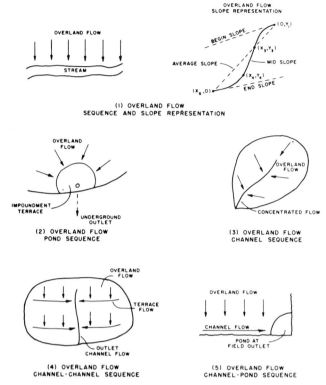

FIG. 1 Schematic representation of typical field systems in the field-scale erosion/sediment yield model.

treated in the model with a specific set of equations for that type of flow. Lateral sediment inflow is from interrill erosion on overland flow areas, or from overland flow (or channels if a set of channels drain into a main channel) for the channel areas. Overland flow or channels, but not both, may drain directly into an impoundment according to the model's structure. Flow in rills on overland flow areas or in channels transports all sediment downstream. Lateral sediment inflow to runoff in rills or channels is assumed regardless of whether the flow is detaching or depositing.

The watershed is represented by selecting a combination of elements from a typical overland flow profile, a main channel, a set of channels draining into a main channel, or a small impoundment as illustrated in Fig. 1. The selected combination of elements depends on the site being analyzed. Overland flow and channel elements are divided into segments along their length. Computations proceed downstream segment by segment and element by element. The computational sequence is shown in Fig. 2.

For an overland flow or channel segment, the model computes a potential sediment load which is the sum of the sediment load from the immediate upslope segment plus that added by lateral inflow within the segment. If the potential load is less than the sediment transport capacity of the flow, detachment occurs either at the detachment capacity of the flow or at the rate that will just fill transport capacity, whichever is less. Sediment

detachment by rainfall or flow adds sediment having a given size and density distribution. No sorting is allowed during detachment.

If potential sediment load is greater than transport capacity, deposition is assumed to occur at the rate of (Foster and Meyer, 1975):

$$D_d = \alpha (T_c - q_s) \quad\quad\quad\quad\quad\quad\quad\quad\quad [2]$$

where $D_d$ = deposition rate (mass/unit area/unit time), $\alpha$ = a first order reaction coefficient (length $^{-1}$), and $T_c$ = transport capacity (mass/unit width/unit time). The coefficient $\alpha$ is given by:

$$\alpha = \xi \, V_s/q_w \quad\quad\quad\quad\quad\quad\quad\quad\quad\quad [3]$$

where $\xi$ = 0.5 for overland flow (Davis, 1978), and 1.0 for channel flow (Einstein, 1968), $V_s$ = particle fall velocity, and $q_w$ = discharge per unit width (volume/unit width/unit time). Fall velocity is computed using standard relationships and drag coefficients for a sphere falling in still water.

The assumption that $dT_c/dx$ is constant over a segment permitted use of analytical solutions to equations [1] and [2] where deposition occurred. Where deposition did not occur, sediment load was calculated from:

$$q_s = (D_u + D_l)\Delta x/2 + D_L\Delta x + q_{su} \quad\quad\quad\quad [4]$$

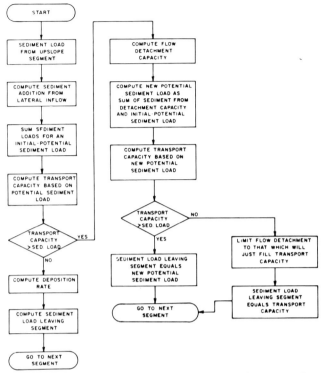

FIG. 2 Flow chart for detachment-transport-deposition computations within a segment of an overland flow or channel element.

where $D_u$ and $D_l$ respectively, rates of detachment by flow at the upper and lower ends of the segment or its portion where detachment by flow is occurring, $\Delta x =$ length of the segment or its portion where detachment by flow is occurring, and $q_{su} =$ sediment load at the upper end of the segment or its portion where detachment by flow is occurring.

The Yalin sediment transport equation was modified (Yalin, 1963; Foster and Meyer, 1972; Davis, 1978; and Khaleel et al., 1979) to describe sediment transport capacity for various particle sizes and densities. A particle type is a class of particles represented by a given diameter and specific gravity. If transport capacity exceeds availability for one particle type while it is less for another, excess transport capacity is shifted from the particle type having the excess to the one having the deficit. Furthermore, simultaneous deposition and detachment of particles by flow is not allowed. Equations [1 - 4] are solved for each particle type within these constraints.

## OVERLAND FLOW ELEMENT

Detachment on interrill and rill areas and transport and deposition by flow in rills are the important erosion-transport processes on overland flow areas. Detachment is described by a modified USLE written as (Foster et al., 1977):

$$D_{Li} = 4.57 \, EI \, (s_0 + 0.014) K\Phi P \, (\sigma_p/V_u). \quad \ldots \ldots \ldots \quad [5]$$

and

$$D_{Fr} = (6.86 \times 10^6) \, \eta \, v_u \sigma_p^{1/3} \, (x/22.1)^{\eta-1} K\Phi P \, (\sigma_p/V_u). \ldots \, [6]$$

where $D_{Li} =$ interrill detachment rate (g/m² of land surface/s), $D_{Fr} =$ capacity rate for rill detachment (g/m² of land surface/s), $EI =$ rainfall erosivity (energy times maximum 30-min intensity) (N/h)*, $x =$ distance downslope (m), $s_o =$ sine of slope angle, $\eta =$ slope length exponent for rill erosion, $K =$ USLE soil erodibility factor* [g h/ (N m²)], $\Phi =$ USLE cover-management soil-loss ratio, $P =$ USLE contouring factor, $V_u =$ runoff amount [volume/unit area (m)], and $\sigma_p =$ peak runoff rate [volume/unit area/unit time (m/s)]. The term $\sigma_p/V_u$ converts a total soil loss for a storm to an average rate for the storm. Only the contouring part of the USLE P factor is used. The model is structured to directly account for other supporting conservation practices like terraces and stripcropping.

For downslope distances less than 50 m, $\eta$ is set to 2.0, but for slopes longer than 50 m, $\eta$ is limited by:

$$\eta = 1.0 + 3.912/\ln x \quad \ldots \ldots \ldots \ldots \ldots \ldots \ldots \ldots \ldots \quad [7]$$

*The units on the K factor from the USLE must be carefully noted. Multiplication of the K in U. S. customary units by 131.7 gives a metric K having units of gh/(Nm²). Foster, G. R., D. K. McCool, K. G. Renard and W. C. Moldenhauer, 1981. Conversion of the Universal Soil-Loss Equation (USLE) to metric SI units. Accepted by J. Soil and Water Conservation. Manuscript available from senior author.

This limit avoids apparent excessive rill erosion for very long slopes (Foster et al., 1977). Equation [7] limits the effective slope length exponent for rill and interrill erosion combined to 1.67 so far as it is a function of length. The effective exponent is also a function of slope and runoff erosivity relative to rainfall erosivity.

The detachment equations [5] and [6] (except for the $\sigma_p/V_u$ term) as originally developed (Foster et al., 1977) were on a storm basis, whereas the transport equation is on an instantaneous rate basis. The two were combined by assuming that the computed sediment concentrations are average concentrations for the runoff event.

Cover and management effects on detachment are described with the USLE soil loss ratios (Wischmeier and Smith, 1978). Cover and management affect transport by their reduction of $\sigma_p$ and $V_u$ (estimated outside of the model) and by reducing the flow's shear stress acting along the soil-water interface. The concept (Graf, 1971) of dividing shear between form roughness (cover like mulch or vegetation) and grain roughness (soil) is used to estimate the proportion of total shear stress acting on the soil. The shear stress acting on the soil, $\tau_{soil}$, is estimated by:

$$\tau_{soil} = \gamma y s_0 \, (n_{bov}/n_{cov})^{9/10} \quad \ldots \ldots \ldots \ldots \ldots \quad [8]$$

where $Y$ = weight density of water, $y$ = flow depth for bare, smooth soil, $n_{bov}$ = Manning's n for bare soil, and $n_{cov}$ = Manning's n for rough, mulch, or vegetative covered soil. Flow depth is estimated by the Manning equation as:

$$y = [q_w \, n_{bov}/s^{1/2}]^{3/5} \quad \ldots \ldots \ldots \ldots \ldots \ldots \ldots \quad [9]$$

where $q_w$ = discharge rate per unit width. Although the Darcy-Weisbach equation with a varying friction factor for laminar flow might be more accurate in some cases for y, most users are better acquainted with estimating Manning's n. Values for Manning's n may be selected from Foster et al. (1980a) and Lane et al. (1975).

Segments along the overland flow profile are established by the model. The overland flow profile may be uniform, convex, concave, or a combination of these shapes. Input data requirements are slope length, average slope steepness, location of the end points of a uniform section at midslope, slope at the upper end of the profile, and slope at the lower end of the profile. A quadratic curve is fitted to curved portions of the slope so that it passes through an end point of the uniform segment at midslope and is tangent to the profile near each of its ends. Convex portions of a profile are divided into three equal length segments while concave portions are divided into ten equal length segments because calculation of deposition on concave slopes is quite sensitive to the number of segments, and accurate computation of the location of the beginning of deposition is important. Uniform portions of a profile are single segments. Additional segment ends are designated by the model where K, Φ, P, or $n_{cov}$ change.

## CHANNEL ELEMENT

The channel element describes detachment, transport, and deposition by flow in terrace channels, diversions, natural waterways, grassed waterways, row middles or graded rows, tailwater ditches, and other similar channels where topography has caused overland flow to con-

verge. The channel element does not describe erosion in gullies or large streams.

The same basic concepts are used in both the channel and overland flow elements. Discharge along the channel is assumed to vary directly with upstream drainage area. An initial discharge is permitted at the upper end of a channel to account for upland contributing areas. Changes in controlling variables like slope and cover along the channel are allowed.

Flow in most channels in fields is spatially varied, with discharge increasing along the channel. The model approximates the energy gradeline along the channel assuming a triangular channel section and steady flow at the characteristic peak discharge from a set of polynomial curves fitted to solutions of the normalized spatially varied flow equation (Chow, 1959). This feature approximates either drawdown or backwater at a channel outlet like at the edge of a field where vegetation may hinder runoff. As an alternative in the model, the slope of the energy gradeline can be assumed equal to the channel slope. After the slope of the energy gradeline is estimated, a triangular, rectangular, or "naturally eroded" section is selected at the user's option to compute flow hydraulics and channel erosion and sediment transport.

In the spring immediately after planting, concentrated flow from intense rains on a freshly prepared seedbed may erode through the finely tilled layer to the depth of secondary tillage. If the soil is susceptible to erosion by flow when tilled, the flow may erode deeper to the depth of primary tillage. Often the soil is much less erodible at this level and downcutting will stop here. Before the channel reaches a nonerodible layer, its width is a function of the flow's shear stress and the soil's critical shear stress. Once the flow reaches a relatively nonerodible layer, the channel widens. As it widens, the erosion rate decreases until it approaches zero as the channel approaches a maximum width. The maximum width depends on the flow's shear stress and the soil's critical shear stress. Data from rill erosion studies (Meyer et al., 1975; Lane and Foster, 1980) suggest that erosion by flow over a tilled, loose seedbed may be described by:

$$D_c = K_{ch} \, (1.35\bar{\tau} - \tau_{cr})^{1.05} \quad \ldots \ldots \ldots \ldots \ldots \quad [10]$$

where $D_c$ = erosion rate in a channel (mass/unit area of wetted perimeter/unit time), $K_{ch}$ = soil erodibility factor for a channel erosion $\bar{\tau}$ = average shear of the flow at a channel section, and $\tau_{cr}$ = a critical shear stress below which erosion is negligible. Critical shear stress of the surface layer of soil seems to increase greatly over the year as the soil consolidates (Graf, 1971; Foster et al., 1980a).

The shear stress acting on the soil is the shear stress used to compute detachment and sediment transport capacity. Grass and mulch reduce this stress. Total shear is divided into that acting on the vegetation, mulch, or large scale roughness and that acting on the soil using sediment transport theory (Graf, 1971).

Shear stress at a channel location varies with time as runoff rises and falls. The model assumes that shear stress is triangularly distributed in time during the runoff event to estimate the time $t_b$ that shear stress exceeds the critical shear stress. Shear stress is assumed constant and equal to shear stress computed from the characteristic peak discharge for this time period. This tends to overestimate total erosion for the storm. The derivation and

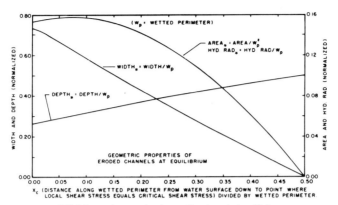

FIG. 3 Geometric properties of an eroding channel at equilibrium.

validation of the equations for channel erosion discussed below were described by Lane and Foster (1980).

Until the channel reaches the nonerodible layer, an active channel is assumed to be rectangular with the width obtained by Fig. 3 and 4 and equations [11] and [12]. The solution requires that a value for $x_c$ be found. Given the discharge Q, Manning's n, and friction slope $S_f$, a value $g(x_c)$ is calculated from:

$$g(x_c) = (Qn/S_f^{1/2})^{3/8} \ (\gamma S_f/\tau_{cr}) \ \ldots \ldots \ldots \ldots \ldots \ [11]$$

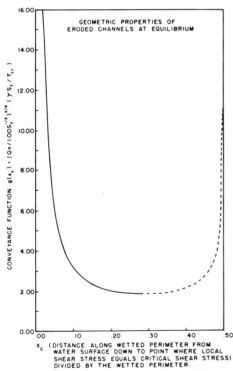

FIG. 4 Function $g(x_c)$ for an eroding channel at equilibrium.

Given a particular value $g(x_c)$, a value of $x_c$ is obtained from Fig. 3. Having determined $x_c$, a value for $R_* =$ hydraulic radius/wetted perimeter and $W_* =$ width /wetted perimeter is read from Fig. 4. The width of the channel before it reaches the nonerodible layer is then calculated from:

$$W_{ac} = (Qn/S_f^{1/2})^{3/8} \ W_*/R^{5/8}) \ \ldots \ldots \ldots \ldots \ldots \ [12]$$

The channel moves downward at the rate $d_{ch}$:

$$d_{ch} = D_c/\rho_{soil} = K_{ch} \ (1.35\tau - \tau_{cr})^{1.05}/\rho_{soil} \ \ldots \ldots \ldots \ [13]$$

where $\rho_{soil} =$ mass density of the soil in place. The erosion rate in the channel is:

$$E_{ch} = W_{ac} \ K_{ch} \ (1.35 \ \tau - \tau_{cr})^{1.05} \ t_b \ \ldots \ldots \ldots \ldots \ [14]$$

where $E_{ch}$ is the soil loss per unit channel length for the storm (mass/unit length).

Erosion rate $e_p$ (mass/unit area of wetted perimeter/unit time) normal to the wetted perimeter at a point is assumed equal to:

$$e_p = K_{ch} \ (\tau_x - \tau_{cr})^{1.05} \ \ldots \ldots \ldots \ldots \ldots \ [15]$$

where $\tau_x =$ the shear stress at a given point along the wetted perimeter. In order for a channel to be eroding downward in an equilibrium shape at an equilibrium rate, the vertical component of the erosion rates, $e_p$, must be equal at all points along the wetted perimeter. Equations [10 - 13] and Figs. 3 and 4 are based on this condition. The 1.35 factor is the ratio of the shear stress in the center of the channel to the average shear stress for the cross section.

Once the channel reaches the nonerodible boundary, erosion rate decreases with time as the channel widens. The rate decreases even if discharge rate remains constant. The width W of the channel at any time after the channel has eroded to the nonerodible layer is estimated by:

$$W_* = [1-\exp(-t_*)] \ \ldots \ldots \ldots \ldots \ldots \ldots \ [16]$$

where:

$$W_* = (W - W_i)/(W_f - W_i) \ \ldots \ldots \ldots \ldots \ldots \ [17]$$

$$t_* = (t - t_i) (dW/dt)_i/(W_f - W_i) \quad \dots \dots \dots \dots \dots [18]$$

where $W$ = width at t, $W_i$ = width at $t_i$, $W_f$ = final eroded width for $t \to \infty$ and the given Q, t = time, and $(dW/dt)_i$ = rate that channel widens at $t = t_i$. The initial widening rate is given by:

$$(dW/dt)_i = 2K_{ch} (\tau_b - \tau_{cr})^{1.05}/\rho_{soil} \dots \dots \dots \dots [19]$$

where $\tau_b$ is given by:

$$f(x_b) = \left(\tau_b/\bar{\tau}\right) = \exp [0.127 - 0.516 \ln x_b - 0.408 (\ln x_b)^2 -$$

$$0.0344 (\ln x_b)^3 ] \qquad x_b \geq 0.02 \quad \dots \dots \dots [20]$$

or

$$f(x_b) = \left(\tau_b/\bar{\tau}\right) = 0.13 \, x_b/0.02 \qquad x_b < 0.02 \quad \dots \dots \dots [21]$$

where $x_b$ = flow depth/wetted perimeter.

The final width $W_f$ is determined by finding the $x_{cf}$ that gives:

$$(Qn/S_f^{1/2})^{3/8} (\gamma S_f/\tau_{cr}) = \left\{[x_{cf}(1 - 2x_{cf})]^{3/8} f(x_{cf})\right\}^{-1} \quad \dots [22]$$

where $f(x_{cf})$ is the function given by equation [20] or [21] and evaluated at $x_{cf}$. The final width is:

$$W_f = \left\{(Qn/S_f^{1/2}) [(1 - 2x_{cf})/x_{cf}^{5/3}]\right\}^{3/8} \quad \dots \dots \dots [23]$$

Equations [16 - 23] are based on the assumption that in a rectangular channel on a nonerodible layer, the channel widens at the rate that the flow erodes the channel wall at the nonerodible layer. Widening ceases when the shear stress at the nonerodible boundary equals the critical shear stress.

Channel erosion after the channel reaches the nonerodible layer is:

$$E_{ch} = \Delta W \, H_{sw}\rho_{soil} \dots \dots \dots \dots \dots \dots [24]$$

where $\Delta W$ = the change in width calculated from equations [16 - 23] and $H_{sw}$ = the height of the channel sidewall.

## IMPOUNDMENT ELEMENT

The impoundment element describes deposition behind impoundment terraces and other small structures that drain between storms through a pipe near the bottom of the impoundment where an orifice controls discharge.

Deposition is the main sedimentation process occurring in impoundments. Since transport capacity in impoundments is essentially nonexistent, the amount of sediment trapped in an impoundment depends primarily on time available for sediment to settle to the bottom of the impoundment before flow can carry the particles from the impoundment. The equations for the impoundment element were developed from regression analyses where relationships were fitted to simulate the data from a more complex model (Laflen et al., 1978). That model had previously been validated with field data (Laflen et al., 1972.

The fraction of a given particle type that passes through the impoundment is:

$$F_{pi} = A_p \exp(B_p \, d_{epi}) \dots \dots \dots \dots \dots [25]$$

where $F_{pi}$ = fraction passing through the impoundment

for particle type i, $A_p$ and $B_p$ = coefficients given below, and $d_{eqi}$ = the equivalent sand diameter in microns of particle tipe i. Equation [25] is integrated over a particle class interval to obtain the total discharge for the particle class.

The coefficients $A_p$ and $B_p$ are given by:

$$A_p = 1.136 \exp(Z_s) \dots \dots \dots \dots \dots \dots [26]$$

$$B_p = -0.152 \exp(Y_s) \dots \dots \dots \dots \dots \dots [27]$$

with $Z_s$ and $Y_s$ in turn given by:

$$Z_s = (-6.68 \times 10^{-6}) f (0.3048)^{B-2} -0.0903B +$$
$$(1.19 \times 10^{-4})C_{or} -(1.21 \times 10^{-4})V_{in} -0.0185I \dots \dots [28]$$

$$Y_s = (3.28 \times 10^{-5}) f (0.3048)^{B-2} + 0.123B - (2.4 \times$$
$$10^{-4})C_{or} + (2.86 \times 10^{-4}) V_{in} -0.0108I \dots \dots \dots [29]$$

where f and B = coefficient and exponent in the power equation relating surface area to depth $S_a = fY_d^B$, $Y_d$ = depth in the impoundment (m), $S_a$ = surface area (m²), $V_{in}$ = volume of runoff reaching the impoundment (m³), and I = infiltration rate in the impoundment (mm/h). The coefficient $C_{or}$ related to the orifice in the pipe outlet is given by:

$$C_{or} = 0.15 \, d_{or}^2 \dots \dots \dots \dots \dots \dots [30]$$

where $d_{or}$ = diameter of the orifice (mm).

Less water leaves the impoundment than entered it because of infiltration through the boundary of the impoundment. The volume leaving is estimated by:

$$V_{out} = 0.95 \, V_{in} \exp(Z_r) \dots \dots \dots \dots \dots [31]$$

where $V_{out}$ = volume of runoff discharged, and $Z_r$ is given by:

$$Z_r = (-9.29 \times 10^{-6}) f (0.3048)^{B-2} + 0.0282B + (1.25 \times$$
$$10^{-4})C_{or} - (1.09 \times 10^{-4}) V_{in} - 0.0304I \dots \dots [32]$$

In addition:

$$\text{If } I = 0.0, V_{out} = V_{in} \dots \dots \dots \dots \dots \dots [33]$$

$$\text{If } V_{out} > V_{in}, V_{out} = V_{in} \dots \dots \dots \dots \dots [34]$$

are additional constraints on $V_{out}$ from equation [31] because 0 and $V_{in}$ are not lower and upper limits for equations [31] and [32].

## ERODED SEDIMENT CHARACTERISTICS

Sediment eroded from field-sized areas is often a mixture of primary particles and aggregates. The size and density distribution of these particles as they are detached is a function of soil properties, soil management, and rainfall and runoff characteristics. If deposition changes the distribution, usually the coarse and dense particles are deposited first, leaving a mixture of finer sediment. The initial particle input to the model is the distribution of the sediment as it is detached; the model calculates a new distribution when it calculates the occurrence of deposition. No selectivity is assumed in detachment of particles.

**246**

TABLE 1. TYPICAL SEDIMENT CHARACTERISTICS OF
DETACHED SEDIMENT BEFORE DEPOSITION FOR A
MIDWESTERN SOIL.*

| Particle type | Diameter | Specific gravity | Fraction of total |
|---|---|---|---|
| | mm | | mass basis |
| Primary clay | 0.002 | 2.60 | 0.05 |
| Primary silt | 0.010 | 2.65 | 0.08 |
| Small aggregate | 0.030 | 1.80 | 0.50 |
| Large aggregate | 0.500 | 1.60 | 0.31 |
| Primary sand | 0.200 | 2.65 | 0.06 |

*Particle distribution in soil mass: Clay = 25%, Silt = 60%, Sand = 15%.

Based on a survey of existing data, values given in Table 1 are typical of some midwestern soils. If the particle distribution is unknown, the model estimates the distribution from the primary particle size distribution of the soil mass using the following equations:

$$PSA = SAO (1.0 - CLO)^{2.49} \quad \dots \dots \dots \dots \dots [35]$$

$$PSI = 0.13 \ SIO \quad \dots \dots \dots \dots \dots \dots \dots [36]$$

$$PCL = 0.2 \ CLO \quad \dots \dots \dots \dots \dots \dots \dots [37]$$

$$SAG = 2 \ CLO \quad CLO < 0.25 \dots \dots \dots \dots \dots 38]$$

$$SAG = 0.28 \ (CLO - 0.25) + 0.5 \quad 0.25 \leqslant CLO \leqslant 0.50 \dots [39]$$

$$SAG = 0.57 \quad 0.5 < CLO \quad \dots \dots \dots \dots \dots [40]$$

$$LAG = 1.0 - PSA - PS - PCL - SAG \quad \dots \dots \dots \dots 41]$$

if LAG < 0.0, multiply PSA, PS, PCL, and SAG by same ratio to make:

$$LAG = 0.0 \quad \dots \dots \dots \dots \dots \dots \dots \dots \dots [42]$$

The variables, CLO, SIO, SAO, PCL, PSI, PSA, SAG, and LAG, are respectively, fractions for primary clay, silt, and sand in the original soil mass, and primary clay, silt, sand, and small and large aggregates in the sediment at the point of detachment. The diameters for the particles are defined as:

$$DPCL = 0.002 \ mm \quad \dots \dots \dots \dots \dots \dots [43]$$

$$DPSI = 0.010 \ mm \quad \dots \dots \dots \dots \dots \dots [44]$$

$$DPSA = 0.200 \ mm \quad \dots \dots \dots \dots \dots \dots [45]$$

$$DSAG = 0.03 \ mm \quad CLO < 0.25 \quad \dots \dots \dots \dots [46]$$

$$DSAG = 0.20 \ (CLO - 0.25) + 0.03 \ mm \quad 0.25 \leqslant CLO \leqslant 0.60$$

$$\dots \dots \dots \dots \dots \dots \dots \dots \dots \dots \dots \dots \dots [47]$$

$$DSAG = 0.1 \ mm \quad 0.60 < CLO \quad \dots \dots \dots \dots [48]$$

$$DLAG = 2 \ CLO \ mm \quad \dots \dots \dots \dots \dots \dots [49]$$

where DPCL, DPSI, DPSA, DSAG, and DLAG are, respectively, the diameters of the primary clay, silt, and sand, and the small and large aggregates in the sediment. The assumed specific gravities are shown in Table 1. The primary particle composition of the aggregates is estimated from:

$$CLSAG = SAG \ [CLO/(CLO + SIO)] \quad \dots \dots \dots \dots [50]$$

$$SISAG = SAG \ [SIO/(CLO + SIO)] \quad \dots \dots \dots \dots [51]$$

$$SASAG = 0.0 \quad \dots \dots \dots \dots \dots \dots \dots \dots \dots [52]$$

$$CLLAG = CLO - PCL - CLSAG \quad \dots \dots \dots \dots \dots [53]$$

$$SILAG = SIO - PSI - SISAG \quad \dots \dots \dots \dots \dots [54]$$

$$SALAG = SAO - PSA \quad \dots \dots \dots \dots \dots \dots \dots [55]$$

where CLSAG, SISAG, and SASAG = fractions of the total for the sediment of, respectively, primary clay, silt, and sand in the small aggregates in the sediment load, and CLLAG, SILAG, and SALAG are corresponding fractions for the large aggregates.

If the fraction of clay in the large aggregate based on the mass of the large aggregate and not on the total mass of sediment is less than 0.5 times CLO, the distribution of the particle types is recomputed. A sum of Γ is computed whereby:

$$\Gamma = PCL + PSI + PSA \quad \dots \dots \dots \dots \dots \dots [56]$$

The fractions PSA, PSI, and PCL are not changed. The new SAG is:

$$SAG = (0.3 + 0.5 \ \Gamma) \ (CLO + SIO)/([1-0.5(CLO + SIO)] \quad \dots [57]$$

Equation [57] is derived given (i) previously determined values for PCL, PSI, and PSA; (ii) the assumption that the sum of primary clay fractions for the total sediment is 1; and (iii) the assumption that the fraction of primary clay in LAG equals one half of the primary clay in the original soil.

The model also computes an enrichment ratio using values for specific surface area of organic matter, clay, silt, and sand. Organic matter is distributed among the particle types based on the proportion of primary clay in each type. The enrichment ratio is the ratio of total specific surface area of the sediment to that for the original soil.

## DISCUSSION

The model gave reasonable results, when compared with data from concave plots under simulated rainfall, single terrace watersheds, small watersheds with impoundment terraces, and a small watershed under conservation tillage. The simulations were made using measured rainfall and runoff values.

**Concave Plots**

Three concave plots 10.7 m long were carefully shaped in uniform soil so that slope along the plots continuously decreased from 18 percent at the upper end to 0 percent at the lower end (Foster et al., 1980b). Simulated rainfall at 64 mm/h was applied to one of the plots and deposi-

247

TABLE 2. COMPARISON OF OBSERVED SOIL LOSS FROM CONCAVE FIELD PLOTS WITH
THAT COMPUTED BY THE MODEL.

| | Plot length | Slope at lower end | Sediment yield | Particle distribution in size class | | | | |
|---|---|---|---|---|---|---|---|---|
| | | | | 0.002 | 0.03 | 0.3 | 0.75 | 1.5 mm |
| | m | % | kg/m | Fraction | | | | |
| Observed* | 7.0 | 6 | 8.6 | 0.05 | 0.36 | 0.15 | 0.17 | 0.27 |
| Observed | 8.8 | 3 | 3.9 | 0.07 | 0.48 | 0.12 | 0.12 | 0.21 |
| Computed | | | 6.5 | 0.08 | 0.58 | 0.24 | 0.10 | 0.01 |
| Observed | 10.7 | 0 | 3.0 | 0.10 | 0.85 | 0.02 | 0.01 | 0.02 |
| Computed | | | 3.0 | 0.19 | 0.80 | 0.01 | 0.00 | 0.00 |

*These data were used to calibrate soil erodibility factor, Manning's n, and particle distribution of sediment reaching deposition area. Source of data: Foster et al. (1980b).

tion began at 7 m from the upper end. Plot ends were installed at 7.0 m and 8.8 m on the other two plots. The measured particle distribution of the sediment entering the deposition area was used, and the soil erodibility factor and Manning's n were adjusted in the model to give the observed soil loss and particle distribution for the 7.0 m plot. The results shown in Table 2 for the 8.8 and 10.7 m plots were obtained using these calibrated values and the approximate slope shape curves in the model rather than the actual slope shape.

### Single Terrace Watersheds

Soil loss was simulated for eight years of data, about 53 runoff producing storms, from small, single terrace watersheds at Guthrie, Oklahoma (Daniel et al., 1943). The simulations were made without calibration using instructions in the user manual for the model (Foster et al., 1980a). Table 3 gives computed and measured results.

### Impoundment Terraces

Soil loss was simulated under a range of rainfall and runoff characteristics for six selected storms at the Charles City, and Guthrie Center, Iowa, and for five storms at Eldora, Iowa. Data were taken from an impoundment terrace study (Laflen et al., 1972). The model was run using the user manual instruction without calibration. Table 4 gives the results.

### Small Watershed

Simulations were run without calibrating for approximately 2½ years of data, about 35 runoff producing storms, from the P2 watershed at Watkinsville, Georgia in conservation tillage systems for corn (Smith et al., 1978). Deposition in the backwater from the flume at the watershed outlet was modeled. Deposition measured in

the flume backwater was about equal to the measured sediment yield on a similar, nearby watershed (Langdale et al., 1979). The computed sediment yield total for the period of record was 1.47 kg/m², while the measured value was 1.95 kg/m².

### Overland Flow Detachment

The relationships for detachment used in the overland flow element gave good results for a watershed at Treynor, Iowa. Estimates were better than those from the USLE using storm EI (Foster et al., 1977) and those obtained using the USLE and runoff volume and peak discharge (Onstad et al., 1977) as measures of erosivity. These results were confirmed by Lombardi (1979) for data from natural rainfall on uniform slopes. On long-term simulation, the model should produce results similar to those of the USLE for uniform slopes.

### Overland Flow Sediment Transport

As the results in Table 2 indicate, estimates of sediment transport by overland flow may be in error by a factor of two. The Yalin equation was selected to describe sediment transport by overland flow after studies showed that it gave better results than did several other widely used equations (Alonso, 1980; Neibling and Foster, 1980). However, overland flow conditions are outside the range of most sediment transport equations developed for stream flow. Many give results greatly in error for overland flow.

### Channel Detachment

The relationships for channel erosion are the ones most likely to be in error, because data for flow concentrations 300 mm wide from the studies (Meyer et al., 1975; Lane and Foster, 1980) where the relationships were derived may not apply to 2 m wide channels. Also, parameter values for channel soil erodibility and critical shear stress are not readily available. Few models except that of Bruce et al. (1975) consider the decay in erosion with time due to previous erosion. This component of the model may require calibration.

TABLE 3. COMPUTED AND OBSERVED SOIL LOSS FOR 8 YEARS OF DATA FROM SINGLE TERRACE WATERSHEDS AT GUTHRIE, OKLAHOMA*.

| Terrace | Total soil loss for period per unit area of watershed | |
|---|---|---|
| | Observed | Computed |
| | kg/m² | kg/m² |
| Uniform grade of 0.0017, 457 m long | 4.8 | 4.6 |
| Uniform grade of 0.005, 457 m long | 12.1 | 10.6 |
| Variable grade, 0.005 at outlet to 0 at upper end, 871 m long | 13.8 | 11.9 |
| Variable grade, 0.0033 at outlet to 0 at upper end, 773 m long | 12.2 | 6.4 |

*Source of data: Daniel et al. (1943)

TABLE 4. SIMULATED AND OBSERVED SOIL LOSS FOR IMPOUNDMENT TERRACES IN IOWA.

| Location | Total soil loss per unit area of watershed for selected storms | |
|---|---|---|
| | Observed | Computed |
| | kg/m² | kg/m² |
| Eldora | 0.115 | 0.069 |
| Charles City | 0.043 | 0.016 |
| Guthrie Center | 0.050 | 0.050 |

## SUMMARY

An erosion-sediment yield model for field-sized areas was developed for use on a storm-by-storm basis. The overall objective was to develop a model incorporating fundamental erosion-sediment transport relationships for use in evaluating best management practices for control of erosion and sediment yield from farm fields. The procedure allows parameters to change along the overland flow profile and along waterways to represent both spatial variability and the variations that occur from storm-to-storm. Many of the model parameters are directly from the Universal Soil-Loss Equation (USLE) and other similar, process-type relationships. For this reason, we feel that the model has immediate applications without extensive calibration.

Individual components of the model were tested using experimental data from studies of overland flow, erodible channels, and impoundments. Testing suggests that the model gives reasonable results and may be a useful tool for analyzing the influence of alternate management practices on erosion and sediment yield from field-sized-areas.

## References

1 Alonso, C. V. 1980. Selecting a formula to estimate sediment transport in nonvegetated channels. In: CREAMS - A Field Scale Model for Chemicals, Runoff, and Erosion from Agricultural Management Systems. Vol III: Supporting Documentation. Conservation Research Report No. 26. USDA, Sci. and Educ. Admin. Chpt. 5. pp. 426-439.

2 Beasley, D. B., E. J. Monke, and L. F. Huggins. 1980. ANSWERS: A model for watershed planning. TRANSACTIONS of the ASAE. 23(4):839-944.

3 Bruce, R. R., L. A. Harper, R. A. Leonard, W. M. Snyder, and A. W. Thomas. 1975. A model for runoff of pesticides from small upland watersheds. J. Environ. Qual. 4(4):541-548.

4 Chow, V. T. 1959. Open-Channel Hydraulics. McGraw-Hill Book Co. New York, NY. 680 pp.

5 Davis, S. S. 1978. Deposition of nonuniform sediment by overland flow on concave slopes. M. S. Thesis. Purdue Univ., W. Lafayette, IN.

6 Daniel, H. A., H. M. Elwell and M. B. Cox. 1943. Investigations in erosion control and reclamation of eroded land at the Red Plains Conservation Experiment Station, Guthrie, OK, 1930-40. USDA-Tech. Bull. No. 837. 94 pp.

7 Donigian, A. S., Jr. and N. H. Crawford. 1976. Modeling nonpoint source pollution from the land surface. EPA-600/3-76-083. U. S. Environ. Protection Agency. 279 pp.

8 Einstein, H. A. 1968. Deposition of suspended particles in a gravel bed. J. Hydr. Div. ASCE. 94(HY5):1197-1205.

9 Ellison, W. D. 1947. Soil erosion studies. AGRICULTURAL ENGINEERING 28:145-146, 127-201, 245-248, 297-300, 349-351, 402-405, 442-444.

10 Foster, G. R., L. J. Lane, and J. D. Nowlin. 1980a. A model to estimate sediment yield from field sized areas: Selection of parameter values: In: CREAMS - A Field Scale Model for Chemicals, Runoff, and Erosion from Agricultural Management Systems. Vol. II: User Manual. Conservation Research Report No. 26. USDA, Sci. and Educ. Admin. Chpt. 2. pp. 193-281.

11 Foster, G. R., W. H. Neibling, S. S. Davis, and E. E. Alberts. 1980b. Modeling particle segregation during deposition by overland flow. In: Proc. Hydrologic Transport Modeling Symposium. ASAE Publ. 4-80. pp. 184-195.

12 Foster, G. R. and L. D. Meyer. 1972. Transport of soil particles by shallow flow. TRANSACTIONS of the ASAE 15(1):99-102.

13 Foster, G. R. and L. D. Meyer. 1975. Mathematical simulation of upland erosion by fundamental erosion mechanics. In: Present and Prospective Technology for Predicting Sediment Yields and Sources. ARS-S-40. USDA, Agric. Research Ser. pp. 190-207.

14 Foster, G. R., L. D. Meyer, and C. A. Onstad. 1977. A runoff erosivity factor and variable slope length exponents for soil loss estimates. TRANSACTIONS of the ASAE 20(4):683-687.

15 Graf, W. H. 1971. Hydraulics of Sediment Transport. McGraw-Hill Book Co. New York, NY. 544 pp.

16 Khaleel, R., G. R. Foster, K. R. Reddy, M. R. Overcash and P. W. Westerman. 1979. A nonpoint source model for land areas receiving animal wastes: III. A conceptual model for sediment and manure transport. TRANSACTIONS of the ASAE 22(5):1353-1361.

17 Laflen, J. M., H. P. Johnson, and R. O. Hartwig. 1978. Sedimentation modeling of impoundment terraces. TRANSACTIONS of the ASAE 21(6):1131-1135.

18 Laflen, J. M., H. P. Johnson, and R. C. Reeves. 1972. Soil loss from tile outlet terraces. J. Soil and Water Conserv. 27(2):74-77.

19 Lane, L. J. and G. R. Foster. 1980. Concentrated flow relationships. In: CREAMS - A Field Scale Model for Chemicals, Runoff, and Erosion from Agricultural Management Systems. Vol. III: Supporting Documentation. Conservation Research Report No. 26. USDA, Sci. and Educ. Admin. Chpt. 10. pp. 474-485.

20 Lane, L. J., D. A. Woolhiser, and V. Yevjevich. 1975. Influence of simplification in watershed geometry in simulation of surface runoff. Hydrology Paper No. 81. Colorado State Univ., Fort Collins. 50 pp.

21 Langdale, G. W., A. P. Barnett, R. A. Leonard, and W. G. Fleming. 1979. Reduction of soil erosion by no-till systems in the Southern Piedmont. TRANSACTIONS of the ASAE 22(1):82-86, 92.

22 Li, R. M. 1977. Water and sediment routing from watersheds. In: Proc. River Mechanics Institute. Colorado State Univ., Fort Collins. Chpt. 9.

23 Lombardi, F. 1979. Universal Soil Loss Equation (USLE), runoff erosivity factor, slope length exponent, and slope steepness exponent for individual storms. Ph.D. Thesis. Purdue Univ., W. Lafayette, IN. 128 pp.

24 Meyer, L. D., G. R. Foster, and S. Nikolov. 1975. Effect of flow rate and canopy on rill erosion. TRANSACTIONS of the ASAE 18(5):905-911.

25 Meyer, L. D. and W. H. Wischmeier. 1969. Mathematical simulation of the process of soil erosion by water. TRANSACTIONS of the ASAE12(6):754-758, 762.

26 Neibling, W. H. and G. R. Foster. 1980. Sediment transport capacity of overland flow. In: CREAMS - A Field Scale Model for Chemicals, Runoff, and Erosion from Agricultural Management Systems. Vol. III: Supporting Documentation. Conservation Research Report No. 26. USDA, Sci. and Educ. Admin. Chpt. 10. pp. 463-473.

27 Onstad, C. A., R. F. Piest, and K. E. Saxton. 1976. Watershed erosion model validation for Southwest Iowa. In: Proc. of the Third Federal Inter-Agency Sedimentation Confer. Water Resources Council. Washington, DC. Chpt. 1. pp. 23-24.

28 Smith, C. N., R. A. Leonard, G. W. Langdale, and G. W. Bailey. 1978. Transport of agricultural chemicals from small upland Piedmont watersheds. EPA-600/3-78-056. U. S. Environ. Protection Agency. 364 pp.

29 U. S. Department of Agriculture (USDA), 1980. CREAMS - A Field Scale Model for Chemicals, Runoff, and Erosion from Agricultural Management Systems. Conservation Research Report No. 26. Sci. and Educ. Admin. 643 pp.

30 Williams, J. R. 1975. Sediment-yield prediction with universal equation using runoff energy factor. In: Present and Prospective Technology for Predicting Sediment Yields and Sources. ARS-S-40. USDA, Agric. Research Ser. pp. 244-253.

31 Wischmeier, W. H. and D. D. Smith. 1978. Predicting Rainfall Erosion Losses. Agricultural Handbook No. 537. USDA, Sci. and Educ. Admin. 58 pp.

32 Yalin, Y. S. 1963. An expression for bed-load transportation. J. Hydr. Div. ASCE. 89(HY3):221-250.

### LIST OF SYMBOLS

| | |
|---|---|
| $A_p$ | Coefficient in equation for deposition in an impoundment |
| B | Exponent in surface area - depth relationship for an impoundment |
| $B_p$ | Exponent in equation for deposition in an impoundment |
| CLLAG | Clay content of large aggregates, fraction of total sediment |
| CLO | Fraction of original soil made up of primary clay |
| CLSAG | Clay content of small aggregates, fraction of total sediment |
| $C_{or}$ | Orifice coefficient for drainage from impoundment |
| $d_{ch}$ | Rate that channel erodes downward, (depth/time) |
| $d_{eq}$ | Equivalent sand diameter of a sediment particle |
| $d_{or}$ | Diameter of orifice in an impoundment drain |
| $D_c$ | Rate of sediment detachment by flow in channels (mass/area/time) |
| $D_d$ | Rate of deposition by flow (mass/area/time) |
| $D_F$ | Rate of detachment or deposition by flow (mass/area/time) |
| $D_{Fr}$ | Rate of sediment detachment by rill erosion, (mass/area/time) |

$D_l$    Rate of detachment by flow at lower end of a segment, (mass/area/time)

$D_L$    Rate of lateral inflow of sediment, (mass/area or length/time)

$D_{Li}$    Rate of sediment from interrill areas (mass/area/time)

$D_u$    Rate of detachment by flow at upper end of a segment, (mass/area/time)

DLAG    Diameter of large aggregate sediment particles

DPCL    Diameter of primary clay sediment particles

DPSA    Diameter of primary sand sediment particles

DPSI    Diameter of primary silt sediment particles

DSAG    Diameter of small aggregate sediment particles

$E_p$    Erosion rate normal to channel boundary, (mass/area/time)

$E_{ch}$    Erosion rate per unit length of channel, (mass/length of channel)

EI    Rainfall erosivity, total storm energy times maximum 30-min intensity

f    Coefficient in surface area-depth relationship for impoundment

$f(x_b)$    Shear stress distribution around a channel

$F_{pi}$    Fraction of a particular particle class deposited in an impoundment

$g(x_c)$    Conveyance function for flow in an eroding channel at equilibrium

$H_{sw}$    Height of channel sidewall

i    Particle class index

I    Infiltration rate through boundary of an impoundment

K    Soil erodibility factor for the USLE

$K_{ch}$    Soil erodibility factor for channel erosion

LAG    Fraction of sediment made up of large aggregates

n    Manning's n

$n_{bov}$    Manning's n for bare, smooth, overland flow surface

$n_{cov}$    Manning's n for a covered or rough overland flow surface

P    Contouring component of USLE supporting practices factor

PCL    Fraction of sediment made up of primary clay

PSA    Fraction of sediment made up of primary sand

PSI    Fraction of sediment made up of primary silt

$q_s$    Sediment load, (mass/width/time)

$q_{su}$    Sediment load at upper end of segment, (mass/width/time)

$q_w$    Rate of runoff discharge per unit width (volume/time/width)

Q    Discharge rate, (volume/time)

$R_*$    Ratio of hydraulic radius to wetted perimeter

$s_o$    Sine of angle of slope

$S_a$    Surface area in an impoundment

$S_f$    Friction slope for flow hydraulics in a channel

SAG    Fraction of sediment made up of small aggregates

SALAG    Fraction of sand in large aggregates, fraction of total sediment

SAO    Fraction of original soil made up of primary sand

SASAG    Fraction of sand in small aggregates, fraction of total sediment

SILAG    Fraction of silt in large aggregates, fraction of total sediment

SIO    Fraction of original soil made up of primary silt

SISAG    Fraction of silt in small aggregates, fraction of total sediment

t    Time

$t_b$    Time that shear stress exceeds critical shear stress

$t_*$    Normalized time for channel erosion

$t_i$    Initial time

$T_c$    Transport capacity, (mass/area/time)

$V_{in}$    Runoff volume into impoundment

$V_{out}$    Runoff volume out of impoundment

$V_s$    Particle fall velocity

$V_u$    Runoff volume per unit area, (depth)

W    Channel width

$W_*$    Normalized channel width

$W_{ac}$    Width of an eroding channel at equilibrium

$W_f$    Final eroded channel width

$W_i$    Initial channel width

x    Distance

$x_b$    Normalized distance around wetted perimeter to nonerodible boundary

$x_c$    Normalized distance around wetted perimeter to location where $\tau = \tau_{cr}$ for an eroding channel at equilibrium

$x_{cf}$    Normalized distance around wetted perimeter to location where $\tau = \tau_{cr}$ at nonerodible boundary

y    Flow depth

$Y_d$    Depth in impoundment

$Y_s$    Exponent in deposition equation for an impoundment

$Z_r$    Exponent in equation for runoff reduction by an impoundment.

$Z_u$    Exponent in equation for deposition in impoundment

$\alpha$    Reaction coefficient for deposition by flow, length$^{-1}$

$\gamma$    Weight density of water

$\Gamma$    Sum of PCL, PSI, and PSA

$\Delta x$    Segment length

$\Delta W$    Change in channel width

$\eta$    Slope length exponent for rill erosion

$\varepsilon$    Coefficient in deposition equation

$\rho_{soil}$    Mass density of soil in place

$\sigma_p$    Peak runoff rate, (depth/time)

$\bar{\tau}$    Average shear stress around wetted perimeter

$\tau_b$    Shear stress in channel at a nonerodible boundary

$\tau_{cr}$    Critical shear stress

$\tau_{soil}$    Shear stress acting on soil

$\tau_x$    The shear stress at a given point along the wetted perimeter.

$\Phi$    Soil loss ratio from USLE

# 17

Copyright © 1984 by the British Society for Research in Agricultural Engineering

Reprinted from *Jour. Agric. Eng. Research* **30**:245–253 (1984)

# A Predictive Model for the Assessment of Soil Erosion Risk

R. P. C. MORGAN*; D. D. V. MORGAN*; H. J. FINNEY*

A model is developed for predicting annual soil loss from field-sized areas on hillslopes. The model comprises a water phase and a sediment phase. In the sediment phase erosion is taken to result from the detachment of soil particles by rainsplash and their transport by runoff. Splash is related to rainfall energy and rainfall interception; runoff transport capacity depends upon the volume of runoff, slope steepness and crop management. Rainfall energy and runoff volume are estimated from annual rainfall amount in the water phase. The predicted rate of soil loss is compared with a topsoil renewal rate to determine changes in topsoil depth over time. Model validation was carried out by comparing predicted and observed values of annual runoff and erosion for 67 sites in 12 countries. As an example of how the model can be used, a 100-year simulation exercise is presented for soil erosion under shifting cultivation in Peninsular Malaysia.

## 1. Introduction

The need to predict the rate of soil erosion, both under existing conditions and those expected to occur following soil conservation practice, has led to the development of the Universal Soil Loss Equation[1] and the CREAMS model.[2] These techniques require considerable inputs of data and are generally too complicated for initial assessments of erosion in reconnaissance surveys. This paper presents an alternative procedure for soil loss prediction, bringing together the results of research by geomorphologists and agricultural engineers into a model which, although empirical, has a stronger physical base than the Universal Soil Loss Equation and is simpler and more flexible than CREAMS. Although it is an amalgam of existing operating functions and, in that sense, is not new, the model was developed with the specific objective of seeing to what extent existing work could be combined in a simple format to predict annual soil loss from field-sized areas on hillslopes.

## 2. Approach

The model[3] separates the soil erosion process into a water phase and a sediment phase (*Fig. 1*). The structure of the sediment phase is a simplification of the soil loss model described by Meyer and Wischmeier.[4] It considers soil erosion to result from the detachment of soil particles from the soil mass by raindrop impact and the transport of those particles by overland flow. The energy of rainfall for splash detachment and the volume of overland flow are estimated in the water phase. Operating functions (Table 1) were selected from the geomorphological and engineering literature according to their predictive ability, simplicity, and ease of determination of their input parameters (Table 2).

In the water phase annual rainfall (*R*) is used to determine the energy of the rainfall for splash detachment and the volume of runoff. Rainfall energy (*E*) is modelled empirically by extending the relationship between energy and intensity (*I*) developed by Wischmeier and Smith[1]

*Silsoe College, Silsoe, Bedford MK45 4DT

251

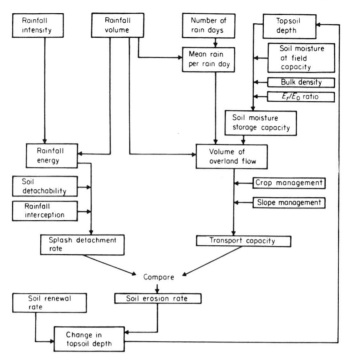

*Fig. 1. Flowchart of the model.*

TABLE 1
**Operating functions**

*Water phase*
$$E = R(11 \cdot 9 + 8 \cdot 7 \log_{10} I), \qquad \ldots (1)$$
$$Q = R \exp(-R_c/R_0), \qquad \ldots (2)$$
where $R_c = 1000 \, M \gamma D_r \, (E_t/E_0)^{0.5}$ and $R_0 = R/R_n$.

*Sediment phase*
$$F = K(Ee^{-aP})^b \times 10^{-3} \qquad \ldots (3)$$
$$G = CQ^d(\sin S) \times 10^{-3} \qquad \ldots (4)$$

$E$ = Kinetic energy of rainfall (in J/m²); $Q$ = volume of overland flow (in mm); $F$ = rate of splash detachment (kg/m²); $G$ = transport capacity of overland flow (kg/m²)
Values of exponents: $a = 0 \cdot 05$;[8,9] $b = 1 \cdot 0$;[7] $d = 2 \cdot 0$[6,10,11]

to an annual basis using the annual rainfall and an estimate of the typical hourly intensity of erosive rain [eqn (1) in Table 1]. Runoff volume is predicted from eqn (2) in Table 1, as given by Carson and Kirkby,[5] which assumes that runoff occurs whenever the daily rainfall exceeds a critical value ($R_c$) representing the moisture storage capacity of the soil–crop complex and that the daily rainfall amounts approximate an exponential frequency distribution. The value of $R_c$ depends upon the soil moisture storage at field capacity ($M$), the bulk density of soil ($\gamma$) and the topsoil rooting depth ($D_r$) (Withers and Vipond[6]) after allowing for the crop cover effect through evapotranspiration ($E_t/E_0$) using the empirical relationship adopted by Kirkby.[7]

The sediment phase is divided into two components: splash detachment and runoff transport. Splash detachment is modelled using the widely accepted power relationship with rainfall energy (Meyer[8]) modified to allow for the rainfall interception effect of the crop. Rainfall energy at the

TABLE 2
**Input parameters**

| | |
|---|---|
| $M$ | Soil moisture content at field capacity or 1/3 bar tension, % w/w |
| $\gamma$ | Bulk density of the top soil layer, Mg/m³ |
| $D_r$ | Topsoil rooting depth (in m), defined as the depth of soil from the surface to an impermeable or stony layer, to the base of the A horizon, to the dominant root base or to 1·0 m, whichever is the shallowest. Reasonable values are 0·05 for grass and cereal crops and 0·1 for trees and tree crops |
| $D_s$ | Total soil depth (in m), defined as the depth of soil from the surface to bedrock |
| $K$ | Soil detachability index (in g/J), defined as the weight of soil detached from the soil mass per unit of rainfall energy |
| $W$ | Rate of increase in soil depth by weathering at the rock–soil interface, mm/a |
| $V$ | Rate of increase of the topsoil rooting layer (mm/a) as a result of crop management practices and the natural breakdown of vegetative matter into humus |
| $S$ | Steepness of the ground slope expressed as the slope angle |
| $R$ | Annual rainfall, mm |
| $R_n$ | Number of rain days in the year |
| $I$ | Typical value for intensity of erosive rain, mm/h |
| $P$ | Percentage rainfall contributing to permanent interception and stemflow |
| $E_t/E_0$ | Ratio of actual ($E_t$) to potential ($E_0$) evapotranspiration |
| $C$ | Crop cover management factor. Combines $C$ and $P$ factors of the Universal Soil Loss Equation to give ratio of soil loss under a given management to that from bare ground with downslope tillage, other conditions being equal |
| $N$ | Number of years for which the model is to operate |

TABLE 3
**Sensitivity analysis**

| *1% change in* | *Percentage change in* | |
|---|---|---|
| | *transport capacity* | *detachment rate* |
| $R$ | $2(1 + R_c/R_0)$ | 1 |
| $M$; $\gamma$; $D_r$; $R_n$ | $-2(R_c/R_0)$ | |
| $E_t/E_0$ | $-R_c/R_0$ | |
| $C$; $\sin S$ | 1 | |
| $K$ | | 1 |
| $I$ | | $(3\cdot1 + 2\cdot3 \log_{10} I)^{-1}$ |
| $P$ (1% absolute change) | | $-5$ |

ground surface is assumed to decrease exponentially with increasing interception.[9,10] The transport capacity of overland flow is determined from eqn (4) in Table 1, developed by Kirkby,[5] with experimental support from erosion plot data,[5,11] and depends upon the volume of overland flow, the slope steepness and the effect of the crop cover. Although Kirkby uses the tangent of the slope angle, the sine value is adopted here. This makes very little difference to predictions at low slope angles, but prevents unacceptably high predictions at high slope angles. Tests of this equation[12] show that $C$ takes values in relation to crop cover which are of the same order of magnitude as the $C$ factor values used in the Universal Soil Loss Equation. These are therefore used at present as a convenient method of determining values of $C$.

The model compares the predicted rate of splash detachment with the transport capacity for overland flow, and equates the rate of soil loss to the lower of the two values, thereby indicating whether detachment or transport is the limiting factor.

The predicted rate of soil loss is compared with an estimated rate of weathering ($W$) at the soil–bedrock interface to determine the annual loss or gain in total soil depth ($D_s$). A similar calculation is made for the change in topsoil rooting depth ($D_r$) by comparing the rate of soil

TABLE 4
**Predicted and measured rates of annual runoff and erosion**

| Site | Runoff, mm | | Soil loss, kg/m² | |
|---|---|---|---|---|
| | Observed | Predicted | Observed | Predicted |
| *Lushoto, Tanzania* | | | | |
| Clay soil, maize and beans intercropping | 0·3–1·0 | 6·85 | 0·0012–0·0013 | 0·0018 |
| (2 years) | 0·2–1·0 | 3·74 | 0·0005–0·0010 | 0·0005 |
| Sandy clay loam, evergreen forest (2 years) | 2·6–5·6 | 2·59 | 0·0006–0·0015 | 0·0000 |
| | 4·2–5·7 | 5·96 | 0·0012–0·0030 | 0·0000 |
| Sandy clay loam, steep slope, evergreen | 8·5–12·7 | 3·25 | 0·0013–0·0057 | 0·0000 |
| forest (2 years) | 10·4–14·8 | 7·31 | 0·0105–0·0129 | 0·0001 |
| Clay, steep slope, maize and beans | 0·4–0·8 | 6·85 | 0·0007–0·0013 | 0·0038 |
| intercropping (2 years) | 0·4–0·8 | 3·74 | 0·0008–0·0010 | 0·0011 |
| *Adiopodoumé, Ivory Coast* | | | | |
| Sandy loam, secondary tropical forest | 15·0 | 85·9 | 0·001–0·02 | 0·003 |
| Sandy loam, bare ground | 707–1415 | 1268·1 | 6·9–15·0 | 15·5† |
| Sandy loam, oil palm | 43–172 | 57·8 | 0·001–0·05 | 0·02 |
| Sandy loam, banana with mulch | 11–86 | 102·4 | 0·004–0·005 | 0·002 |
| Sandy loam, maize | 643–1608 | 355·0 | 3·5–13·1 | 3·53 |
| Sandy loam, groundnut | 579–1565 | 452·7 | 5·9–12·0 | 7·17 |
| *Sefa, Senegal* | | | | |
| Loam, secondary tropical forest | 1·6–19·2 | 154·8 | 0·002–0·02 | 0·0009 |
| Loam, groundnut | 130–699 | 370·8 | 0·29–1·63 | 1·38 |
| Loam, cotton | 15–699 | 429·0 | 0·05–1·85 | 1·84 |
| Loam, maize with mechanization | 504 | 420·3 | 1·03 | 0·71 |
| Loam, sorghum | 390–683 | 442·5 | 0·33–1·24 | 1·57 |
| *Pong Khrai, Thailand* | | | | |
| Clay loam, upland rice | 22–32 | 34·4 | 1·40–2·40 | 0·07 |
| Clay loam, upland rice, bench terraces | 16–53 | 34·4 | 1·1–1·3 | 0·011 |
| *Marchiazza Basin, Italy* | | | | |
| Loamy sand, bare soil with tufted grass | 201–261 | 186·7 | 2·7–3·1 | 3·25 |
| Loamy sand, Molinia moor grass | 51–58 | 56·9 | 0·05–0·09 | 0·0062 |
| Loamy sand, chestnut and oak trees | 36–38 | 48·3 | 0·009–0·018 | 0·0014 |
| *Hesbaye, Belgium* | | | | |
| Sandy soil, sugar beet | n/a | 60·7 | 0·13–2·95 | 0·10 |
| Sandy soil, winter wheat | n/a | 78·6 | 0·045–0·10 | 0·10 |
| Sandy soil, bare ground | n/a | 415·6 | 0·6–8·25 | 12·01† |
| *Trier, West Germany* | | | | |
| Sandy loam, vines | n/a | 5·8 | 0·0027–0·0044 | 0·0046 |
| *Taiwan* | | | | |
| Clay loam, citrus, clean cultivation | 1268 | 580·2 | 15·64 | 10·25† |
| Clay loam, citrus, bench terracing | 344 | 580·2 | 0·50 | 4·71 |
| Clay loam, citrus with mulch | 109 | 517·9 | 0·094–0·28 | 0·22 |
| Clay loam, banana, clean cultivation | 1113–1449 | 279·7 | 3·94–6·37 | 5·40 |
| Clay loam, banana with mulch | 189 | 245·6 | 0·009 | 0·042 |
| Clay loam, banana with contour bunds | 483–1029 | 279·7 | 0·11–0·39 | 0·54 |
| *Henderson, Zimbabwe* | | | | |
| Clay, maize | 8–61 | 26·7 | 0·2–0·3 | 0·013 |
| Clay, cropped grass | 8–26 | 18·6 | 0·05–0·1 | 0·0000 |
| *Mpwapwa, Tanzania* | | | | |
| Clay loam, bare ground | 446 | 212·7 | 14·7 | 2·78 |
| Clay loam, sorghum and millet | 80–259 | 5·72 | 5·5–9·0 | 0·0007 |
| Clay loam, tufted grass | 8–65 | 2·9 | 0·0–0·07 | 0·0000 |
| Clay loam, savanna grass | 3–4 | 2·04 | 0·0 | 0·0000 |
| *Lyamungu, Tanzania* | | | | |
| Clay loam, coffee, clean cultivation | 15–232 | 28·18 | 4·3 | 0·04 |
| Clay loam, coffee with cover crops | 10–98 | 9·57 | 0·4 | 0·002 |
| Clay loam, coffee with contour ridges | 36 | 28·18 | 0·3 | 0·004 |
| Clay loam, coffee with cover crops and | | | | |
| contour ridges | 27 | 9·57 | 0·1 | 0·0001 |

TABLE 4—*continued*

| Site | Runoff, mm | | Soil loss, kg/m² | |
|---|---|---|---|---|
| | Observed | Predicted | Observed | Predicted |
| *Tuanshuangou, China* | | | | |
| Sandy soil, millet/mungbean; potato; | n/a | 278·3 | 0·1 | 7·89† |
| millet/mungbean (2 years); alfalfa | n/a | 250·3 | 43·9 | 7·59† |
| rotation (5 years) | n/a | 278·3 | 6·3 | 7·89† |
| | n/a | 278·3 | 23·4 | 7·89† |
| | n/a | 238·5 | 4·4 | 0·23 |
| Mean values | n/a | 264·7 | 13·1 | 6·29 |
| *Malaysia* | | | | |
| Sandy loam, oil palm | 263 | 294 | 0·77 | 0·29 |
| (2 years) | 657 | 757 | 0·89 | 0·76 |
| Sandy loam, bare soil | 532–642 | 827 | 2·93–3·39 | 3·73 |
| Sandy loam, groundnut | 273–328 | 273 | 0·64–1·01 | 1·04 |
| Sandy loam, maize | 365–378 | 340 | 0·56–0·81 | 1·07 |
| Sandy loam, maize with mulch | 73–80 | 298 | 0·04–0·06 | 0·06 |
| Sandy clay loam, bare soil | 688–941 | 829 | 2·44–3·92 | 3·93 |
| Sandy clay loam, groundnut | 241–388 | 266 | 0·51–0·97 | 0·99 |
| Sandy clay loam, cowpea | 260–302 | 291 | 0·59–0·61 | 1·18 |
| Clay loam, primary rain forest | n/a | 181 | 0·004–0·024 | 0·027 |
| *United Kingdom* | | | | |
| Sandy soil, bare ground | 66 | 341 | 3·9 | 8·0† |
| Sandy soil, grass | 17 | 28 | 2·3 | 0·001 |
| Sandy loam, woodland | 9 | 2 | 0·001 | 0·0000 |
| Clay, spring barley | 1 | 9 | 0·07 | 0·003 |
| Clay, winter wheat and spring barley | 6 | 5 | 0·05 | 0·001 |
| Sandy loam, oats, wheat, beans | 11 | 11 | 0·06 | 0·005 |
| Chalk, winter wheat | 5 | 5 | 0·07 | 0·002 |

*Sites where data on soil properties are based on field measurements
†Detachment-limited erosion
Sources of measured data are given in Morgan and Finney [14]

loss with an estimated rate for top soil renewal ($V$).[13] The new values of $D_s$ and $D_r$ are used as inputs to the following year of simulation. In this way the model is able to simulate positive feedback in the erosion system, where reductions in the topsoil rooting depth lead to reduced soil moisture storage and result in greater runoff and an increased rate of erosion. The process continues until first the topsoil and, finally, the subsoil disappear.

## 3. Sensitivity analysis

The extent to which predictions of soil loss by the model are affected by small changes in the values of the input parameters was assessed using partial differentiation. The results (Table 3) show that predictions are most sensitive to changes in annual rainfall and the soil properties when erosion is transport-limited and in rainfall interception and annual rainfall when erosion is detachment-limited. The most sensitive parameters need to be assessed with the greatest accuracy.

## 4. Validation

Validation of the model was carried out using published data for 67 sites in 12 countries where agricultural soil erosion has been measured on hillside erosion plots (Table 4).[14]

The model generally predicted erosion as being limited by the transport capacity of the runoff,

TABLE 5
**Input data for simulation of soil erosion with shifting cultivation in a 16-year cycle on an 8° slope**

| M | $\gamma$ | $D_r$ | $D_s$ | K | W | V | sin S | N |
|---|---|---|---|---|---|---|---|---|
| 0·26 | 1·28 | 0·15 | 1·00 | 0·30 | 0·02 | 0·19 | 0·140 | 100 |

| Year | R | $R_n$ | T | P | $E_t/E_o$ | C |
|---|---|---|---|---|---|---|
| 1 | 2029 | 181 | 25·0 | 35·0 | 0·90 | 0·002 |
| 2 | 2425 | 195 | 25·0 | 20·0 | 0·60 | 0·15 |
| 3 | 2413 | 180 | 25·0 | 20·0 | 0·60 | 0·20 |
| 4 | 2209 | 181 | 25·0 | 25·0 | 0·90 | 0·10 |
| 5 | 2161 | 183 | 25·0 | 30·0 | 0·90 | 0·05 |
| 6 | 2479 | 193 | 25·0 | 35·0 | 0·95 | 0·002 |
| 7 | 2305 | 181 | 25·0 | 35·0 | 0·95 | 0·002 |
| 8 | 2163 | 181 | 25·0 | 35·0 | 1·00 | 0·001 |
| 9 | 2152 | 171 | 25·0 | 35·0 | 1·00 | 0·001 |

Data shown for the first nine years only. Rainfall data are a synthetic sequence generated by Monte Carlo method from a log-normal distribution with the same mean and standard deviation as sample data from the study area. Selection of the parameter values is discussed in Morgan *et al.*[16]

but where runoff rates were very high erosion was detachment-limited. The soil loss predictions were worst at very low annual rates of soil erosion ($<0·1$ kg/m$^2$), when they were often an order of magnitude or more out, and at very high rates of erosion ($>20$ kg/m$^2$). Runoff predictions were often poor where mechanical soil conservation measures, such as ridging and terracing, were used or where mulching was adopted. This can be explained by the inability of the model to allow for surface depression storage of rainfall in the water phase. The soil conservation practices are accounted for in the sediment phase, however, through the values selected for the crop management factor so that soil loss predictions are often satisfactory when the runoff prediction is not.

In comparing predicted values ($Y$) with measured data ($X$), reduced major-axis lines[15,16] were fitted in preference to regression lines because of the likelihood of errors in both the predicted and measured values. The following relationships were obtained:

$$Y = 19·776 + 0·775X \text{ for runoff}, \quad r = 0·735, n = 56, \qquad ...(5)$$
$$Y = 0·472 + 0·503X \text{ for soil loss}, r = 0·583, n = 67. \qquad ...(6)$$

The lower value of the correlation coefficient ($r$) for soil loss is partly explained by the failure of the model to predict sufficiently closely two extremely high soil erosion rates in China. If these two cases are omitted, the relationship becomes

$$Y = -0·090 + 0·896X \text{ for soil loss}, r = 0·671, n = 65, \qquad ...(7)$$

where the slope of the regression line is not significantly different from unity ($P>0·05$).

Since correlation analysis gives equal weight to the differences between predicted and observed values regardless of their magnitude, an alternative validating procedure was sought which allowed much greater differences to be acceptable at very high and, in particular, at very low values of soil loss. Predictions were viewed as successful if (1) the annual predicted and observed values were both less than 0·1 kg/m$^2$, the model therefore correctly predicting that soil erosion was unlikely to be a problem, or otherwise (2) the ratio of the predicted value to either a single observed value or to the mid-point of a range of observed values was between 0·5 and 2·0. No threshold value was applied to the runoff predictions and their success was judged solely in terms of whether the ratio of predicted to observed values was in the range 0·5–2·0.

TABLE 6
Changes in topsoil rooting depth (in mm) for 100-year simulations of soil erosion

| Shifting cultivation cycle | Slope steepness, degrees | | | | |
|---|---|---|---|---|---|
| | 2 | 4 | 8 | 16 | 34 |
| 16 years | + 18·3 | + 17·6 | + 16·2 | + 13·1 | + 5·4 |
| 11 years | + 17·1 | + 16·1 | + 14·1 | + 9·4 | − 3·4 |
| 4 years | + 12·4 | + 9·4 | + 2·2 | − 26·0 | − 150·0 |

Initial topsoil rooting depth was 150 mm. For the four-year cycle on the 34° slope, the topsoil disappears in Year 73

TABLE 7
Selected output for the simulation of a four-year shifting cultivation cycle on a 34° slope

| Year | Rain, mm | Soil loss, $kg/m^2$ | Rooting depth, mm | Soil depth, mm |
|---|---|---|---|---|
| 5 | 2161 | 0·0475 | 148·2 | 997·6 |
| 9 | 2152 | 0·0839 | 146·4 | 995·3 |
| 13 | 2282 | 0·0808 | 142·0 | 990·3 |
| 17 | 2214 | 0·0791 | 140·7 | 988·5 |
| 37 | 2669 | 1·0112 | 122·1 | 967·3 |
| 41 | 2389 | 0·3579 | 115·9 | 960·5 |
| 45 | 2483 | 0·9312 | 106·4 | 950·5 |
| 49 | 2406 | 1·9267 | 93·5 | 937·2 |
| 65 | 2473 | 3·9908* | 32·0 | 873·6 |
| 69 | 2106 | 3·3986* | 15·5 | 856·6 |
| 73 | 2305 | 3·7197* | 0·0 | 840·3 |
| 89 | 2499 | 4·0328* | 0·0 | 775·5 |
| 93 | 2021 | 3·2614* | 0·0 | 759·2 |
| 97 | 2128 | 3·4341* | 0·0 | 743·0 |

Data are given for fourth year of each cycle. Values for rooting depth ($D_r$) and soil depth ($D_s$) are for the end of the year of simulation allowing for the rate of erosion, the bulk density of the soil, a topsoil renewal rate of 0·19 mm/a and a weathering rate of 0·02 mm/a. Initial values were $D_r = 150$ mm and $D_s = 1000$ mm.
*Detachment-limited erosion rate
From Morgan *et al.*[16]

Applying these criteria, the model successfully predicted runoff for 33 out of 56 test sites and soil loss for 47 out of 67 test sites, giving respective success rates of 59% and 70%. If only those sites are considered, 31 in all, where data on soil properties were obtained in the field instead of being estimated, the success rates are 57% and 90%, respectively.

## 5. Applications

The availability of data restricted the validation of the model to single years. Whilst it can be used to predict annual runoff and erosion, the design of strategies for soil conservation also demands information on trends in rates of erosion in the longer term. An attempt was therefore made to see how well the model could simulate erosion problems over time with a view to providing an understanding and indication of trends rather than details of absolute quantities.

It was decided to examine the effects of shifting cultivation on soil erosion in a tropical rainfall forest environment using data from Malaysia. Simulations were carried out for a typical area of undulating hill country to the south and west of Kuala Lumpur with shifting cultivation following traditional practice. One crop per year is obtained, usually of upland rice, with the land being cleared in April and burned in May and the crop being planted in June and harvested in December. The land is farmed for two years and then allowed to revert to fallow for 14 years, giving a 16-year cycle. The cycle is simulated by making appropriate changes to the parameter values of $I$, $P$, $E_t/E_0$ and $C$ according to the guidelines set out in Reference (3) (Table 5).

To assess the effect of population pressure on the land reducing the length of the cycle, simulations were also carried out for 11- and four-year cycles.

The results of the simulations[17] (Table 6) show that erosion is not a problem on slopes of 2° and 4°, as evidenced by the net increases in topsoil rooting depth over the 100-year period ranging from 9·4 mm to 18·3 mm. This is also the case for the 11- and 16-year cycles on an 8° slope where the topsoil rooting depth increases by 14·1 mm and 16·2 mm, respectively. The four-year cycle on the 8° slope reveals an approximate balance between the rates of erosion and soil renewal.

The four-year cycle on a 34° slope (Table 7) illustrates the worst effects of erosion. In the early years of simulation, topsoil rooting depth is reduced by between 1·3 mm and 4·4 mm over a four-year cycle, but between years 38 and 45 the rate increases to between 6·2 mm and 12·9 mm. This change in rate is associated with an increase in the number of years when soil loss is detachment-limited until it becomes detachment-limited every year. The high rate of decrease in rooting depth is maintained until all the topsoil is removed after 73 years.

## 6. Conclusion

The model predicts annual soil loss from hillslopes. The results of a validation exercise show that, except for very low and very high rates of erosion, the model gives realistic predictions over a wide range of conditions. In addition, it is reasonably efficient in terms of data input requirements, ease of computation and time. The case study of shifting cultivation in Malaysia indicates that the model provides a reasonable simulation of runoff and sediment production on a hillside over a long sequence of years in such a way that the effects of the various factors influencing the soil erosion system, including the adoption of soil conservation practices, can be readily understood and that, in general trend, accords with reality. Only the 16-year cultivation cycle is acceptable on all the slopes tested, which compares well with the minimum 10-year fallow period recommended by Hurni[18] in Thailand. All the cycles tested are acceptable up to 8° slope, implying that continuous cropping with agronomic soil conservation measures would also be possible; this fits well with the 7° slope recommended by Sheng[19] as the maximum slope steepness for the tillage without bench terracing in humid tropical areas.

## Acknowledgements

Research on the model was carried out with the aid of grants from the United Kingdom Natural Environment Research Council and the International Institute for Applied Systems Analysis, Laxenburg, Austria. The encouragement of Mr V. Svetlosanov (IIASA) is appreciated.

REFERENCES

[1] **Wischmeier, W. H.; Smith, D. D.** *Predicting rainfall erosion losses. A guide to conservation planning.* USDA Agric. Hdbk No. 537, 1978

[2] **Knisel, W. G., Jr (Ed.)** *CREAMS: a field scale model for chemicals, runoff and erosion from agricultural management systems.* USDA Conserv. Res. Rep. No. 26, 1980

[3] **Morgan, R. P. C.; Morgan, D. D. V.; Finney, H. J.** *Stability of agricultural ecosystems: documentation of a simple model for soil erosion assessment.* Int. Inst. Applied Systems Analysis, Collab. Paper No. CP-82-59, 1982

[4] **Meyer, L. D.; Wischmeier, W. H.** *Mathematical simulation of the process of soil erosion by water.* Trans. ASAE, 1969 **12** 754–758, 762

[5] **Carson, M. A.; Kirkby, M. J.** *Hillslope Form and Process.* Cambridge University Press, 1972

[6] **Withers, B.; Vipond, S.** *Irrigation: Design and Practice.* London: Batsford, 1974

[7] **Kirkby, M. J.** *Hydrological slope models: the influence of climate.* In *Geomorphology and Climate* (Derbyshire, E., Ed.). London: Wiley, 1976 247–267

[8] **Meyer, L. D.** *How rain intensity affects interill erosion.* Trans. ASAE, 1981 **24** 1472–1475

[9] **Elwell, H. A.** A soil loss estimation technique for southern Africa. In *Soil Conservation: Problems and Prospects* (Morgan, R. P. C., Ed.). Chichester: Wiley, 1981 281–292

[10] **Laflen, J. M.; Colvin, T. S.** *Effect of crop residue on soil loss from continuous row cropping.* Trans. ASAE, 1981 **24** 605–609

[11] **Mou, J.; Xiong, G.** *Prediction of sediment yield and evaluation of silt detention by measures of soil conservation in small watersheds of north Shaanxi.* Preprint, Int. Symp. River Sedimentation, Beijing, 1980 [in Chinese with English summary]

[12] **Morgan, R. P. C.; Hatch, T.; Sulaiman, W.** *A simple procedure for assessing soil erosion risk: a case study for Malaysia.* Z. Geomorph. Suppl. 1982 **44** 69–89

[13] **McCormack, D. E.; Young, K. K.** Technical and societal implications of soil loss tolerance. In *Soil Conservation: Problems and Prospects* (Morgan, R. P. C., Ed.). Chichester: Wiley, 1981 365–376

[14] **Morgan, R. P. C.; Finney, H. J.** *Stability of agricultural ecosystems: validation of a simple model for soil erosion assessment.* Int. Inst. Applied Systems Analysis, Collab. Paper No. CP-82-76, 1982

[15] **Kermack, K. A.; Haldane, J. B. S.** *Organic correlation and allometry.* Biometrika, 1950 **37** 30–41

[16] **Till, R.** *The use of linear regression in geomorphology.* Area, 1973 **5** 303–308

[17] **Morgan, R. P. C.; Finney, H. J.; Morgan, D. D. V.** *Stability of agricultural ecosystems: application of a simple model for soil erosion assessment.* Int. Inst. Applied Systems Analysis, Collab. Paper No. CP-82-90, 1982

[18] **Hurni, H.** *Soil erosion in Huai Thung Choa, northern Thailand: concerns and constraints.* United Nations University/University of Bern, Highland–Lowland Interactive Systems Project, Scient. Rep., 1981

[19] **Sheng, T. C.** *A treatment-oriented land capability classification scheme for hilly marginal lands in the humid tropics.* J. Scient. Res. Coun. Jamaica, 1972 **3** 93–112

Part V

# SOCIAL AND ECONOMIC ASPECTS

# Editor's Comments
# on Papers 18, 19, and 20

**18**  **HUDSON**
*Non Technical Constraints on Soil Conservation*

**19**  **TEMPLE**
*Soil and Water Conservation Policies in the Uluguru Mountains, Tanzania*

**20**  **RUNOLFSSON**
*Soil Conservation in Iceland*

Compared with the treatment afforded to soil erosion processes and techniques of soil conservation in numerous standard texts since the 1940s, little attention has been paid to the social, economic, and political factors that act as constraints on the implementation of erosion control measures. While continuing to recognize the importance of technical research, Hudson (Paper 18) presents a systematic review of the cultural environment stressing those aspects that often determine why well-intentioned and technologically sound conservation schemes fail.

The basis of successful soil conservation is wise use of the land. It is important first to distinguish between areas where erosion rates are naturally high and those where the rates are high because the erosion is man-induced. Second, for the latter case, a distinction must be made between erosion associated with agriculture on land that is fundamentally unsuitable for that purpose and erosion associated with agriculture on land that is basically suitable. Only in this last instance where erosion is the result of poor management of the agricultural system can soil conservation measures be applied with any expectation of success. Where erosion is related to misuse of the land, there is little that conservation technology can offer to improve the position. Thus, as stated by Hudson, encouraging people to move into steeply sloping and mountainous land in the tropics and expecting them to settle and grow food is a recipe for social and ecological disaster. Since the land is marginal, investing large sums of money in an attempt to reduce erosion will produce an economic disaster as well. High pressure of population on the land, which forces people to migrate into these marginal areas, is a social problem and should be dealt with as such. Soil conservation should not and cannot be used as a surrogate for a lack of social policy.

Although it is generally desirable to conserve ecologically sensitive areas by restricting access and giving them the status of reserved land, this approach can create pressure on the unreserved areas. Temple, in his study of the Uluguru Mountains (Paper 19), describes how the German administration, through an enactment in 1909, forced people onto unreserved land which then came under such pressure that the traditional system of shifting cultivation broke down. The result was a loss of soil fertility and an increase in soil erosion. As erosion, in turn, causes land to be abandoned, even greater pressure is placed on the land remaining in use which becomes an ever-diminishing resource. According to Runolfsson in Paper 20, this is basically the history of Icelandic agriculture.

Iceland provides a case study that illustrates many of the points made in Hudson's paper. As in most areas of the world, soil erosion in Iceland dates from the time of settlement. According to historical documents, when this occurred in the period 875 to 930 A.D., much of the country was covered with low-growing birch woodland and a dense understory of forbs and grasses. This probably occupied about 39% of the land area (Thorgilsson, c. 1100; Einarsson, 1962; Fridriksson, 1972). Today only about 20% of the country is covered with soil and vegetation. The loss of half of the original plant cover in just 11,000 years is attributable to climatic changes, particularly the effects of cold periods between 1740 and 1840 and between 1880 and 1890; ashfalls from volcanic eruptions, especially those of Mount Hekla; wood cutting for fuel, which has resulted in a 95% reduction in the area under trees so that they now occupy only 1% of the total land area; overgrazing, particularly by sheep which have increased in numbers from 450,000 in 1947 to 800,000 today; and highly erodible soils derived from volcanic ash, glacial outwash, and loess. By 1890 it was clear that wind erosion was threatening the existence of most farms in the south-central part of the country and this provided the stimulus for action. In Paper 20 Runolfsson explains how, against this background, the State Soil Conservation Service (Landgraedsla Ríkisins) was formed, initially starting life in 1907 as the Sand Reclamation Service.

Successful soil conservation requires the development of appropirate technology. Runolfsson describes the history of the research and trials on the techniques of sand reclamation and outlines the methods in use today. A major problem is that reclamation is expensive because most of the grass seeds have to be imported. Even the varieties of grasses of proven value in stabilizing and revegetating dunes do not produce a reservoir of local seed because of the severe climate. Research is therefore in progress on using *Lupinus nootkatensis,* which seems well adapted to Icelandic conditions and may provide a good supply of seed (Arnalds, 1980). The Icelandic experience is a good example of developing local technology and contrasts with the view that, according to both Hudson and Temple, frequently prevails elsewhere and that emphasizes the direct transfer of North American technology. This view is epitomized by the statement made with reference to the Uluguru, "What was good enough for the Ohio Valley was good enough for the Ruvu."

Field trials of soil conservation techniques need to be supported by basic research. For example, it was recognized (Ólafsson, 1973) that the cause of the deterioration of the vegetation cover and subsequent soil erosion in Iceland was largely the result of a lack of knowledge of the range resources and their carrying capacity. Accordingly, the Agricultural Research Institute (Rannsóknastofnun Landbúnadarins), which works very closely with the State Soil Conservation Service, initiated a detailed vegetation survey of the country (Gudbergsson, 1980; Steindórsson, 1980). Through studies of the botanical composition of the plant communities and a knowledge of the palatibility of each plant species for sheep, the carrying capacity can be estimated (Thorsteinsson, 1980a; 1980b; 1980c), assuming that no more than 50% of the annual growth of the most palatable species is removed by grazing. These studies have confirmed the importance of the upland summer grazings. Although the dry matter production of the lowland pastures is greater than those of the uplands, sheep grazed on the latter have higher growth rates (Gudmunsson, 1980), a fact known by the local farmers for some considerable time (Sturluson, c. 1230).

Unless the erosion control measures proposed by a soil conservation service or similar body are based on sound research, local trials, and field demonstrations of their advantages to farmers, they are unlikely to be taken up. In Temple's view the attempts to introduce bench terracing in the Uluguru Mountains showed a disregard for these matters. The measures were ill conceived, based on inaccurate data and subjective assessments of the problem. It was assumed that sheet erosion was the process that needed to be controlled, whereas more recent research has indicated the importance of landslides. The labor costs of terrace installation were ignored, as were the effects on crop yield where the terraces resulted in exposure of the subsoil. When designing soil conservation strategies the local knowledge of the farmers was neglected even though they understood their environment better than the ill-trained field officers. Thus the role that the traditional practices of ladder terraces, intercropping, and tree planting might play was never considered. The soil conservation program had no planned objectives and no adequate extension backup. No attempt was made to demonstrate the value of bench terracing.

Hudson argues that implementing soil conservation through legislation does not work and the case study of the Uluguru Mountains presented by Temple supports this. Indeed, Temple shows how coercion and legislation can be counterproductive, resulting in all soil conservation work being discredited and thereby making it more difficult for future governments to initiate alternative schemes. On achieving independence, the Tanzanian government was forced into a policy of noninterference because it was not politically expedient to relate to policies with a tarnished reputation, policies that were also associated with the former colonial administration. Unfortunately, non-intervention has been as disastrous as coercion and there has been a massive increase in soil erosion.

That legislation only works when community support for soil conservation is so strong that it is not needed is illustrated by the Icelandic experience. The State Soil Conservation Service has the power to take over any land and protect it for as long a period as necessary to reclaim it and make it fit for controlled grazing. In practice, compulsory acquisition has never been used. The farmers voluntarily make a legal contract with the Service which holds the land for up to 30 years and then gives the farmer first option on taking it back into private use. Through this policy the Service has fenced and reclaimed about 2% of the country.

Hudson stresses that most farmers have a very good appreciation of their land and its value. They are generally experienced and efficient managers and respond well to new ideas that they can identify with. Runolfsson describes how the Icelandic farmers changed their attitudes from skepticism to willing acceptance when the saw that land could be successfully reclaimed. The Gunnarsholt experiments where 1300 ha of land were turned from bare, moving sands and gravels in 1948 to pasture for hay and silage production today were a clear demonstration to farmers that soil conservation and land reclamation could work.

The commitment to soil conservation in Iceland is long-term, and the Service is provided by the government. There is also a community commitment—one example of this being the free assistance provided by pilots of Iceland Air to help the Service with aerial seeding during the summer months. The Icelandic system is therefore in line with the principles proposed by Hudson, which acknowledge that some form of financial assistance to the farmer is both required and justified for soil conservation work. This is because the benefits of such work extend beyond the individual farmer. They affect neighboring farmers, particularly those downstream, and—in terms of the ability of the agricultural sector of the population to produce food—the whole community. Within this wider context, Hudson cites several studies showing that soil conservation programs can be worthwhile investments on the basis of cost-benefit analysis. Unfortunately, the Icelandic program is at present hampered by the failure of the government to develop a policy that will reduce the number of sheep. Until this is done, much of the reclamation work will have limited impact.

The papers reviewed here show that farmers will take up soil conservation measures given the right conditions. These are that the measures must be well researched, properly demonstrated, and implemented through a well-trained and enthusiastic extension service. In addition, suitable financial incentives must be made available. Unfortunately, the provision of monetary incentives by the state within a basically captitalist society is politically unacceptable in many countries.

The challenge is not simply better technology. Nor is it merely overcoming the nontechnical constraints. It is in devising strategies that are compatible with the farming system and relevant to the erosion system. In making this match, a much greater understanding of and sensitivity to the cultural environment is required along with a proper analysis of erosion to identify the critical processes. These elements are essential if mistakes are not to be repeated and the foundation is to be laid for successful soil conservation practice.

## REFERENCES

Arnalds, A., 1980, Lúpínurannsóknir, *Rannsóknastofnun Landbúnaðarins Report No. 59.*

Einarsson, Th., 1962, Vitnisburður frjógreiningar um gróður, veðurfar og landnám á Íslandi, *Saga* **1962**:442–469.

Friðriksson, S., 1972, Grass and grass utilization in Iceland, *Ecology* **53**:785–796.

Guðbergsson, G. M., 1980, Gróðurkortagerð, *Íslenzkar Landbúnaðar Rannsóknir* **12**(2):59–83.

Guðmunsson, Ó., 1980, Approaches, materials and methods, in *Consultancy Reports for the Project on Utilization and Conservation of Grassland Resources in Iceland,* Guðmunsson, Ó., and A. Arnalds, (eds.), *Rannsóknastofnun Landbúnaðarins Report No.* 61: 3–33.

Ólafsson, G., 1973, Nutritional studies of range plants in Iceland. I. Range problems in Iceland, *Íslenzkar Landbúnaðar Rannsóknir* **5**(1-2):3–8.

Steindórsson, S., 1980, Flokkun gróðurs í gróðurfélög, *Ísenzkar Landbúnaðar Rannsóknir* **12**(2):11–52.

Sturluson, Snorri., c. 1230, *Egils saga Skalla-Grímssonar,* English transl., H. Pálsson, and P. Edwards, 1976, Egil's Saga, Penguin, Harmondsworth, p. 76.

Thorgilsson, Ari fróđi., c. 1100. *Íslendingabók*

Thorsteinsson, I., 1980a, Gróđurskilyrdi, gróđurfar, uppskera gróđurlenda og plöntuval búfjár, *Íslenzkar Landbúnađar Rannsóknir* **12**(2):85–99.

Thorsteinsson, I., 1980b, Nýting úthaga-beitarthungi, *Íslenzkar Landbúnađar Rannsóknir* **12**(2):113–122.

Thorsteinsson, I., 1980c, Beitagildi gróđurlenda, *Íslenzkar Landbúnađar Rannsóknir* **12**(2):123–125.

# 18

Reprinted from *South-East Asian Regional Symposium on Problems of Soil Erosion
and Sedimentation,* T. Tingsanchali and H. Eggers, eds., Asian Institute of
Technology, Bangkok, 1981, pp. 15-26.

## NON TECHNICAL CONSTRAINTS ON SOIL CONSERVATION

NORMAN W. HUDSON
*Professor of Field Engineering*
*National College of Agricultural Engineering*
*Silsoe, Bedford*
*England*

ABSTRACT

The theme of this paper is that the knowledge of soil conservation
techniques should be improved, but the most pressing need is to put into
practice what is already known. Past experience of implementing soil
conservation programmes has been disappointing, and the paper tries to
identify the more important reasons.

Under Political Aspects, there is a discussion of the lack of political
will at several different levels to implement policies of wise land-use. The
use of legislation to regulate and control land use is discussed but
considered to be impractical.

Among the social difficulties discussed are land ownership, land tenure,
the fragmentation of land holdings, and uneven population distributions. The
significance of livestock is discussed, and the reasons are analysed for the
apparent reluctance of the peasant farmer to adopt new farming methods.

Economic constraints include the inability of the peasant farmer to take
risks, and the difficulty of balancing the long-term benefits of soil conservation against the costs of construction. The allocation of costs is
discussed, and the conclusion is that the practice in developed countries of
the landowner bearing a portion of the cost is not applicable in developing
countries.

## 1. INTRODUCTION

The hypothesis presented in this paper is that there is much to be
learnt about developing soil conservation techniques which are applicable to
South-East Asia, but there is an even greater challenge in coming to terms
with the non-technical constraints. Practical soil conservation methods have
developed historically through several phases. During the first half of this
century, the emphasis was on the traditional North-American style of

conservation, mainly manipulation of the land to control surface run-off by
the use of graded channel terraces. During the 1950's, the identification
by Ellison and his colleagues of rainsplash as a major component in the
erosion process, led to a decade of development of agronomic conservation
practices using rotations, mulches, and cover. By the sixties a successful
conservation package had been developed in the US, and we rather tended to
assume that all we had to do in the Third World was to modify it slightly
for local conditions before putting it into practice. In the 1970's research
in many countries, and at the international centers IITA and ICRISAT, led to
the realisation that minor modification is not sufficient. There has to be a
new approach, and this has now led to a new surge of effort to develop solutions appropriate to the very different conditions in developing countries in
the tropics.

But the development of technical solutions is only one part of solving
the soil erosion problem. It is also necessary to apply the solutions on the
ground. There have been several approaches to achieving this objective:

1.   The concept that soil conservation is such a "good thing" that it is
only necessary to show the people what to do and they will start to do it.
In this scheme, once the technical knowledge is available all that is
required is education and extension. This approach has been actively
practised in the US for several decades using the full armoury of extension
methods, such as grass-roots Soil Conservation Districts, youth involvement
through 4H Clubs and Young Farmers Clubs,  cooperative programs, and saturation extension programs.

2.   Another approach is to use financial incentives to encourage the
implementation of conservation works by the government paying a substantial
part of the cost of the works and the rest being a charge to the landowner.

3.   The third possibility is to use coercion. This may range from low-key
pressure, a kind of moral blackmail to make people feel bad if they do not
conform, to the outright enforcement of legislation.

Various combinations of these three elements are possible. The US has
been for three years debating the right mix in connexion with the Soil and
Water Resources Conservation Act of 1977 referred to as RCA. (Cook, 1980).
In a useful contribution to this discussion Timmons (1980) refers to the
"green ticket strategy" which is a combination of the first two approaches,
and the "red ticket strategy" which is the third alternative with an element
of financial assistance as a sweetener.

What seems to me to be a most significant development is that in the
seventies we first had the realisation that new technical solutions are
required, and now we are seeing a parallel realisation that putting soil
conservation programs into operation is extremely difficult, and that
programs are seldom as effective as their authors expect them to be. This
is not entirely new:  there have been several realistic appraisals of the
soil conservation movement in the US which have showed that success does not
automatically follow good intentions. Two examples of these early reviews

are Held & Clawson (1965), and Simms (1970). More recent studies suggest that the concept of good stewardship of the soil is wearing thin in some parts of the US in spite of the soil conservation program. Timmons (1980) says that the Iowa Department of Soil Conservation report that at the present time "only one third of Iowa's cropland is adequately treated with erosion control measures". In South-East Asia soil conservation programs are much less intensive than those of the US, and so it is not surprising that it is possible to identify many cases of well-planned and well-intentioned conservation measures being a great disappointment. Taking just one example, the "Greening Programme" in Indonesia is well conceived and planned, and has strong political and financial support but has nevertheless been disappointing in operation. At the International Conference on Soil Conservation at NCAE in July 1980 it became apparent that the problem of implementation of soil conservation programmes is widespread. The problem usually arises from a complicated interaction between a large number of factors. For convenience I will consider these factors under the three main categories of political aspects, social difficulties, and economic problems.

## 2. POLITICAL ASPECTS

### 2.1 Political policy

Most politicians and most political parties pay lip service to the ideal of good husbandry and the conservation of natural resources but, in practice, soil conservation does not win votes. Government policies are not translated into action programs unless there is the political will to make them work, so the situation in many countries today is that plans are made for the conservation of natural resources but they have little practical effect.

At the level of the individual politician, in both developed and undeveloped countries it is almost expected that an elected representative will attempt to divert funds into his own constituency - frequently disrupting the planned priorities by doing this, and often diverting resources from long-term conservation works into measures like roads and schools which have a more immediate impact on the voters.

At the level of National policies there can also be a conflict between the short-term objective of self-sufficiency and the long-term maintenance of the country's land resource base. The understandable desire of national governments to increase food production can lead to excessive pressure on the land. (Dudal 1980).

At the International level too there is room for improvement. The present situation is analysed in the report of the Brandt Commission (1980 Chapter 5) which calls for a much more realistic dialogue between the North and the South.

## 2.2 State land and state forests

Many developing countries have a historical legacy of land reserved for the use of some agency of power. In territories previously under British Colonial rule all the land not otherwise allocated was "Crown Land". Other Colonial regimes had similar systems. In Sovereign states and feudal systems the land automatically belonged to the ruler, except where some form of title was granted in return for services rendered. Whatever the historical background, many developing countries have a sizeable proportion of land which was previously reserved.

As authoritarian management has declined, population pressure and land hunger have increased. So has the chance of getting away with unauthorised or illegal encroachment on reserved land. There are many examples where encroachment has been tolerated because it was politically expedient.

The point is that the restriction or reservation of land was often to preserve the income or power of the ruling elite, but there are also many examples where the land was deliberately witheld from settlement because it was ecologically unsuitable. This is particularly true of steep mountainous land in the humid tropics, and terrifying destruction is occuring as the result of the spread of small-scale peasant farming into land which cannot possibly support this use in the long-term. Examples are the steady climb of cultivation up the mountain slopes in Central Java, in Kenya in the Machakos and Aberdare districts, in the lower Himalayas of India and Nepal, in the Andes in South America.

Some political leaders neither know nor care what is happening: some know but do not care: some even know and worry about it: but none are going to do anything about it. Even if the political will were there, it is not clear what technical solution could be offered to this problem.

## 2.3 Land allocation

A corollary to tolerating encroachment is the deliberate allocation of land to non-landowners. In India the declared policy is "land for the landless" and by government legislation 50% of all state or public land must be made available for allocation to people who do not own any land. This policy is neither realistic nor practical in a country like India which has been intensively farmed and densely settled for centuries and has no reserve of spare land suitable for cultivation. The common land is usually worn out, or eroded, or so steep that it is only suitable for forestry. But recently I saw examples of allocations of 1 ha of land on a $40^{\circ}$ slope with a soil depth of less than 5 cm. An Indian colleague summed it up - "They ask for land for food, and we give them stones to lick". Whether practical or not this policy of land allocation will almost certainly continue. It is not itself a political issue, but it would be political suicide for any party to suggest changing it.

India is by no means an isolated case. The political desirability of breaking up large land holdings into small units is widespread. Iran had a major land reallocation policy nearly 20 years ago, Kenya has since independence bought many of the large 'white-settler' estates for resale in small parcels, and in Sri Lanka the tea estates were first nationalised and then later subdivided. There are some cases where reallocation of this kind has led to improved productivity, but there are more cases where the total production is lower after the reallocation.

The Brandt report points out that in what they call the Asian poverty belt reallocation is not likely to solve the problems of the rural poor, and quotes the case of Bangladesh where "land reform can only provide small relief for these people, since large holdings account for only 0.2% of the total land" (Brandt, 1980, p 86).

## 2.4 Legislation

The question is "should government try to enforce its policies for land use by legislation?" It is a tempting thought. After all, surely the experts, the technocrats, the management, must know best what is best for the country. The record on legislation was reviewed some years ago by FAO (1971) and more recently in a series of articles "Soil Conservation - Persuasion or Regulation?" (Soil Conservation Society of America 1974).

My personal opinion is that legislation cannot provide the solution for two reasons. First, it is morally and ethically wrong to force on the populace theories and practices whose validity has not been proved (this theme is forcefully explored by Shaxson, 1980a) but secondly and more importantly, the fact is that legislation does not work.

In the old days of colonialism and imperialism the approach "Daddy knows what's best for you" worked to some extent. In British territories in Africa, a great deal of practical field conservation work was carried out, two examples being Kenya and Malawi. With independence two possibilities arose. One was that the newly-independent government could discard conservation control as part of the process of throwing off colonial impositions. Alternatively the government might say "now that the land is really ours we will look after it", and increase the conservation effort. In practice, neither has happened - the concept of regulation and control remains, but it has no truth.

My conclusion is that asking any reasonably democratic country to pass legislation to control land use is rather like going to the bank and asking for credit. You can only get it when you have so many assets that you do not really need it. Some countries have quite sweeping powers to control land use and enforce soil conservation practices, but in every case it has been possible to pass the legislation because there is a widespread acceptance that the misuse of land is an offence against society and socially unacceptable. When this approach to land use has been accepted by the community as a whole it is not necessary to invoke the legislation because incurring the displeasure of the rest of the community is a more effective regulator of behaviour.

Soil conservation programmes can only be effective when they are "moved from below i.e. by full involvement of the rural population" (Dudal, 1980). Programmes which are imposed "from above" will not succeed even if they are technically correct.

## 3.  SOCIAL FEATURES

### 3.1  Land ownership and land tenure

Among the factors which lead to an overstressing of land resources, we must put high on the list the cultural ethic that everyone has an automatic right to own land.  This is not spelled out with the other jargon phrases like "the right to freedom of speech", or "the right to freedom of worship", but it is there at the heart of land management in most developing countries, and with some justification.  Partly it is historical;  in the past there was enough land for everyone to have some, and an increase in population just meant bringing more land into use.  It is also partly because in dominantly agricultural economies there are no alternatives to working on the land.

It is easy for us in the North to say that the ethic is no longer tenable as an increasing population seeks a share of a static or dwindling land resource.  But then it was easy for us in the North to stop being farmers because there were other jobs to do.  But it is very different when there is no other employment for those who cannot be farmers.  The Brandt Commission concludes that only the expansion of industry can ease the problems of those who have no opportunities on the land.

The situation might be temporarily eased if those who take up non-farming work transferred their land to others who stay in farming.  At the moment they seldom do.  The University lecturer or civil servant or trader in the developing countries nearly always retains his land - perhaps to retire to, or to give him security which is lacking in his job, or often just because he cannot bear the thought of parting with it.  Even so the number of people who have this option is so small that even if they all gave up or sold all of their holdings it would hardly affect the overall position.

Undesirable pressure on the land can also come from some forms of land tenure.  Communal ownership can lead to mismanagement, particularly over-grazing by cattle, or the over-enthusiastic removal of firewood.  Sometimes the existing system is difficult to change because there are vested interests at stake such as the Chief having special rights.  Where this is not a problem it should be possible to move the land management towards a system which will bring greater rewards to the community as a whole.  Shaxson (1980a) quotes a remarkable case of this in Central India.  After many years of patient groundwork by extension workers, the village decided to rest and restore their main hill-grazing area and operated a self-imposed and self-regulated programme of zero grazing and intensive planting of fuelwood trees.  If this example continues to work it will be a classical demonstration of how this problem can be overcome.

Other aspects of land tenure which can lead to soil depletion are short-term cultivation rights, share-cropping tenancies, and absentee land-lords. These need no elaboration here, but it is worth pointing out that it should not be assumed that these problems only arise in developing countries. They are not usually associated with farming in the American mid-West but Timmons (1980) includes tenant farming among the significant causes of poor land use in Iowa.

Alongside the questions arising from ownership and tenure, there are other historical rights which do not fit into today's land use pattern. Throughout India for example there are entrenched rights of villagers to graze certain areas, to cut fuelwood, and to take timber for making agricult-ural implements. The story of the rights is so ancient and so complex that Muslim, Moghul, and British rulers have in turn accepted the situation as too difficult to change, and so has the present-day government. The problem is that these rights make it extremely difficult to manage efficiently even the so-called "reserved forests".

## 3.2 Fragmentation

Another consequence of the concept of a universal right to own land is that it leads to the progressive subdividion of land holdings. Extreme examples of this are seen in Indian villages (Shaxson 1980a). In some States this has led to the "consolidation programme" which in effect puts all the land into a common pool and redistributes it in more manageable parcels. A similar consolidation programme was applied some years ago in Malawi, but successful and long-lasting consolidation schemes are much less common than examples of the wish or intention to implement such schemes. Again, the problem of fragmentation is by no means confined to the developing world and we also have examples from Europe and North America.

There are some self-regulating mechanisms which do something to lessen the problem. One is that when it becomes impractical to further divide a field, it may remain under the management of one of the brothers who claim part of it. But he still has to share the produce with the others, so it does not alter the fact that an increasing number of people are drawing on the same land. In India another regulator is that although the law has for some years decreed that the father's estate should now be shared among the daughters as well as the sons, this is seldom applied to land. And for a very sensible reason. The sons usually stay in their village, so their holdings are fragmented but within the village. The daughters marry outside the village so if they were to take land rights with them then after only two generations a man could own parcels of land in four different villages.

However, neither these modifying influences, nor the changes in the laws of inheritance are likely to alter the basic fact that fragmentation is a serious constraint on optimal land use.

## 3.3 The social significance of cattle

The part played by livestock in adding to pressures on the land resource base varies a great deal from one country to another. In many parts

of Africa and India it is an important part of the problem:  in the humid
tropics it may be less severe.  In Africa, cattle are symbols of status, and
evidence of wealth.  In India both of these aspects apply and there is the
added complication of the religious significance of the cow.  All these
social factors lead towards quantity rather than quality, and more livestock
than are necessary or desirable.  Associated problems are low standards of
livestock management and low levels of production.  The total effect of all
these combined factors is a high, and unnecessarily high, stress on the
ecological system.

## 3.4  Reluctance to move

It is a fact of life that humans are gregarious and do not like to move
from their accustomed environment.  There are cases where large scale resettlement has been effective - usually either in irrigation schemes (The
Ghezira) or in relatively high technology farming (the mid-west in America).
However, it does not appear to be a starter as a solution to the problem of
over-development of mountainous regions of the humid tropics.

If it could work anywhere it should be in Indonesia.  Java is desperately crowded, and erosion in the mountains is very serious, but the outer
islands have vast areas of undeveloped land, much of it with good soil and
good rainfall.  The Transmigration Plan aimed to achieve a massive transfer
of population but has been very disappointingly ineffective.

Another example is Venezuela.  The Andes range is steep and erodible,
but the population increases, and the cultivation climbs higher and higher
while there are millions of hectares of unused land in the southern plains.
Sri Lanka is a further example.  Even a surplus of irrigated land in the
plains is not easing the increasing pressure on the steep hill areas.

India has the worst problem of all.  The over-crowded, over-grazed,
over-cultivated foothills of the Himalayas are disintegrating so fast that
the sediment load of the main rivers is higher by an order of magnitude than
comparable rivers on other continents.  But in this case there is no room
for manoeuvre and no opportunity for resettlement, for the plains are already
settled to capacity.  It is difficult to see any practical solution to this
problem.  People do not like to move, and I see no way to persuade, cajole,
or force them to do so.

## 3.5  Reluctance to change

It is commonplace to refer to the conservation of the peasant farmer,
or to his inertia, and the difficulty of changing existing patterns.  There
is no doubt that change occurs infrequently and slowly, but I wonder if this
is attributed to the real reasons.  The subsistence farmer is too often
thought of as uneducated, ignorant, and doing what he does because he does
not know any better.  The corollary to this assumption is that it is only
necessary for the expert to tell him about hybrid varieties of crop, or
fertiliser, or new implements, and his problems will be solved overnight.
But as we know from experience it does not work out like that.  My belief is

that the small-scale peasant farmer is usually an experienced and efficient practitioner of his craft - but we forget that this craft is not the business of farming as we know it in the North, but rather the art of survival. He is not unwilling to change, but he is locked into an economic prison from which he is unable to escape - and this leads to the next constraint.

## 4.   ECONOMIC CONSTRAINTS

### 4.1  The element of risk

The economic prison holding the subsistence farmer is his inability to take risks. The essence of farming is trying to improve the odds in the gamble against weather, pest, and disease. The crunch is that the peasant has no risk capital to gamble with, so his whole strategy is geared to safety. He would rather use a low-yielding variety which gives some yield every year than an improved variety which will give an increased yield most years, but none at all in the bad year. Even if the chance of an increased yield is nine years out of ten, it is still not an acceptable gamble for the small subsistence farmer. In the tenth year, the year of failure, his family will starve. It is neither stupidity nor lethargy when he sticks to his old variety, it is accepting the realities of life. Fortunately there is a solution to this dilemma, and one which is being tentatively tried out in some development projects. This is to inject money into the production system in the form of risk capital. In the past the practice has tended to be to inject money by subsidising the price of seed or fertiliser or by offering support prices, but all these only help the man who already has some capital of his own to invest. Instead the money should be used to underwrite the risk of failure. When a Department of Agriculture promotes a new variety with promises of improved yield four years out of five let it also promise to give grain free in the fifth year when the improved variety fails. Only by this kind of approach will the subsistence farmer be able to change.

### 4.2  The time scale of soil conservation

In order to measure the benefit of conserving the soil, one has to use a long time scale. What is the cash value today of conserving the land so that our grandchildren may also use it? Considering the land resource as a national asset the development, use, or exploitation of the resource should be viewed in the context of something like 50 or 100 years. It is not necessary to look for infinite use because at some time in the future it will be possible to manufacture synthetic foods in the same way that synthetic clothing is manufactured today.

However, the managers of our national land resources are also our political leaders, and their time-scale seldom extends beyond the date of the next election, so on the whole they are not interested in long-term conservation. The farmer's economic cycle is even shorter. He is probably working on cashflows over 12 months, so it is unreasonable to expect him to pay now for preserving the land for posterity. That is a luxury he cannot afford. Two quotations from Dudal (1980) are appropriate. "The first requirement for effective soil conservation is that the income from the farm

is large enough to provide a sufficient proportion for the maintenance of
the soil capital". Obviously for a very large number of peasant farmers the
immediate problems of today override any consideration of the future.

### 4.3  Who benefits? and who should pay?

Apart from the on-site benefits of erosion control, there may be other
reasons why erosion control is required. The down-stream damage caused by
sedimentation - increased flooding, reduced hydro-electric power, interference with irrigation - may be more important than the loss of soil. This
is certainly the case in some parts of Northern India where the economic
consequences of siltation are extremely serious, but the main source of
sediment is "moonscape land" of negligible agricultural value.

The peasant farmer in the Himalayas cannot reasonably be expected to
care very much about the silting of an irrigation reservoir one thousand
kilometres away in the Gangetic Plain, and he certainly should not be
expected to pay for the required remedial measures. If sedimentation is
increasing the cost of hydro-electric power and irrigation water, then the
increased cost, or the cost of corrective measures, should be passed on to
the consumer or borne by the State. The peasant farmer in the hills cannot
afford to do anything about it.

However, the money to meet the cost of conservation works must come from
somewhere, and the question is how to distribute the costs fairly. My
suggestion is that the division should be very different in different
countries. In a wealthy developed country such as the US it is reasonable to
adopt the approach suggested by Libby (1980) who suggests that while there
should continue to be an element of government support there should be a
shift towards more responsibility by the landowners or land users.

In countries where the farmers are generally poor their ability to contribute towards the cost is less or non-existent. In this case the answer
has to be that since it is the long-term interest of the State, nation, or
community which requires that the soil should be conserved, then it is the
State which should pay. This does not necessarily have to be in the form of
an uncontrolled handout. Injection of money through subsidies on works which
are carried out by the landowner achieves simultaneously the objectives of
injecting money into the agricultural economy and also ensuring that the
farmer retains a personal interest in the works. This is often a problem in
the case of works which are constructed by the government at government
expense.

When considering the economics of soil conservation a constantly
recurring theme is the difficulty of establishing the costs and the benefits.
Programmes which involve the expenditure of funds whether national, bi-
lateral, or international have to decide how to use the available funds to
the best advantage. But it is not easy to decide on relative priorities
solely in economic terms nor is it simple to convert all the benefits into
cash values.

It is fairly easy to calculate costs of soil conservation: more difficult to determine the benefits: even more difficult to quantify the benefits in cash terms. With some highly subjective guesswork one can make an estimate and put cash values on increased production. With more dubious calculations one can compute a value for not losing the soil. Secondary benefits like job creation and injecting money into the economy can be included, but there will always be at the end some intangible benefits which cannot be measured in cash values. There is no way of putting a cash value on the improved quality of life which comes from better health, a more balanced diet, or reduced drudgery among the rural population.

I think it is fair to say that the necessity for a much greater effort in studying this problem is now generally accepted (e.g. Marsh 1980), and there is now a good base for future research. Seckler has produced useful studies from India (1977) and for Lesotho (Nobe and Seckler 1979). The US SCS has undertaken some serious heart-searching about how much erosion should be tolerated (Mannering 1980) and a model for calculating sensible tolerance losses has been produced on the basis of his work in India by Shaxson (1980b). As a result of studies of erosion in El Salvador, Wiggins (1980) has produced an interesting model which suggests that at least in those circumstances a soil conservation programme can be justified in straightforward economic terms over a relatively short time scale.

Clearly a great deal more work requires to be done on this subject because the decision makers need to have more reliable evidence on which to make their judgements, but progress is interesting and encouraging.

## 5. SUMMARY AND CONCLUSIONS

I have tried to point out some of the difficulties which hinder the adoption of soil conservation measures. The division into social, political, and economic problems is rather arbitrary, but the three aspects are strongly interactive so it does not matter which heading we discuss them under. What is important is that we should recognise these constraints and try to understand them. Then we can start looking for solutions, because in few cases do I see a solution available now. The hypothesis which I started with is worth repeating. There are undoubtedly gaps in our technical knowledge, and many other cases where the existing technology needs to be improved, but I believe that the most immediate task should be to get into operation the knowledge we already have, and we should therefore address ourselves to the problem of understanding and overcoming the non-technical constraints which at the moment inhibit the adoption of soil conservation practices.

## REFERENCES

BENBROOK C (1978). Integrating soil conservation and commodity programs. Journal of Soil and Water Conservation. Vol. No. 34 1979.

BRANDT W (1980). North-South: A programme for survival. London. Pan Books.

COOK K A (1980). On the horizon for RCA. Journal of Soil and Water Conservation. Vol. No. 35 1980 pp 51-53.

DUDAL R (1980). An evaluation of conservation needs. In Soil Conservation: Problems and Prospects. Ed. R P C Morgan. Chichester. Wiley.

FAO (1971). Legislative principles of soil conservation. Rome. FAO Soils Bulletin No. 15.

HELD R B & CLAWSON M (1965). Soil conservation in perspective. Princeton. Johns Hopkins.

HUDSON N W (1980). Social, political and economic aspects of soil conservation. In Soil Conservation: Problems and Prospects. Ed. R P C Morgan. Chichester. Wiley.

LIBBY L W (1980). Who should pay for soil conservation? (Guest Editorial). Journal of Soil and Water Conservation. Vol. No. 35 1980 pp 155-157.

MANNERING J V (1980). The use of soil loss tolerance as a strategy for soil conservation. In Soil Conservation: Problems and Prospects. Ed. R P C Morgan. Chichester. Wiley.

MARSH B a'B (1980). Economics of Soil Loss: A Top Priority Research Need. In Soil Conservation: Problems and Prospects. Ed. R P C Morgan. Chichester. Wiley.

NOBE K C & SECKLER D W (1979). An economic and policy analysis of soil-water problems and conservation programs in the Kingdom of Lesotho. Lesotho Agricultural Sector Analysis Project. Department of Economics, Colorado State University, and Ministry of Agriculture, Maseru, Lesotho. LASA Research Report No. 3.

SECKLER D W (1977). A Guide to the Economic Evaluation of Soil and Water Conservation Projects. Ford Foundation and Central Soil and Water Conservation Institute, Dehra Dun, India.

SHAXSON T F S (1980a). Reconciling social and Technical Needs in conservation work on Village Farm Lands. In Soil Conservation: Problems and Prospects. Ed. R P C Morgan. Chichester. Wiley.

SHAXSON T F S (1980b). Developing concepts of Land Husbandry for the Tropics. In Soil Conservation: Problems and Prospects. Ed. R P C Morgan. Chichester. Wiley.

SIMMS H D (1970). The soil conservation service. Praeger Library of US Government Departments and Agencies.

SOIL CONSERVATION SOCIETY OF AMERICA. Proceedings of 29th Annual Meeting, August 1974, Syracuse, New York.

TIMMONS J F (1980). Protecting agriculture's natural resource base. Journal of Soil and Water Conservation. Vol. No. 35 1980 pp 5 - 11.

WIGGINS S (1980). The Economics of Soil Conservation in the Acelhuate River Basin, El Salvador. In Soil Conservation: Problems and Prospects. Ed. R P C Morgan. Chichester. Wiley.

# 19

Reprinted from *Geog. Annaler* **54-A:**110-123 (1972)

# SOIL AND WATER CONSERVATION POLICIES IN THE ULUGURU MOUNTAINS, TANZANIA

BY PAUL H. TEMPLE

Department of Geography, University of Dar es Salaam

## Introduction

The Uluguru mountains form one of the major stream source areas of Tanzania. The rivers Ruvu, Ngerengere and Mgeta all rise in these mountains (Fig. 1). The Ruvu basin (18,389 km²) drained by these rivers, is recognized as an area having considerable development potential.

Heijnen (1970) provides a bibliography of the development projects, some of which are already constructed, others being under way and yet others under active consideration. The Ruvu river supplies the water demands of urban and industrial users in Dar es Salaam. The Ngerengere river is the major water source for the numerous sisal estates in its catchment while the Morogoro river is the main water source for Morogoro town designated, under the Second Five Year Plan, for expanded industrial and urban development.

Yet the mountain sources region of these perennial streams has long been identified as prone to severe soil erosion and consequent deterioration of water yields in terms of either quantity or quality. Large areas of the mountains have been deforested over the last century and a half in the course of expansion of peasant agriculture. The rural economy is essentially subsistence-orientated, relying heavily on annual crops which provide little protection against sudden heavy rainfall on the generally steep slopes.

The agricultural economy of the mountains sustains a fairly-high population density, though out-migration is widespread. There is thus an urgent need to prevent resource deterioration within the area to sustain the existing popula-tion at least at their present level and a need for awareness of the possible repercussions such a deterioration could have on the developmental prospects of the whole Ruvu Basin.

There is indeed a long and intricate history of conservation policy and practice in the Ulugurus. It was in the Ulugurus, for example, that the British colonial government attempted one of its most ambitious and comprehensive schemes of soil conservation and land use improvement in the early 1950s.

This paper presents a critical review of past conservation policies and their impact. The paper covers environmental appraisal, policy recommendations and administrative action. It also examines the reactions of the local agricultural population to the measures adopted. It will be demonstrated that many of these measures were based upon inaccurate data and subjective assessments. While this situation was to some extent inevitable, it is certainly a major explanation of the failure of many of the conservation schemes. It will further be demonstrated that, in one classic case, conservation measures actually accelerated resource wastage.

The purpose of this review is to draw lessons for the present from the frequently painful and costly experiences of the past and to highlight the need for objective, factual technical data as the only sure basis for proper decision-taking.

## The natural environment and its early modification

The topography and geology of the Uluguru mountains is described by Sampson and Wright

(1964). Average annual rainfall, rainfall reliability and monthly rainfall data for the Ruvu basin are discussed by Jackson (1970). Hydrological data on stream flow is available for all the major rivers but this information has not yet been fully analysed (WD & ID, Yearbooks 1963 & 1967). Analysis of population density (Fig. 2) and land use has been made by Thomas (1970). Work on land systems is in progress (B. A. Datoo, personal communication).

Contrasts in natural vegetation in the Ulugurus reflect local differences in altitude, slope, soils and rainfall, but it is probably variations in the latter which are most important in explaining the pattern. Generally average annual rainfall exceeds 150 cm; on the eastern slopes annual rainfall rises to over 285 cm as at Tegetero, but on the western slopes conditions are drier due to a pronounced rain shadow effect. The main rainy season extends from late February to early May at most stations, though some show an important secondary peak from late November to early December (e.g. Bunduki and Mfumbwe).

Under natural conditions and before the heavy impact of man, the Uluguru mountains were mainly covered by forest and woodland. Only on the upper levels of the Lukwangule plateau at altitudes of above 2600 m does forest give way to a grassy scrub climax (Hill 1930). Below this summit level, montane forest extended downslope to varying elevations in response to rainfall amounts. On the northern Ulugurus, where all the detailed study areas are located, the natural lower limit of forest varied over an altitude range of 1000 m from 800 m a.s.l. on the wetter eastern slopes to 1300—1400 m above Morogoro and to 1800 m above Kienzema. On the southern Ulugurus, which are drier, the natural lower limit of forest varied between 1200 m on the eastern, seaward-facing slopes and 1800 m on the western slopes (T. Pocs, personal communication). On steeper slopes and at higher elevations, particularly in the west, a part of this forest cover still remains and is protected as Forest Reserve. Below the forest limits, woodland of various types covered the remaining slopes; most of this has been cleared for agriculture.

Various dates have been advanced for the beginnings of woodland and forest clearance and the intensive settlement of the mountains. These range between 1884 (Savile 1947), 1800 (Bagshawe 1930, Cory undated) and 13—15 generations or up to 300 years ago (Young & Fosbrooke 1960). The account of Bagshawe and Cory is the most plausible and is followed here.

According to these authors, the Luguru, who form the bulk of the African cultivators, migrated to the Ulugurus from Ubena in Iringa Region. They were originally plains people who knew little about the cultivation of steep slopes and carried their lowland crops with them into the mountains, where they were driven through pressure from the stronger pastoral Ngoni. They were a pastoral people but their cattle were decimated in this area by East Coast Fever and could not flourish on the insufficient pastures in the mountains. The Luguru settled initially in the open woodlands of the western and south-western slopes, where relics of abandoned *shambas* (small cultivated plots) can be discerned in areas of open grassland, the original vegetation of which has been almost entirely destroyed by fire and cultivation. These areas were progressively abandoned as they became impoverished (the upper Mlali valley is an example) and the tribe expanded into the more densely wooded upper slopes around Kienzema and Bunduki, and later still into the wetter eastern forested areas. It is unlikely that any conservation practices were followed in this initial phase of exploitation of a virgin environment. The land was cleared and cultivated until it became impoverished and was then abandoned.

## History of conservation policy and practice

### Pre-1945: Piecemeal measures

This shifting cultivation was interrupted by the enforcement of the first conservation measures in 1909. These resulted from the advocacy of forest and watershed protection by Bruchhausen (reported by Cory *op. cit.*) and a report on rapid forest clearing by Stuhlmann (1895). An area of 277 km² was declared Forest Reserve and its boundary demarcated by the German colonial administration. In many areas the reserve boundary lay well within the woodland and forest zone (Cory *op. cit.*) but in some areas cultivated or cleared land was included within the Reserve boundary and

**281**

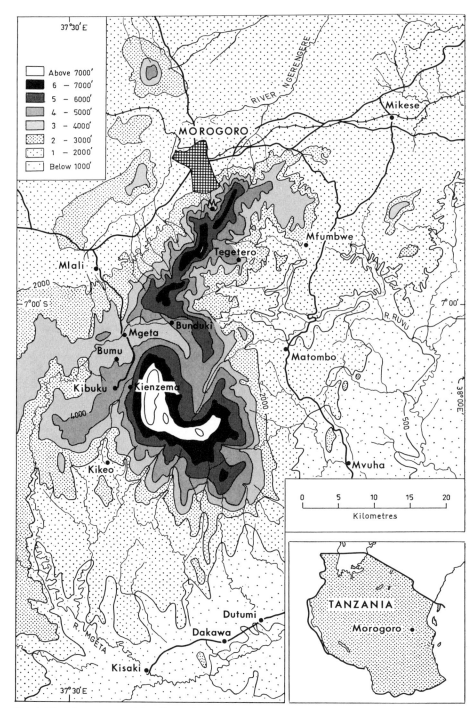

Fig. 1. The Uluguru mountains, Tanzania.
M = Morningside in the Morogoro river catchment.

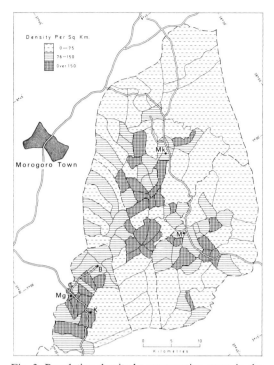

Fig. 2. Population density by enumeration areas in the Uluguru mountains (after Thomas).
B = Bunduki; K = Kienzema; M = Matombo; Mf = Mfumbwe; Mg = Mgeta; Mk = Mkuyuni.

occupants of these plots were expelled and compensated (Platt, Huggins, Savile & Parry 1945). The aim of this step was to safeguard perennial stream flow in the surrounding lowlands. In many areas the present Forest Reserve boundary follows the original demarcation but in some areas woodland cover has been pushed back beyond it as near Kienzema. The effect of this conservation measure according to Savile (1945—1946) was to intensify the exploitation of remaining non-reserved land and to accelerate its deterioration; land was cleared before vegetation regeneration had fully restored the soil fertility and a greater proportion of steep slopes were cultivated leading to increased soil erosion by sheet wash. Cultivated land of these types showed a decreased fertility and "in order to compensate . . . the natives increased the area under cultivation which in turn caused a decrease in the resting period which the land received before being reopened for cultivation. Thus we have the start of a vicious circle in which sheet erosion

has encouraged the natives to steadily increase the area under cultivation, thereby in turn, increasing the amount of sheet erosion that took place" (Savile *op. cit.*). This view almost certainly exaggerated the rate of soil loss through wash (for which there was no reliable data), underestimated the ability of exposed soils in this area, even on steep slopes, to resist erosion and overestimated the recovery period of soils under fallow.

During the period of the First World War, a number of those expelled from the Forest Reserve area by the Germans returned to their original *shambas* and were not expelled by the less efficient early British administration, but it is clear from later reports that these incursions were limited (Platt, Huggins, Savile & Parry *op. cit.*).

During the 1920s limited conservation measures were put into operation both by local action and through legal pressure. Around Mgeta in 1923 the local people began planting trees (mainly black wattle) in a number of small areas "with the double idea of obtaining firewood and preventing erosion" (Bagshawe *op. cit.*) though wood was also in demand for pit props for the local mica mines (Duff, personal communication). Furthermore the "Mgeta system" of laying down grass and weeds in ridges along contours as a method of counteracting sheet wash probably evolved during this period of agricultural intensification (D.O., Morogoro to P.C. 1930). This method of conservation was replaced by improved practices in the late 1930s in the Mgeta area but is still common in other areas, as for example in the Morogoro catchment. In 1929 regulations were laid down (Native Authority Orders, 1929) that no one was allowed to burn grass or bush on land other than his own without permission.

In 1930 the administration became alarmed by the Land Development Commissioner's Report (Bagshawe *op. cit.*). This report which, besides indicating that there was no remaining land in the mountains suitable for alienation, added that "owing to approaching congestion on the higher slopes it is necessary to stop further alienation within an area below", presumably in order to allow room for future migration of a labour force and settlers into the surrounding plains, then being developed as sisal estates. A spate of reports and recom-

283

mendations for improvements followed. Among the measures proposed was the encouragement of coffee planting (D.O., Morogoro to P.C., 1930 & P.C. to D.O., Morogoro *op. cit.*), with the aim of encouraging more permanent cropping and thereby reducing sheet wash. Encouragement of the indigenous trash contour ridges was also advocated.

Reafforestation of denuded areas was proposed and improved protection of residual forest areas advocated but nothing was done. A report on local fuel and pole requirements (probably for the local mica mines) in the western and southern areas of the mountain (Baldock & Hutt 1931) noted that there were marked shortages of both commodities, and mentioned the difficulty of acquiring land for forest plantations due to population pressure. 100 hectares of forest were nonetheless set aside as Clan Forest Reserve, presumably additional to the existing Crown Forest Reserve. The report added that lines of wattle should be planted along the tops and bottoms of shambas in Mgeta, but no steps were taken to implement this sensible proposal.

The first agricultural conservation work of significance came in 1936—1937 with the establishment of trial plots and experimental bench terraces for vegetable and potatoe growing at Mgeta (Page-Jones & Soper 1955). The plots were adjudged successful and were described as the model for the ladder or step terrace. This is doubtful as ladder terraces are a completely natural and economical way of cultivating steep hillsides with a hoe, and probably evolved earlier. They also differ fundamentally from bench terraces.

The ladder or step terrace (*matuta ya ngazi*) merits some discussion here as it is still currently the most widely employed and generally accepted soil conservation method on the western side of the mountains. "When cultivating, vegetation and crop residues are spread on the top of the terraces and covered with soil cut from the face of the terrace above. This is also done when weeding and so there is a slow mechanical movement of soil downwards. But these terraces give better yields as little subsoil is dug out and organic matter is incorporated regularly". (Grant 1956). Furthermore soils cultivated in this way acquired an open and free-draining structure. "Only rarely does a terrace system of this nature break down under

heavy rain and cause gully formation" (Page-Jones & Soper *op. cit.*, p. 4). Another conservation method tested at this time was the planting of live grass barriers. Stronger regulations against indiscriminate burning were put into operation, involving a metre-wide firebreak around all *shambas* in an attempt to prevent one man setting his neighbours plot alight (Native Authority Orders 1936). As a demonstration of the benefits to be derived from protection from annual fires, a small area of uncultivated hillside near Morogoro was set aside and protected (Page-Jones & Soper *op. cit.*, p. 4).

A lull in direct conservation efforts followed between 1937 and 1943 apparently due to a policy decision to concentrate on education and demonstration and not to undertake large schemes to check or prevent erosion (Harrison 1937, p. 5). This policy was based on a 6 year review of available information and lengthy discussions. But others pressed for direct action (Gillman 1938). Gillman argued that "in the wage of devegetation follows very pronounced soil erosion and a change of stream regime from permanence to intermittance which means excessive and often destructive floods in the wet and dearth of water in the dry season. With the shrinking of the forest oases ever wider regions of the surrounding lowlands are threatened with desiccation ... and all for the sake of feeding at best a few generations". (*op. cit.*, p. 20). He reinforced this statement later when he wrote ... "The chief function of our highlands is *not* to provide a shortlived subsistence or profit for their excessive exploiters ... but to maintain a regular run-off from the climatically—favoured more humid heights into the thirsty surrounding arid lowlands. By fulfilling this function they will not only permit a *limited* population to continue its highland life in safety ... but will enable an increasing number of people to till the lowlands" (1943, p. 103). Gillman's reports were a decisive influence on later policy, but they were based on inadequate quantitative data and they ignored the importance of public opinion and public reaction.

The administration responded in 1943 with new territorial regulations to prevent large-scale burning while large-scale re-afforestation and compulsory movement of population out of the Ulugurus were mooted and seriously considered in 1944 (Page-Jones & Soper *op. cit.*). Demonstrations of storm-draining, terracing and tie-ridging were set up in August 1944 at Kienzema, Kibuku and Mgeta and these plots were said to have achieved their purpose, which was identified as the prevention of runoff and increased crop yields (Platt, Huggins, Savile & Parry *op. cit.*). How rigorously these experiments were conducted is not known nor is it possible to find out whether runoff control and yield increases were significantly different from those measured on control plots (if there were any). The report suggests that many local peasants turned over to such methods without further pressure, an assertion contradicted by subsequent instructions to the Native Authority to enforce their adoption. This move set the stage for the larger-scale interference which followed.

### 1945—1955: Genesis and implementation of the Uluguru Land Usage Scheme (ULUS)

A new approach to the problems of the area was initiated with the establishment of the Committee on the rehabilitation of eroded areas in the Ulugurus in 1945. This committee made recommendations for soil and water conservation together with comments on forestry policy and administration, prefacing their report with the following remarks: "The existing advanced state of soil erosion and detereoration of water supplies warrants the immediate adoption of such physical methods of control as will effectively halt the present rate of destruction" (Platt, Huggins, Savile & Parry *op. cit.*). No evidence of either the extent or speed of soil erosion or the type of water supply detereoration observed was presented; it was presumed that soil erosion posed a serious threat and that previous conservation practice had been inadequate. No mention was made of the ladder terrace and the efficacy of live grass barriers was questioned on all but the most gentle slopes.

The administration believed that the agricultural situation in the mountains was so grave that the people were no longer self-supporting; indeed food shortages had been experienced on the northern and western slopes (Duff, personal communication). This explains their support of the conservation proposals. The alternative must have appeared to be long-term subsidies and assistance or large-scale

investments in resettlement schemes. Their first step was to re-enforce the regulations against burning; "except when breaking or preparing new land, burning of grass and bush in hill country is at all times forbidden and is permitted on flat country only to destroy weeds" (Native Authority Orders 1945).

The Department of Agriculture believed that the Luguru were destroying their land due to inadequate fallowing practice and excessive burning and by exposing cultivated soil on steep slopes to sheet wash. Savile wrote that a stage had been reached "when a family has to cultivate four or five times as much land as was necessary thirty years ago" and that "sheet erosion has advanced so far that, in many cases, the natives are endeavouring to produce crops on clayey subsoil underlying a surface layer, of 2 or 3 inches of immature surface soil " (Savile 1947). He also claimed that large areas of abandonned land were common and that some did not recover even after 40 years under fallow. Another section of this report covered the supposed repercussions of these adverse changes upon the headwaters of the Ruvu system and the resulting effects on the plain. Savile's statement "was forceful and persuasive, and it formed the official rationale for establishing the Uluguru scheme" (ULUS) (Young & Fosbrooke 1960, p. 143).

The remedial action proposed by the Committee was wide-ranging as it represented a balance of departmental interests. Specific recommendations were to (a) redemarcate the forest boundary as it had been during later German times and to restore forest and woodland down to that line, (b) plant trees outside these limits to provide fuel and poles, with the aim of reducing cutting and fire damage on the margins of the reserve itself; (c) control annual burning of grass, weeds and trash in order to allow the regeneration of fallow land; (d) adopt contour tie-ridging on a large scale in order to control sheet wash. These last two measures were meant to increase infiltration and reduce surface runoff. Should these measures prove inadequate after trials, it was recommended that the whole area should be turned over to forest reserve and the people forcibly resettled on the plains. (Platt, Huggins, Savile & Parry *op. cit.*).

However there are reasons to believe that

several of the arguments put forward by the committee were erroneous. Certainly the Luguru possessed a greater knowledge of their own environment and soils than the officials were prepared to concede. Witness the ladder terracing, tree planting and intercropping practiced especially around Mgeta. Furthermore both pressure on land and population growth were almost certainly overestimated due to inadequate information. Population density in the Mgeta area, according to the 1967 census reached 118.5 per km² (Tanzania: 1967 Population Census, 1969); it was estimated at 290 per km² in 1945. It was admitted in 1945 (Platt, Huggins, Savile & Pàrry, *op. cit.*) that the Forest Reserve, demarcated by the Germans, had not been seriously depleted though isolated forest stands were probably being rapidly reduced. The seriousness of sheet erosion was probably overestimated, for no accurate data was available, and soil regeneration in some areas at least is very rapid (see below). In assessing the changes of the hydrological regime of the Ruvu Savile made no detailed analysis either of the rainfall records or of the hydrological data but contented himself with a discussion of the increasing difficulties of paddy rice cultivation around Bagamoyo, an argument which was far from proving his case.

Nonetheless a sum of £50,000 was made available in 1947 for a 10 years rehabilitation scheme. There is however no record of any formal policy decision about the type of operations necessary (Duff 1960), presumably because it was assumed in most quarters that these would follow the 1945 recommendations. It was soon to appear otherwise for the Provincial Administration decided to concentrate effort and available money on building bench terraces and, as an after-thought, to set up demonstration plots to show how effective these bench terraces were. Removal of barren land from cultivation, tree planting, demarcation of special reserves for stream headwater protection formed the remainder of the operation. While the last two measures represented a consistent and agreed policy, the decision to concentrate primarily upon bench terracing represented a complete reversal of agreed agricultural policy which was to have tragic consequences for the whole conservation effort.

The history of the Uluguru Land Usage Scheme (ULUS) has been described by Young and Fosbrooke (*op. cit.* pp. 141—167) and Duff (*op. cit.* & 1961) and can thus be treated in summary form.

The scheme began full-scale operation in 1950 by which time the finance available had been raised to £68,000 (Notes for 1955/56 ULUS estimates, 1956). When it was abruptly dropped in July 1955 after serious riots and general disorder, over £47,000 had been spent and 5 years of the scheme were left to run (ULUS Estimates for 1953 & 1954). "By the end of 1950 monthly reports disclose that the main line of policy had somehow been established. This was to be compulsory bench-terracing in *shambas* of medium gradients, with reafforestation of steeper slopes, involving their closure to cultivation of annual crops". (Duff 1960). A basic presumption behind the whole policy was that sheet wash and flash runoff were the major problems and that large terraces were therefore the most effective method of control. No thorough investigations were conducted in advance to ascertain the rate of sheet wash or the effect of bench terracing on crop yields. On the experimental bench terraces at Kienzema some increase of yield was claimed but no firm evidence was available. Experimental terraces established in the Morningside valley had proved barren, resulting in either total or partial crop failure. No data had been collected on the heavy labour inputs necessary to build large bench terraces. Nor was any consideration given to the probability that, by holding heavy rainfall on such steep slopes by means of large terraces, other forms of soil erosion, notably landslides, might be encouraged.

Official bench terracing began at Matombo in 1950, at Mkuyuni in 1952 and later in other areas (at Mgeta in 1953). It was backed to the full by the Colonial Administration and the Native Authority. At first the Luguru appeared to accept bench terracing. At least there was initially no resistance, but the people were apathetic and had been antagonised from the start by the closing of steep land, the prohibition of burning and of the rumour of enforced migration to the plain. In mid-1952, the District Commissioner summed up progress as follows: "£20,000 spent, four field officers established in houses in the northern Ulugurus, a few areas of trees planted and out of 400 square miles of mountain under native cultivation, perhaps 5 had been put under some form of ULUS control" (quoted by Duff 1960). A lot of money was spent on a road, never properly surveyed, from Kibuku to Kisaki. Furthermore the field officers were ill-trained and uninterested. Amidst a largely Roman Catholic population, the influencial Holy Ghost Mission had been antagonised and the Fathers openly preached against the new policies (Duff 1960). When the new District Commissioner, who had no say in the operation of the most important project in his district and who came fresh from the experiences of the failing Western Usambara Land Development Scheme, dared to question the wisdom of universal bench-terracing, he was told by the Provincial Commissioner to get on with his job for the policies were "a proven success". The Provincial Commissioner, who was an influential member of the Legislative Council, played a major role in the direction of the scheme and its policies, and was much impressed by the enormous conservation projects then proving successful in the Appalachian mountains of the southeastern U.S.A. Duff categorised his attitude with exactness; "what was good enough for the Ohio valley was good enough for the Ruvu" (1960).

From late-1952 the scheme was intensified. Competent field officers were brought in and bush-schools set up in an intensive campaign to try to secure the Scheme's acceptance by the people. Under this new impetus the programme was diversified. Fish farming, stall-feeding of cattle, soft wood and coffee planting and experiments with cloves, citrus and coconuts were introduced, though compulsory bench terracing remained the core (Hill & Moffett 1955). The administration set goals which accelerated temporarily the rate of terracing. Each household was forced to bench terrace 500 m$^2$ per year. This coersion apparently led to a seven-fold increase in bench terrace construction between 1953 and 1954, and nearly four times as many terraces were reported to have been dug in the first 6 months of 1955 as were dug in the equivalent period of 1954 (Grant op. cit.). The·validity of these data is however in serious doubt (Kunambi, personal communication).

Other conservation methods also seemed to be succeeding. "By the end of 1954 trash

bunding, live barriers, and cored ridging had become almost standard practice in fields which were not terraced, over most of the hills. *Most of these measures were successful in holding up wash during storms of rain*" (Grant, op. cit.) (author's italics).

A stringent control on bush fires was enforced and this was particularly effective in 1954: "the effect of this protection over a number of years was to be seen in the flow of the streams along the northern face of the Ulugurus. During the heavy rains experienced in February and April the streams feeding the Ngerengere did not flood as in former years and the water remained comparatively clear" (Grant op. cit., p. 6).

But "as time went on, a hostile reaction developed to the new agricultural techniques, or to their implications; the hostility was expressed through unauthorised political channels and eventually led to intimidation, open violence and the·use of force" (Young & Fosbrook, op. cit., p. 146). Except in the Mgeta area, where the ladder terrace system largely obviated the burning of trash, the people complained bitterly over the prohibition of fires. In the east, where terracing of any type was unknown before ULUS, they also complained about terracing. The amount of work done on bench terrace construction declined rapidly and a successful strike was organised in June 1955; within a fortnight all bench terracing work had ceased throughout the area. Fires were lit all over the mountain in July to signal defiance of the land use rules. ULUS collapsed, a classic example of attempted innovation and its resistance.

Various reasons were advanced to explain the scheme's failure. Page-Jones, the Provincial Commissioner, and Soper (op. cit.) in their departmental enquiry stressed the general multiplicity of orders and rules, the widespread corruption of headmen and instructors, who could ignore breaches of rules and permit burning if bribed. On the northern and eastern slopes the heavy labour of terracing was considered a main factor, while in the east in particular, yields were bad, and field officers and instructors incompetent.

Young and Fosbrooke (op. cit.) identified much more fundamental causes, pointing out that ULUS had threatened the traditional social and cultural system of Lurugu particularly over the authority to allocate the land. They also reemphasised the deficiencies of administration. They also drew attention to the inadequacy of the technical recommendations, a fact inexcusably glossed over in the Page-Jones-Soper report, and showed how discontent engendered by all these deficiencies was used by emerging national political groups for their own purposes, which were not essentially the undermining of ULUS but the discrediting of the conservative Native Authority, and its replacement (Cliffe 1970).

The technical shortcomings were identified by Young and Fosbrooke as the incomplete nature of the initial plans which lacked clear and agreed objectives, and the lack of experimental work, demonstration plots, or adequate extension services. They noted the fact that no study of the possible impact of, and reaction to, the proposals had been made. Duff (1961) largely supported these judgements, but he argued that "these technical short-comings were less important than a general opposition to a lot of hard work under the direction of outsiders" and the disorganisation of the traditional cultivation patterns and institutions. He identified the chief fault as the unequivocal administrative directive from the Provincial Commissioner which forced the local officials into an impossible position just as it did the peasant cultivators. Bench terraces were dug to order in the knowledge that they were useless; the Luguru deliberately chose sterile sites for bench terraces in order to avoid damaging fertile land (Duff 1960).

As bench terraces were again proposed as an appropriate measure to prevent erosion in the Mgeta area in 1970, some further comment on the technical problems involved seems desirable. "An easy method of digging in crop residues and grass with a hoe without pulling down soil from the face of the terrace above was not found and this is the chief difficulty involved in cultivating with a hoe . . . when the grass is left on the terrace faces and crop residues and vegetation must not be burnt" (Grant op. cit., p. 5). Secondly bench terraces construction on steep slopes with a thin soil cover invariably exposed infertile subsoil: this was a serious drawback when it was important to show improved results to justify the extra labour involved. Only on fair depths of good soil where slopes were not too steep (i.e. on

areas not seriously endangered by soil wash) had maize done better on terraced land. In most areas the terraces had proved barren. "The best that could be hoped for was no diminution of yield but the chances were that there would be a loss" (Page-Jones & Soper op. cit., p. 11).

*1955—present: Non-interference*
The civil disturbances were brought to an end at the price of scuttling ULUS. New regulations were brought out in July 1955 and cultivators were given the choice of what conservation method they wished to use (Grant op. cit., p. 6). They mostly chose to employ none at all. Bench terracing was abandoned. People complained about barrier hedges and they were abandoned. Contour ridges and trash bunds had never been fully accepted and their use was not enforced. Permission was given to burn fields with "safeguards" against fire spreading (Grant op. cit., p. 3). What these safeguards were and how they were to be enforced was discretely glossed over; in fact they were never put into practice because of opposition. All pressure towards soil conservation measures was withdrawn as a matter of policy for two years following the advice of Page-Jones and Soper (op. cit., p. 34).

Only in the Mgeta area did anti-erosion methods continue while "on the eastern and southern parts of the Uluguru no soil conservation measures of any sort have been observed . . . Long established live barriers have actually been dug up. There has been a general tilling of the banks of water courses with annual crops and a re-opening of steep land which had been closed (Grant op. cit., p. 6). ULUS had discredited all conservation efforts and its longterm effect was to reduce the area over which any form of conservation was practiced.

With their conservation policy in ruins, the colonial administration began belatedly to establish a series of experimental trials. Experimental plots were set up to measure sheet-wash and to observe crop yields under different managements (Mfumbwe) (Temple & Murray-Rust 1972) and experimental trials were made with a wide range of crops. Other projects concentrated on methods of improving the productivity of individual holdings, on catchment area experiments, and on establishing an irrigation scheme on the plain (Mlali). This enforced change was financed by the saving which the failure of ULUS had brought about. By 1958 enthusiasm for direct action had waned. In most cases the experimental plots and crop trials were allowed to revert to bush. Mfumbwe was abandoned in 1960 and none of the results obtained were properly analysed or made available (see below).

After independence in 1961, only the Mlali resettlement and irrigation project continued while in the mountains a new phase of forest clearance began. The more accessible parts of the Forest Reserve were plundered of timber and in some areas planted forests were felled and burned to provide new agricultural land (e.g. at Bunduki). In other areas relic rain forest was cleared by fire and cultivation extended into areas never before under agricultural use e.g. south of Kienzema. The total extent of these incursions is difficult to assess but they may be seen as a logical extension of the systematic resource wastage set in motion by the failure of ULUS.

Repercussions of this apparently increasing devegetation, though locally variable, were soon remarked upon. By 1963 flood damage, bank erosion and silting had become serious problems even within the township of Morogoro itself due to the condition of catchments on the northern face of the mountains (Little 1963). Large differences between wet season high flows and dry season low flows were said to be increasingly apparent while severe short-duration flash floods of high sediment content were causing considerable damage. Little argued from an analysis of 10 years of hydrological records that the situation was deteriorating.

According to Little's analysis, the river Ngerengere was drying up completely in the dry season with increasing frequency. First recorded in 1930, this phenomenon was repeated in 1934, 1943, 1949, 1953, 1955, 1958 and 1960. For the first time in 1960, and again in 1966, sisal production on estates dependent upon the river for water came to an enforced halt for 2 months. The cost of ensuring alternative dry season supply for these enterprises from dams was placed at £260,000 while such construction would flood over 400 hectares of developed plantation land. Flood erosion and silting within Morogoro township were

estimated to have done well over £24,000 of damage in 10 years, and the Morogoro river had been too low to be recorded in October 1958 and 1960, a serious problem in view of the fact that it provided the urban water supply.

In the late 1960s a new and ominous phenomena was recorded for the first time in the mountains namely catastrophic landslide damage. In Matombo in 1968 an unusually violent rainstorm triggered a large but unrecorded number of landslides. Several people were killed and several houses damaged. Landslides as an erosion danger had never been recorded before though they had undoubtedly occurred. On 23/24 March, 1969 a sudden storm devastated the area south of Kienzema and over 200 slides occurred in one night, but because the area was rather isolated, the catastrophe passed largely unnoticed. Then on 23 February, 1970 a further storm devastated the area around Mgeta, causing a loss of 6 lives, affecting almost 14 % of the households of the area, killing stock and damaging nearly 1600 *shambas* in the space of 3 hours. A detailed account of this event is presented below (Temple & Rapp 1972). Damage conservatively estimated at over Shs. 600,000 was sustained without evaluation of land damage and soil loss.

No earlier reports on soil erosion mention landslides as a menace, an indication either of their lack of thoroughness or that this was a new phenomenon. The latter hypothesis cannot be sustained though it is interesting to note that severe landsliding elsewhere has been interpreted as the terminal phase of man-induced accelerated erosion eg. by Sternberg (1949) and Tricart and Cailleux (1965).

## Future conservation policy

The purpose of the review presented above was to evaluate past experience in conservation policy and practice. That soil erosion of various types is actually a serious problem in the Ulugurus has already been indicated and will be demonstrated and quantified in detailed case studies presented below.

It remains, however, to make some comments on possible future policy. Because slope, rainfall, soil, cultivation methods and cropping varies from one locality to another, generalisation is difficult and much more technical data needs to be accumulated before definitive recommendations can be made.

Conservation measures to protect soil and water resources in such environments as the Ulugurus, to be acceptable, must bring (a) a demonstrable advantage to individual farmers in the short-run and (b) long-term advantages to the community. Past measures placed excessive emphasis on the latter point and were therefore rejected (Cliffe 1970, Temple 1971). Measures advocated must take account of what is feasible given technical and financial constraints and local opinion. The implementation of conservation measures will remain a more critical problem than their design. Excessive dependence on rules and regulations is no answer while the legacy of past policies and experience is a major barrier in the way of future ameleorative measures.

Certain recommendations can be advanced nonetheless. The existing *Forest Reserve* needs strict protection (Pereira 1969). As the current active encroachment in certain areas indicates, this is not the simple and routine task that Parry (1963) suggests. Forest cover, though a major user of water itself, ensures the maintenance of even stream flow of good quality during the dry season and a slower release of runoff during the wet season. Large-scale expansion of the Forest Reserve has been proposed on occasions in the past; it would have the following disadvantages: (a) by removing presently cultivated land from agricultural use, it would put further pressure on remaining cultivated areas, thereby reducing fallow periods and accelerating the depletion of the soil nutrient status; (b) it would antagonize the local people and thus be political unacceptable; (c) expansion of forest or woodland on a large scale would decrease the total volume of stream flow as EAAFRO's data from the Mbeya catchments (Pereira 1962, Pereira & Hosegood 1962) demonstrate; (d) it would disrupt the subsistence economy of the area.

Conservation measures on *cleared* or *cultivated land* should be adapted to local conditions of slope, soil and cultivation practice. There is still a very great need for detailed research and experimentation of the type begun at Mfumbwe before these can be recommended with confidence. There are however indications that landslides are the most serious erosional threat to cultivation on steep slopes in the Ulugurus. Landslides in this area are primarily caused by the build-up of pore water pressures within the regolith. The remedial or conservation measure normally recommended to remedy such sub-surface water pressure build-up from a geological viewpoint is drainage (ditches, tile drains or other conduits). Drainage increases the internal friction of the regolith by lowering its water content and prevents the entrance of excessive amounts of water into the ground. Drainage cannot however be recommended for this area because (a) it could not be justified economically as its cost would be vastly in excess of the value of the land so protected; (b) it is almost certainly financially impossible even if technically feasible; (c) it would accelerate channel erosion and downstream sedimentation by accelerating surface run-off. Haldemann's conclusions (1956) relating to an almost identical problem in the Rungwe mountains of southern Tanzania are thus confirmed.

Bench terracing is a most unsatisfactory remedial measure for this area. Its disadvantages have already been discussed. Bench terracing (a) holds water on the slopes thus accelerating the build-up of high pore water pressures which are the major cause of landsliding; (b) it involves vast labour inputs with no advantage because (c) it invariably exposes infertile subsoil. Where soils are thin this effect is unavoidable in the short-run; (d) it would certainly be rejected by the people.

The Mgeta landslides study (Temple and Rapp 1972 below) indicates a possible alternative to the measures rejected above, for it confirms the value of a tree cover and even isolated trees in curtailing landslide damage during severe water saturation of the soil. Reasons for this are well known and need not be recapitulated here. Most important is certainly the physical binding of soil and regolith to the underlying rock slope by deep-penetrating and sturdy roots. From the geomorphological data on slide locations and an evaluation of the costs of damage sustained, it is possible to identify specific sites where tree planting would be most beneficial:

1. On the upslope margins of main roads where the natural slope is steep (over 30$^\circ$) or where road construction has created artificially-steepened cuts. The direct cost of clearing slide debris from roads and the

effects of the rupture of communications figure importantly in the storm damage evaluation around Mgeta.

2. In lines, along the contour some distance (approximately 30 m) *below* ridge crests as a counter measure to the build-up of pore water pressure in this zone.

3. Upslope of villages and individual dwellings to protect dwellings and lives.

4. Along stream lines to restrict bank erosion by floods, excessive sediment transport, deterioration of water quality and downstream sedimentation on cultivated land, in culverts and in dams. Sedimentation downstream caused the major quantifiable damage in the Mgeta area in February 1960.

Further advantages of such a policy to those advanced above are: (a) it would ensure a supply of building timber and firewood at hand; (b) it would decrease the illegal depletion of the Forest Reserve; (c) it would generate local employment; (d) it is already recognised locally as an effective remedial measure against landslides; (e) it is cheap and feasible.

The introduction of new tree crops and the expansion of the area covered by tree crops would have the following advantages; (a) it would directly curtail erosion of all types; (b) it *could* be a means of introducing better husbandry methods if the provision of seedlings at low cost was made conditional upon eg. better control of burning or trash-barrier maintenance; (c) it would provide supplementary income and improved diet for the farmers—a tangible short-term incentive. Several types of fruit trees flourish in this area (eg. guava, apricot etc.) and a market (eg. Dar es Salaam, Morogoro) and marketing organization for vegetable export already exists, though it needs radical overhaul.

The proportion of land currently under tree crops in the Ulugurus is very low (5.7 %) and significantly lower than in the Usambara and Pare mountains (11.8 %) (Thomas *op. cit.*) and on Rungwe, Kilimanjaro and Meru. This probably results from clan ownership of land. Clan heads discourage tree planting because it restricts their right to reallocate land and takes arable land out of cultivation (Duff, personal communication).

Any encouragement of perennial crops to reduce the excessive dependence on annual crops like maize on the western slopes and hill-rice on the eastern slopes would also be advantageous as a means of improving soil structure and infiltration capacity. This would retard sheet wash. Though proposed in general terms on numerous occasions in the past, it has not been adopted. Local custom and traditional land ownship patterns appear to be the main constraints.

Conservation practices could be readily improved on land used for annual crops. Local farmers are prepared to adopt simple measures if these are *demonstrated* to provide better crop yields. The widespread employment of ladder terraces in the Mgeta area is an indication of this. The Mfumbwe data (Temple & Murray-Rust, 1972) confirms the value of this practice on steep slopes.

Few of these proposals are in themselves new. The criteria employed in identifying appropriate measures have been feasibility, low cost, the likelihood of local acceptance and technical considerations.

## Conclusions

Past conservation policies were often unsoundly based and unwisely implemented. Before 1945 efforts were piecemeal and uncoordinated. The major attempt at a comprehensive rehabilitation scheme foundered as a result of its own technical inadequacy and the resistance it provoked. The subsequent "laissez faire" policy has contributed little to the improvement of the situation.

Experience from this area indicates clearly that certain conservation measures are both desirable and necessary. The most practicable of these measures are briefly discussed.

## Acknowledgements

Financial support for this investigation was provided by the Research Committee of the University of Dar es Salaam. The author acknowledges helpful comments and data on specific points from Dr. T. Pocs, Dr. J. E. G. Sutton, and Mr. I. D. Thomas. He is indebted to the Director of the Tanzania National Archives (cited below as TNA) for access to material in his keeping in Dar es Salaam. Mr. P. C. Duff, formerly District Commissioner, Morogoro, now of the Overseas Development Administration, and Chief Patrick Kunambi both made numerous most valuable criticisms of the text and provided the author with new insights into many of the events described in which both played key roles. Neither are responsible for the opinions expressed above (except where directly quoted).

# References

Bagshawe, F. J., 1930: A report by the Land Development Commissioner on the Uluguru hills, *Land Development Survey Rept.*, 3, TNA; 61/378/4.

Baldock, W. F. & Hutt, A. M. B., 1931: Native fuel and pole supplies in the western and southern Ulugurus, Unpub. rept., Tanganyika Forest Dept., University microfilm, MF/1/8.

Cory, H., Undated: History of native settlement—Uluguru, Ch. 7; Manuscript, Cory collection 430, University of Dar es Salaam.

Cliffe, L. R., 1970: Nationalism and the reaction to enforced agricultural change in Tanganyika during the colonial period, *Taamuli*, 1, 3—15.

District Officer, Morogoro to Provincial Commissioner, 1930: Untitled letter 21/7/30, TNA; 61/378/4.

Duff, P. C., 1960: Uluguru Land Use Scheme, entry in Morogoro District book, University microfilm, MF/1/8.

— 1961: Land and politics among the Luguru of Tanganyika; a letter, *Tanganyika Notes Rec.*, 57, 111—114.

Gillman, C., 1938: Problems of land utilisation in Tanganyika Territory. *S. Afr. geogr. J.*, 20, 12—20.

— 1943: A reconnaissance survey of the hydrology of Tanganyika Territory in its geographical setting, *Water Consultant's Rept.*, 6 (1940), 136 pp., Govt. Printer, Dar es Salaam.

Grant, H. St. J., 1956: Uluguru Land Usage Scheme; Annual report for 1955, Unpub. rept., TNA; 61/D/3/9.

Haldemann, E. G., 1956: Recent landslide phenomena in the Rungwe volcanic area, Tanganyika, *Tanganyika Notes Rec.*, 45, 3—14.

Harrison, E., 1937: Soil erosion: a memorandum, *Govt. of Tanganyika Territory, Crown Agents*, 1—22: also 1938: Soil erosion memorandum, Conference of Governors of B.E.A.T., Govt. Printer, Nairobi, 9—11.

Heijnen, J. D., 1970: The river basins in Tanzania: a bibliography, *BRALUP Res. Notes*, 5e, 3—5.

Hill, W. J., 1930: Notes on the forest types of the district, Sheet 3, entry in Morogoro District book, University microfilm, MF/1/8.

Hill, J. F. R. & Moffett, J. P., 1955: Tanganyika: a review of its resources and their development, Govt. of Tanganyika, Dar es Salaam (part. 365—376 & 522—525).

Jackson, I. J., 1970: Rainfall over the Ruvu basin and surrounding area, *BRALUP Res. Rept.*, 9, 1—11.

Little, B. G., 1963: Report on the condition of rivers rising in the Uluguru mountains and their catchments, Unpub. rept, W.D. & I.D. Tanzania.

Native Authority Orders, 1929, 1936, 1945 and 1951 (Native Authority Land Usage-Morogoro-Rules), Section 16, Cap. 72, TNA 26/76.

Page-Jones, F. H. & Soper, J. R. P., 1955: A departmental enquiry into the disturbed situation in the Uluguru Chiefdom, Morogoro District, June—September, 1955, Unpub. rept., Dept. of Agricult., Tanganyika; Cory collection 364, University of Dar es Salaam.

Parry, M. S., 1962: Progress in the protection of stream source areas in Tanganyika, in "Effects of peasant cultivation practices in steep stream source valleys", *E. Afr. agric. for J.*, 28, 104—106.

Pereira, H. C., 1962: The research project (4a), in Hydrological effects of changes in land use in some East African catchment areas, *E. Afr. agric. for. J.*, 28, 107—109.

— 1969: Influence of man on the hydrological cycle: guidelines to the safe development of land and water resources. Unpub. paper, F.A.O., Rome.

Pereira, H. C. & Hosegood, P. H., 1962: Suspended sediment and bed-load sampling in the Mbeya range catchments, *E. Afr. agric. for J.*, 27, 123—125.

Platt, S. A., Huggins, P. M., Savile, A. H. & Parry, M. S., 1945: Recommendations for soil and water conservation in the Uluguru mountains. Unpub. rept. of Committee on the rehabilitation of eroded areas in the Ulugurus, TNA/SMP/22446.

Provincial Commissioner, Eastern Province to District Officer, Morogoro, 1952: untitled letter 51/21, TNA 3797/4/4.

Sampson, D. N. & Wright, A. E., 1964: The geology of the Uluguru mountains. *Bull. Geol. Surv. Tanganyika* 37, 1—9.

Savile, A. H., 1945—46: A study of recent alterations in the flood regimes of three important rivers in Tanganyika, *E. Afr. agric. for. J.*, 11, 69—74.

— 1947: Soil erosion in the Uluguru mountains, Unpub. rept. Dept. of Agricult., Tanganyika, University microfilm, MF/1/8.

Sternberg, H. O'Reilly, 1949: Floods and landslides in the Paraiba valley, December 1948. Influence of destructive exploitation of the land, *Int. geogr. Congr.*, 3, 335—364.

Stuhlmann, F., 1895: Über die Uluguru berge in Deutsch-Ost-Afrikas, *Danckelmanns Mitteilungen. Deutsch. Schutzgebieten*, 8, 209.

Tanzania, 1969: 1967 Population census: Volume 1, Statistics for enumeration areas, Central Stat. Bur., Dar es Salaam.

Temple, P. H., 1971: Conservation policies in the Uluguru mountains, Unpub. paper, Second Conference on land use in Tanzania, Morogoro 1971.

Temple, P. H. & Rapp, A., 1972: Landslides in the Mgeta area, western Uluguru mountains; geomorphological effects of sudden heavy rainfall, *Geogr. Ann.*, 54A, 3—4.

Thomas, I. D., 1970: Some notes on population and land use in the more densely populated parts of the Uluguru mountains of Morogoro District, *BRALUP Res. Notes*, 8, 1—51.

Tricart, J. & Cailleux, A., 1965: *Introduction à la géomorphologié climatique* SEDES, Paris (part. 226—253).

Uluguru Land Use Scheme, 1956: estimates for 1953 and 1954 and notes for 1955—56 ULUS estimates, TNA: 61/D/3/13.

W.D. & I.D., 1963: Hydrological year-book 1950—1959, Govt. Printer, Dar es Salaam.

— 1967: Hydrological year-book 1960—1965, U.C.D. Library, Dar es Salaam.

Young, R. & Fosbrooke, H., 1960: *Land and politics among the Luguru of Tanganyika*, Routledge & Kegan Paul, London.

# 20

SOIL CONSERVATION IN ICELAND

S. Runolfsson

Landgraedsla Rikisins

Gunnarsholt, Rang, Iceland

### THE ESTABLISHMENT OF THE ICELANDIC
### SOIL CONSERVATION SERVICE

At the close of the cold period from 1860 to 1890 a good deal was
written and a lively discussion maintained on the problem of wind
erosion in Iceland. The great damage it had caused, particularly
in the previous decades, seemed bound to continue on an ever
escalating scale if protective measures were not undertaken. The
drift-sand was closing in on most farms in the middle of the South
and a great many had already been evacuated.

About the beginning of the 20th century the Agricultural Society
of Iceland showed a growing concern for the problem of erosion in
general and the drift-sand in particular. On its initiative a
number of specialists, including Danes, were sent to undertake a
field study of the problem. The Icelandic Althing granted a small
sum of money for fighting the sand-drift. Some farmers took some
protective measures, which consisted mainly of erecting walls of
stones or timber right across the direction of the erosion at the
junction of the eroded areas and the vegetated land. The results
of these individual and public attempts to stop the erosion were
however negligible.

The State Soil Conservation Service was previously named the
Icelandic Sand Reclamation and the oldest legislation on sand reclamation dates back to an 1895 "Act for Resolution on Sand Erosion
and Reclamation". This contained authority for District Commissions,
but the Act proved a dead letter with no implementation so it made
no mark in its field.

In 1907 an Act on "Forestry and Prevention of the Erosion of Land" was enacted and a special Sand Reclamation representative was engaged. He was Gunnlaugur Kristmundsson of Hafnarfjord and for 40 years he fought a serious shortage of funds and a disbelief in the importance of his function.

The link between sand reclamation and forestry, however, remained very weak and was abolished by a new "Act on Sand Reclamation" in 1914 whereby the Governor was charged with the administrative supervision of sand reclamation with the Agricultural Society of Iceland which was to look after sand reclamation affairs.

Since then there has at all times been some legal connection between the Agricultural Society of Iceland and Sand Reclamation. The act on Sand Reclamation was thereupon augmented and amended in 1923 and again in 1941. Then again 1965 marks the end of an era with the passing of the Act on Land Reclamation. The name of the Institute was then amended to the State Soil Conservation Service and the operation and task of this Institute was vastly extended and is now as follows:

1) To stop and prevent destruction of vegetated areas and soil and further the reclamation of eroded areas;

2) Protection of growth, which is achieved by obstructing the excessive utilisation of plant growth anywhere in the country;

3) Supervision of all grazing areas in Iceland;

The organisation began its activities by importing barbed wire for fencing in areas which were being eroded and protecting them completely from grazing. Sowing of sand lymegrass or "Melgras", in Icelandic, (Elymus arenarius) was undertaken on a very small scale, mainly on a trial and error type basis and it was even tried to plough down the seed.

The farmers living in the neighbourhood of the eroding areas did not have much faith in the experiments and considered their usefulness of a very limited value. Most of the farmers accepted erosion as a fact of life and would not have anything to do with interfering with God's will. According to the Act of 1975 the Service can take any land and protect it as long as is needed in order to reclaim it and make it fit for controlled grazing (under the supervision of the S.C.S.). However, rather than enforcing the previously mentioned law legal contracts have been made between the landowners and the Service.

It soon became evident that some regrowth took place within

the protected areas, even if the plants grew slowly to begin with.
Elymus arenarius took  root first, then came creeping red fescue,
Agrostis spp. and Poa. Systematic gathering of Elymus seed was
undertaken fairly early on, for the spring sowing.  Simultaneously
with the sowing, erection of windbreak fences was undertaken.  Such
were the main activities of the organisation until 1920.  A considerable number of small desert reclamation fences had been erected
in many places in the country.  People were beginning to believe
that wind erosion and drift-sand could be stopped by protecting
the land from grazing, by windbreaks and by sowing of Elymus in
the worst drifts.

After 1920 the organisation began to expand, more money was
allocated in the budget for this purpose, and a number of larger
soil conservation fences were erected.  The greatest effort was
made in Gunnarsholt which is in the middle of the volcanic area
of the South.

In 1975 over 190,000 hectares (nearly 2% of the total area of
Iceland) were enclosed by reclamation fences, but a number of older
areas had been given back to the farmers for controlled grazing.
The activities of the organisation have been confined almost
entirely to the volcanic areas in Iceland, which indicates that the
basic reason for the incredible erosion which has taken place in
the past is the volcanic origin of the soil (discussed later).
The main activities of the S.C.S. will be discussed independently
below.


## SAND DUNE STABILISATION

The total area of the so called inland dunes has been estimated to
have covered in the past up to $2,000km^2$.  They were only to be
found in the volcanic areas.  The texture of the volcanic loess on
the palagonite formation consists mainly of silt and fine sand plus
some considerable percentage of coarser sand, but very little clay
material.  When the vegetative cover on these areas was weakened,
e.g. by overgrazing so that the roots no longer managed to keep
the topsoil in place, the finer soil fractions (silt, fine sand
and organic matter) were carried away by wind, leaving the coarser
fraction behind.  This sorting action removed the most important
material from the standpoint of plant productivity and water
retention.

Eventually a soil condition was created in which plant growth
was minimised and in extreme situations the sand began to drift and
form unstable dunes.  These encroached on better surrounding lands
and in the last two centuries seriously threatened and even caused
the evacuation of many farming areas.

The only way this problem could be tackled at the time was
to fence them off and protect them completely from grazing. Extensive sowings of the Icelandic Elymus arenarius had to be done as
soon as possible. This plant grew wild in many places and its seeds
were formerly used to some extent as a human food. In severe cases
the establishment of the melgras had to be aided by the erection
of windbreaks. These were either made of lumber or stone walls and
seedlings of Elymus managed to survive on the leeward side of the
walls. Elymus does not thrive well except in drift sand and then
only when there is a fresh annual addition of loose sand. In
general Elymus does not form a continuous vegetative cover, however,
but collects the sand around its roots in small heaps which gradually
grow into hills or stable dunes. Hardy grass species like red
fescues (Festuca rubra) and Agrostis spp. were able to start
growing in the sand once it had stopped drifting. Nowadays the
areas in between the dunes are seeded down with red fescue and the
whole lot top dressed with fertilizer by aircraft and the areas
can be exposed to strictly controlled grazing. One area in particular which was reclaimed in this manner 20 years ago and has been
completely protected from grazing ever since, shows a very good
example of plant succession. The area which was tackled first
is now covered by low growing birches and willows, showing the
tendency towards a climax vegetation.

Today most of these moving inland dunes are now being brought
under control by the Icelandic S.C.S.

Coastal sand dunes are very common in Iceland - the total area
of them being approximately 1,500 km$^2$. The origin of these dunes
is quite different from those discussed previously and they are
also found on the coastal basalt formations in addition to the
palagonite formations.

The sand which is washed ashore by wave action is blown inland
on drying and deposited in mounds or small dunes. The initial
product of this process is the formation of a foredune just beyond
the high-tide mark. Halophytic (those plants which are able to
withstand periodically high concentration of salt in the sand)
grass species, mainly Elymus, gain foothold in these but the dunes
are by no means stable especially if not protected from grazing,
since halophytic plants are usually succulent and sought after by
livestock, and particularly sheep. As the size of the foredunes
increases, the sand at the crest is blown further inland where it
accumulates into another dune etc.

The way in which the reclamation of the coastal sand dunes has
been tackled by the Icelandic S.C.S. is first of all by fencing the
areas off, then sowing of Elymus in long continuous strips just
beyond the scouring action of the tides. Gradual build-up of dunes

amongst the shoots of <u>Elymus</u> then takes place, the upward part of the plant being renewed repeatedly after burial. The whole ridge is bound together by the buried stems of the <u>Elymus</u>, but at first the surface sand is loose, giving the name "Mobile Dunes". Later however colonisation of the surface takes place, often initially by mosses, later on by red creeping fescue and this plant community is later invaded by <u>Agrostis</u> and <u>Poa</u> spp.

Because of the continuous supply of sand from the sea and in some cases sediment from river banks, these areas are extremely vulnerable to grazing, and those within a S.C.S. fence are in most cases not grazed at all.

These areas are a considerable threat to the farming communities situated inland from the sand dunes, and quite often large amounts of sand are deposited on grass fields and blown into farm buildings etc. As is often the case with such sand dune areas, a dune slack or very wet area with a constantly very high watertable forms behind the dunes, being fed by rain and drainage from inland. Farmers have drained this land by open ditches and got good grazing pastures out of it; however there have been cases where these ditches have been filled up by the sand blown from the sand dunes.

Once the sand dunes have been reclaimed or stabilised it is usually necessary to protect them from grazing for a long time, and probably grazing should never be allowed there. Experience has shown these areas to be extremely vulnerable and probably not even fit for recreation purposes.

The reclamation work on the dunes is often hampered by the partial flooding of the sand, especially in the winter, with subsequent killing of the seedlings.

## RECLAMATION OF ERODED AREAS

The disastrous erosion which has taken place in Iceland left a great many scars in the volcanic areas where all the vegetation and most of the topsoil have been completely stripped off, leaving in many instances level fine gravel or fluvioglacial sands. Old lavas were often uncovered in a similar manner, but usually quite a lot of aeolian sand is moving within the lava. Loss of vegetated areas in the rangelands did not result in a lowering of the total number of sheep in this century, so this in fact meant increased grazing pressure on the remaining rangelands.

The staff of the Sand Reclamation Service realised this, and in order to relieve the grazing pressure on the most vulnerable grazing areas the Service started reclaiming eroded areas, even

though they were past the stage in the process of erosion where
they did harm to farmland.  This work was begun in Gunnarsholt in
1946 with various grass seeds imported mostly from the U.S. and
Canada.  Later on varieties from Norway, Denmark and Alaska were
also tried.  Over 50 different grass species and a great many
varieties have been tried, both for reclaiming eroded land and also
for stabilising moving sand and fixing the fine volcanic loess.

Of the grasses tested Festuca rubra, Poa macranthe, Poa
pratensis, Bromus inermis (varieties from Canada, Minnesota and
Alaska), Phleum pratensis and Agrostis tenuis, have given best
results.  Other varieties have often managed to survive for one or
two years and then died or were overtaken by low yielding Icelandic
creeping red fescue.  The grasses mentioned above usually last for
5 or 6 years and then the native creeping red fescue dominates in
most cases.

Some legumes were tried without any promising results although
lupins will survive and grow extremely well where there is no sand-
drift and wild white clover grows in lawns and sheltered areas.

In the year 1948 the cultivation of grasses for agricultural
use was started on the sands at Gunnarsholt.  This cultivation has
been kept on and expanded on a growing scale and today there are
over 1,200 hectares of fields utilised for hay and silage on this
farm where before had been practically bare sands and fine gravel.
To begin with these fields required more fertiliser in order to
yield the same as other meadows, but the incidence of winter kill
has been far less than on other soils in Iceland.  The difference
may lie in the higher organic matter of many Icelandic soils,
(average 10%) and the fact that these "sands" drain more easily.

The results of this reclamation of the "sands" at Gunnarsholt
were so promising that some farmers recognised this and in 1954
the S.C.S. assisted the farmers from a district in a volcanic area
of the South to reclaim 300 hectares of eroded land.  This is all
in one field and the farmers manage the fertilisation, haymaking
etc. co-operatively.  Since then the Service has assisted farmers
in other parts of the country to reclaim for hay making purposes
over 3,000 hectares of level eroded areas.  These grass fields
are grazed in the spring, then cut for hay in the summer and grazed
again in the autumn, thus reducing greatly the grazing pressure
on the rangelands.

The reclamation of the older lavas, once covered by vegetation,
did not become a reality on a big scale until the Service bought
its first fertiliser aeroplane in 1957.  Usually there is quite a
lot of volcanic loess and sand on top of these lavas but very little
vegetation.  By sowing Danish creeping red fescue from the air and

applying fertilisers for the first three years these lavas could
be turned into valuable grazing lands. However, this land had to
be fenced off to control the grazing pressure. Hopefully this
land will be an important factor in reducing the grazing pressure
on the commons in the volcanic areas in the future.

In the past few years vast areas on the boundary between common
grazing areas and homeland have been reclaimed. The farmers take
active part in this work and once the vegetation has established
itself these areas are used for grazing in the autumn. This re-
lieves the grazing pressures on the commons.

## IMPROVEMENTS OF RANGELANDS

Range research has revealed that most of the commons in the volcanic
areas are being overgrazed, with consequent changes in botanical
composition to unpalatable and lower yielding species. Many of
these rangelands are already badly damaged by wind erosion. Prolonged overgrazing has left the soil impoverished, but fertilisation
of the rangelands results in rapid, extensive changes in the
vegetation. Grasses become dominant and plant density increases.
Generally an application of 70 kg nitrogen and 70 kg phosphate per
hectare for two successive years has proved satisfactory. As yet
it is somewhat uncertain how stable these changes are, and whether
it is possible to maintain, without further fertilisation, the new
composition and increased herbage yield under proper grazing.

In the Soil Conservation Act of 1965 provisions were made for
enforcing the farmers to reduce the number of grazing livestock
according to the grazing capacity of the commons (a difficult thing
to accomplish in reality) and to assist the farmers to improve their
rangelands by aerial topdressing of fertilisers. This method of
applying fertilisers is especially applicable for the hummocky
surface in Iceland and the relative inaccessibility of Icelandic
Highlands.

The use of aircraft in soil conservation started here in 1957
and has continued on an ever increasing scale. The organisation
now owns two fertiliser aeroplanes. A Piper Pawnee with carrying
capacity of 500 kg of fertiliser and grass seeds and a DC-3 which
carries 4,000 kg. The DC-3 was given to the S.C.S. by the Iceland
Air and the distributing mechanism was imported from New Zealand.
The pilots from Flugleidir h/f fly this aeroplane completely free
of charge which indicates the tremendous interest and willingness
amongst Icelanders to help the reclamation work.

## OTHER SOIL CONSERVATION SERVICE OPERATIONS

In addition to the main activities of the S.C.S. already described, some of its other tasks should be mentioned. The service runs the Gunnarsholt farm, which is the Service's headquarters and main experimental farm. The livestock there has provided the Service with valuable information on the grazing capacity of reclaimed areas.

The S.C.S. always has some fertiliser observations and seeding trials going, although the Agricultural Research Institute at Keldnaholt does the required research work.

Other aspects of the S.C.S. include erection of dams in order to get some sort of a surface irrigation. This is particularly important on the very porous pumice soils, where the establishment of the first vegetation is especially difficult but once some vegetated cover is achieved it helps retain the moisture and the area can eventually be reclaimed for grazing purposes.

## FUTURE OUTLOOK

Today one half of the forage consumed by large herbivorous animals comes from the rangelands. Unfortunately range research in Iceland has revealed the fact that about one half of the rangelands are overgrazed, and what is even more serious is that most of this land is on the volcanic areas. Furthermore a quarter of the range-lands are perhaps properly utilised and one fourth underused. Increasing population places increasing demands upon the vegetated areas for food production. Therefore, there is an urgent need for land reclamation and range improvement by aerial topdressing of fertilisers. Proper allocation of grazing livestock according to grazing capacity of the land is of particular importance.

On the 1,100th anniversary of Icelandic settlement, the Icelandic parliament provided 1,000 million Icelandic kronur for reclamation purposes i.e. soil conservation, forestry and research. This was to be spent according to a five years programme and has vastly increased the activities of the Soil Conservation Service and the aim is to stop all accelerated erosion in the inhabited areas of Iceland before 1979.

Even if we are gaining slightly in our fight against soil erosion, it is a meagre fact in the face of the blatant truth that Icelanders still owe their country over 20,000 $km^2$ of vegetated land.

# AUTHOR CITATION INDEX

Abdul Rashid, F. D., 216
Acland, J. D., 200
Agricultural Land Service, 179
Agricultural Research Service, 239
Ahmad, E., 147
Ahuja, L. D., 148
Aina, P. O., 218
Al-Ansari, N. A., 179
Al-Jabbari, M., 179
Alberts, E. E., 249
Alonso, C. V., 249
Amein, M., 239
Anderson, G. D., 200
Arivastava, M. M., 149
Arnalds, A., 266
Arnoldus, H. M. J., 179
Axelsson, V., 85

Babu, R., 148
Bagnold, R. A., 239
Bagshawe, F. J., 292
Bailey, G. W., 249
Baker, C. F., 179
Baldock, W. F., 292
Bali, J. S., 148, 150
Bali, Y. P., 148
Ballal, D. K., 148
Banerjee, S., 148
Barayev, A., 44
Barnes, F. A., 179
Barnes, K. K., 182
Barnett, A. P., 232, 249
Barrow, G., 180
Battawar, H. B., 74
Bawden, M. G., 85
Bayfield, N. G., 180
Beasley, D. B., 249
Becher, H. H., 157
Benbrook, C., 278
Bennett, H. H., 51, 180
Bennett, J. P., 239
Bergman, J. M., 226, 239
Bermanakusumah, R., 157

Berndt, H. D., 240
Bernhardt, H., 157
Berry, L., 85, 201
Bhattacharya, J. C., 147
Bhimaya, C. P., 148
Bhumbla, D. R., 148
Bibby, J. S., 180
Binnie & Partners, 85
Blake, G. J., 219
Bollinne, A., 93, 158
Bols, P., 216
Bork, H. R., 157
Bormann, F. H., 85
Bower, M. M., 180
Brade-Birks, S. G., 180
Brandt, W., 278
Bridges, E. M., 180
Brook, T. R., 200
Brotherton, D. I., 180
Broughton, W., 183
Brown, I. W., 179, 180
Brown, K. J., 200
Bruce, R. R., 249
Brune, G. M., 74
Bryan, R. B., 216

Caborn, J. M., 180
Cailleux, A., 292
Carroll, K. D., 85
Carson, M. A., 200, 258
Chakela, Q., 85
Chandrasekhar, K., 148
Changlin, C., 117
Chaudhry, S. P., 149
Cheah, K. F., 216
Chee, B. W., 217
Chee, S. K., 218
Chellapah, K., 218
Chin Fatt, 219
Chong, C. P., 217
Choudhury, S. L., 149
Chow, V. T., 232, 239, 249
Christiansson, C., 85, 201

Christy, L. C., 180
Clawson, M., 279
Clayton, K. M., 180
Cliffe, L. R., 292
Coleman, R. A., 180
Colvin, T. S., 259
Conway, V. M., 180
Cook, K. A., 278
Cory, H., 292
Council for Agricultural Science and
    Technology, 44
Countryside Commission, 180
Cox, M. B., 249
Crawford, N. H., 239, 249
Cross, B. V., 159, 217
Curtis, L. F., 180

Dalal, S. S., 149
Dando, W. A., 44
Daniel, H. A., 249
Daniel, J. B., 217
Danz, W., 158
Das, D. C., 93, 148, 149, 150
Davies, D. B., 180
Davis, S. S., 249
Dawdy, D. R., 226, 239
Dayal, R. B., 149
De Ploey, J., 51, 93
Deivasigamnym, J., 148
Demek, J., 181
Dixon, R. M., 93
Donahue, R. L., 8
Donigian, A. S., Jr., 249
Douglas, I., 74, 85, 181, 215
Dudal, R., 279
Duff, P. C., 292
Dunne, T., 74
Durgin, F. A., 44

Eagle, D. J., 180
Eardley, A. J., 74
East Africa Meteorological Department, 200
Edminster, T. W., 182
Einarsson, T. H., 266
Einstein, H. A., 239, 249
Ekern, P. C., 232
Ellison, W. D., 226, 232, 249
Elwell, H. A., 217, 259
Elwell, H. M., 249
Emmett, W. W., 200
Erh, K. T., 216
Evans, R., 94, 181, 200

Fang, C. S., 239
Farbrother, H. G., 200
Fermor, L. L., 215

Finney, H. J., 94, 258, 259
Finney, J. B., 180
Flaxman, E. M., 239
Flegel, R., 157
Fleming, W. G., 249
Flohr, E. F., 157
Food and Agriculture Organization, 279
Fosbrooke, H., 292
Foster, G. R., 51, 239, 243, 249
Fournier, F., 7, 74, 181, 217, 218
Free, G. R., 232
Frevert, R. K., 182
Fridriksson, S., 7, 266
Fu, R., 117

Gabriels, D., 51
Gandhy, D. J., 149
Gardiner, V., 51
Gegenwart, W., 157
Geikie, A., 181
Geol, K. N., 74
George, C. J., 148
Gerlach, T., 200
Gethin-Jones, G. H., 200
Ghulam, M. H., 216
Gillman, C., 292
Gong, S., 7, 8, 117
Graf, W. H., 239, 249
Grant, H. St. J., 292
Grant, K. E., 44
Greenland, D. J., 219
Gregory, K. J., 51
Greig-Smith, P., 182
Grosse, B., 157
Grottenthaler, W., 157
Guðbergsson, G. M., 266
Guðmunsson, O., 266
Guha, D. P., 75
Gupta, G. P., 149
Gupta, R. N., 74
Gupta, R. S., 149
Gupta, S. K., 94, 150

Hadley, R. F., 85, 200
Hahn, W. G., 44
Haldane, J. B. S., 259
Haldemann, E. G., 292
Hall, D. G., 181
Hard, G., 157
Harding, D. M., 180
Hare, F. K., 8
Harper, L. A., 249
Harrison, E., 292
Harrod, M. F., 180
Hartley, C. W. S., 215
Hartwig, R. O., 249

Harvey, A. M., 181
Hashino, M., 240
Hassenpflug, W., 158
Hatch, T., 259
Hawkes, D. E., 179
Heijnen, J. D., 292
Held, R. B., 279
Hempel, L., 158
Henderson, F. M., 239
Hertke, W., 157
Herz, K., 158
Hill, J. F. R., 292
Hill, W. J., 292
Hillier, F. S., 239
Hobbs, J. A., 8
Holtan, H. N., 239
Hosegood, P. H., 292
Hoskins, W. G., 181
Hudson, N. W., 74, 85, 94, 181, 200, 217,
    218, 279
Huggins, L. F., 249
Huggins, P. M., 292
Hupta, R. P., 148
Hurni, H., 259
Hutchinson, Sir J., 200
Hutt, A. M. B., 292

Imeson, A. C., 181
Iveronova, M. I., 51
Iwersen, I., 158

Jacks, G. V., 181
Jackson, D. C., 94, 200, 217
Jackson, I. J., 292
Jackson, W. A. D., 44
Jacot-Guillarmod, A., 85
Jiang, D., 7, 117
Johannsen, H. H., 158
Johnson, C. B., 159, 217
Johnson, H. P., 249
Johnson, R. H., 181
Johnson, V., 44
Jonker, P. J., 51
Judson, S., 74
Jung, L., 158

Kabysh, S. S., 44
Kaith, D. C., 149
Kakshmanan, V., 150
Kalamkara, R. J., 149
Kamnavar, H. K., 149
Kampen, J., 150
Kapur, O. P., 149
Karale, R. L., 148
Karl, J., 158
Käubler, R., 158

Kaul, R. N., 148, 149
Kennedy, J. F., 240
Kermack, K. A., 259
Kern, H., 158
Khaleel, R., 249
Khan, M. H., 149
Khanna, M. L., 74
Khosla, A. M., 149
Kilian, J., 183
King, J. G. M., 201
King, N. J., 200
Kirkby, M. J., 8, 181, 200, 258, 259
Klug, H., 158
Knisel, W. G., 226
Kramer, L. A., 232
Krishnamoorthy, C. H., 149
Kugler, H., 158
Kulasingam, A., 217
Kurian, K., 148, 149

Laatsch, W., 157, 158
Laflen, J. M., 249, 259
Lai, A. L., 218
Lake, H. M., 215
Lakshmanan, V., 149
Lakshmipathy, B. M., 149
Lal, R., 218, 219
Lal, V. B., 148
Lane, L. J., 249
Langbein, W. B., 74, 215
Langdale, G. W., 249
Laurant, A., 158
Laursen, E. M., 232
Le Mare, P. H., 200
Ledger, D. C., 181
Leonard, R. A., 249
Leopold, L. B., 74, 200
Leser, H., 158
Li, R. M., 249
Liang, H., 94
Libby, L. W., 279
Lichty, R. W., 226, 239
Liddle, M. J., 181, 182
Lieberman, G. J., 239
Liggett, J. A., 240
Linde, Y., 117
Ling, A. H., 218
Linke, M., 158
Linsley, R. K., 239
Little, B. G., 292
Lloyd, R. J., 180
Lombardi, F., 249
Lopez, N. C., 239
Lovell, J. P. B., 181
Low, K. S., 217
Lowdermilk, W. C., 8, 44

Luk, S. H., 51
Lundgren, L., 85

McCool, D. K., 243
McCormack, D. E., 259
McDonald, A. T., 181
McManus, J., 179
Mackney, D., 180
Maene, L. M., 216, 217, 218
Maher, C., 200
Maner, S. B., 74
Mannering, J. V., 8, 51, 232, 279
Manning, M. L., 200
Manokaran, N., 219
Marsh, B. A., 279
Martin, L., 182
Masuch, K., 158
Mathur, C. M., 149
Mathur, H. N., 94, 149, 150
Maurice, O. C., 180
Mehta, K. M., 149
Menne, T. C., 200
Meyer, B., 158, 159
Meyer, L. D., 232, 239, 240, 249, 259
Millar, A., 180
Miller, C. R., 240
Milne, G., 200
Misra, V. C., 149
Mitchell, H. W., 200
Moeyersons, J., 93
Moffett, J. P., 292
Mok, C. K., 216, 218
Mokhtaruddin, A. M., 216, 218
Moldenhauer, W. C., 8, 51, 243
Mollenhauer, K., 159
Monke, E. J., 240, 249
Moore, K. G., 182
Morgan, D. D. V., 258, 259
Morgan, R. P. C., 8, 51, 94, 179, 181, 182,
    215, 216, 258, 259
Mortensen, H., 158
Moshra, P. R., 150
Mou, J., 8, 259
Mückenhausen, E., 159
Muckle, T. B., 183
Müller, M. J., 158
Murota, A., 240
Murray-Rust, D. H., 85, 201
Murthy, V. V. N., 149

Narayanan, R., 217
Narayanswamy, 149
Native Authority Orders, 292
Natural Environment Research Council, 182
Negendank, J. F. W., 158
Negev, M., 240

Neibling, W. H., 249
Nikolov, S., 249
Nobe, K. C., 279
Nordin, C. F., 240
Nortcliff, S., 181
Nowlin, J. D., 249

Oakley, K. P., 182
Olafsson, G., 266
Ong, T. S., 216
Onstad, C. A., 8, 249
Overcash, M. R., 249

Page-Jones, F. H., 292
Palit, B. K., 149
Parker-Sutton, J., 183
Parry, M. S., 292
Patel, G. A., 150
Pathak, S., 149
Peat, J. E., 200
Pidgeon, J. D., 182
Penman, H. L., 200
Pereira, H. C., 292
Phillips, J. F. V., 201
Piest, R. F., 249
Pim, A. W., 85
Platt, S. A., 292
Poornachandran, G., 148, 149
Pushparajah, E., 218

Qian, N., 117
Quansah, C., 51

Radley, J., 182
Ragg, J. M., 182
Raghunath, B., 148, 149
Rama Rao, N. S. V., 150
Rao, V. P., 74
Rapp, A., 51, 85, 201, 292
Ratcliffe, J. B., 179
Reddy, K. R., 249
Reed, A. H., 182
Reeves, R. C., 249
Rege, N. D., 150
Renard, K. G., 243
Rensburg, H. J. V., 201
Rentala, G. S., 148
Riaz, A. G., 149
Richardson, E. G., 182
Richardson, E. V., 240
Richter, G., 8, 182, 158
Rickard, P. C., 182
Rigby, P., 201
Robinson, D. N., 182
Rodda, J. C., 182

Roehl, J. W., 74
Roels, J. M., 51
Rogers, J. S., 232
Rohdenburg, H., 157
Roose, E. J., 94
Rounce, N. V., 201
Roy, K., 150
Ruppert, K., 157, 158

Saarinen, T. F., 44
Sabol, G. V., 239
Sampson, D. N., 292
Sastri, R., 148
Satpute, R. V., 150
Savat, J., 93
Savile, A. H., 292
Saxton, K. E., 249
Schäer, R., 157
Schauer, T., 159
Scheffer, F., 159
Schick, P. A., 201
Schmidt, F., 157
Schmidt, R. G., 159
Schultze, J. H., 159
Schumm, S. A., 74, 85, 201, 215
Schwab, G. O., 182
Schwertmann, U., 157
Schwille, F., 159
Schwing, J. F., 182
Sears, P., 44
Seckler, D. W., 279
Seginer, I., 201
Sen, A. T., 149
Shah, R. L., 150
Shallow, P. G. D., 217
Shamra, P. N., 150
Shankarnarayana, H. S., 149
Shaxson, T. F. S., 279
Shen, H. W., 240
Sheng, T. C., 259
Shri, N., 150
Simanton, J. R., 93
Simms, C., 182
Simms, H. D., 278
Simons, D. B., 240
Singh, A., 149
Singh, B., 150
Singh, D. P., 150
Singh, R. L., 150
Singh, S., 93
Singh Teotia, S. P., 75
Sivasundaram, T., 150
Slaymaker, H. O., 183
Smith, C. N., 249
Smith, D. D., 183, 159, 219, 226, 232, 240,
    249, 258

Smith, F. E., 44
Snyder, W. M., 249
Soil Conservation Society of America, 279
Sokollek, V., 158
Soong, N. K., 218, 219
Soper, J. R. P., 215, 292
Spence, M. T., 183
Sperling, W., 158
Spoor, G., 183
Spratt, E. D., 149
Sreenathan, A., 148, 149
Srivastava, R. P., 150
Stallings, J. H., 8, 183
Stamp, L. D., 183
Staples, R. R., 201
Stehlik, O., 159
Steindorsson, S., 266
Sternberg, H. O., 292
Stocking, M. A., 217
Streeter, D. T., 183
Strelkoff, T., 240
Stuhlmann, F., 292
Sturluson, S., 266
Stuttard, M. J., 94
Subba Rao, K. V., 150
Such, A., 157
Sulaiman, W., 259
Süssmann, W., 158
Swaminathan, M. S., 150
Syed Sofi, 218

Tajudin, A., 216
Tan, E. H., 219
Tan, K. H., 218
Tan, K. Y., 218
Tan, P. Y., 218
Tang, H. T., 219
Tanganyika Territory Department of
    Agriculture, 201
Taylor, G. S., 218
Tejwani, K. G., 94, 148, 149, 150
Temple, P. H., 85, 201, 292
Teoh, T. S., 217
Thomas, A. W., 249
Thomas, I. D., 292
Thomas, P. K., 148, 149
Thomas, T. M., 183
Thomas, W. A., 240
Thong, K. C., 216
Thorgilsson, A., 267
Thornton, D., 201
Thorsteinsson, I., 267
Timmons, J. F., 279
Todorovic, P., 240
Toffaleti, F. B., 240
Tricart, J., 183, 292

Troeh, F. R., 8
Tuckfield, C. G., 183

U.S. Department of Agriculture, 249
Uluguru, L. U., 292

Van Zinderen Bakker, E. M., 85
Van Zuidam, R. A., 183
Vandersypen, D. R., 150
Vanoni, V. A., 240
Varnes, D. J., 85
Vasudevaiah, R. D., 75
Verstappen, H. T., 183
Verstraten, J. M., 159
Vipond, S., 258
Vogt, J., 159
Vohra, B. B., 150
Von Gehren, R., 157
Vorndram, G., 159

Wadia, F. K., 150
Wagner, G., 159
Walling, D. E., 8, 183
Walther, W., 159
Wandel, G., 159
Wang, K., 117
Warren, A., 183
Watson, G. A., 217
Webb, B. W., 8
Wen, Y., 94
Wendl, U., 159
Wenzel, V., 159
Werger, M. J. A., 85
Westerman, P. W., 249
White, G., 183
Whyte, R. O., 181
Wiechmann, H., 159
Wiggins, S., 279

Wildhagen, H., 159
Wilhelms, A., 157
Wilkinson, B., 183
Willerding, U., 158
Williams, A. M., 183
Williams, J. R., 240, 249
Willis, J. C., 240
Wilson, L., 75
Wischmeier, W. H., 183, 159, 217, 219, 226, 232, 240, 249, 258
Withers, B., 258
Wittmann, O., 157
Wohlrab, B., 158, 159
Wong, I. F. T., 216
Wong, P. W., 217
Woolhiser, D. A., 240, 249
Wright, A. E., 292
Wubao Central Hydrometric Station, 117

Xiong, G., 117, 259

Yadav, Y. P., 150
Yalin, Y. S., 249
Yang, C. T., 240
Yap, W. C., 218
Yash, P., 150
Yeoh, C. S., 219
Yevjevich, V., 249
Young, A., 75, 183, 201
Young, K. K., 259
Young, R., 292
Yulin Soil and Water Conservation Station, 117

Zachar, D., 51
Zhao, C., 117
Zingg, A. W., 201, 232

# SUBJECT INDEX

Aerial photographs, 49–50, 79
Afforestation, 7, 108, 110, 146, 290–291
Aircraft, use in soil conservation, 299
Arid zone development, 145

Carrying capacity, 264
Cattle, social significance, 274–275
Central Asia, 12. *See also* Soviet Union
Channel sediment production, 233, 236–238, 244–246
Check dams, 109–110, 138
China, 6, 14–16, 24–25. *See also* Huang He; Wuding Valley
Climate, 63, 203
Climatic change, 10, 12
Coercion, 269, 287
Community commitment, 265, 272, 299
Contour cultivation, 130, 176, 178
Contour plowing, 37
Contour ridging, 194–195, 199
Cost-benefit analysis, 92, 266, 277–278
Cover crops, 208
CREAMS model, 224, 241, 251
Crop cover. *See* Plant cover
Crop residue, 37, 98
Crop yield, 153

Dachai commune, 92
Deforestation, 17, 26–27, 80, 99–100
Desertification, 5, 9–10, 19, 21, 36
Drought, 122
Dust Bowl, 18–19, 32–39

Economic constraints, 28, 30–31, 34, 38, 175, 276–278
Economic return on investment, 92, 138
Enrichment ratio, 247
Eroded sediment characteristics, 246–247
Erosion, downstream effects, 14, 277
Erosion equilibrium. *See* Soil stability
Erosion experiments, 48, 50, 52–53, 58–60, 63
Erosion inventory, 49–51

Erosion mapping, 156
Erosion measurement, 46–48, 50, 62–63, 156, 184, 192, 226. *See also* Erosion pins; Tree roots
Erosion models
    accuracy required, 225–226
    algorithms, 238
    approaches, 223, 233–238, 241–243
    calibration, 225, 238
    errors, 225, 239
    formulation, 227–228, 233, 243–246
    physically-based models, 223
    sensitivity analysis, 225, 255
    simulation models, 223–224, 227, 241–249
    simple annual models, 224, 251–258
    stochastic models, 233
    verification, 225, 238, 247–248, 255–257
Erosion monitoring, 50, 214
Erosion pins, 48, 63
Erosion plots, 46–51, 52–53, 63, 71, 155, 184–197, 214, 288
    collecting tanks, 55–56, 59–60
    collecting troughs, 54
    divisors, 56, 59–61
    flumes, 55
    plot boundaries, 53–54
    sills, 54–55, 192
    surplus runoff disposal, 57, 59
    tank screens, 56
Erosion rates
    effect of catchment size, 48, 71
    estimation from fluvial deposits, 156–157
    seasonal variability, 104, 178
    spatial variability, 4–5, 23, 49–50, 76–77, 88–89, 197
Erosion risk assessment, 170–174, 208–211
Erosion survey, 78, 84, 97, 128, 141, 175, 209–210
Errors
    measurement, 47–48, 192, 226
    models, 225, 239
Europe, 27, 90, 151
Extension services, 96, 265, 269

Factor interactions, 49, 66
Fallowing, 37, 40
Farmers
   appreciation of land, 265, 276, 286
   inability to take risks, 276
   initiative in soil conservation, 93
   use of local knowledge, 264
Financial incentives, 266, 269
Fire, 165, 284, 287
First-order reaction law, 235, 239, 242
Foster and Meyer model, 235–236, 242

Gerlach troughs, 48
Germany, 89, 151–153
Geomorphological mapping, 49
Government policies, 90, 96
Grass, 189, 198
Grass strips, 192, 199
Great Plains, 33–35, 38
Greening program, 270
Gully erosion, 14, 77, 133–136, 164, 234, 236
Gully erosion control, 92, 108–112, 136–138
Gunnarsholt experiments, 265, 298

Huang He. *See also* Wuding Valley
   erosion, 14–16, 25, 89, 99
   soil conservation, 7, 112–114, 116

Iceland, 263–264, 293–300
   overgrazing, 7, 263
   State Soil Conservation Service, 263, 265, 293, 298
Impoundments, 246
India, 92, 118–146
   cattle, 275
   erosion inventory, 128
   land allocation, 271
   land tenure, 272, 274
   Soil Conservation Regions, 90, 122–127
   soil groups, 119–121
Inertia, 275–276
Intercropping, 198
Iron Age, 151

Land
   allocation, 271–272
   capability, 174–175, 210
   fragmentation, 274
   inheritance, 27
   reform, 272
   tenure, 273–274
   use, 49, 64–66, 90, 119, 151, 156, 168, 197–199
Landslides, 77, 80, 286, 289–290
Legislation, 265, 272, 290, 294

Legumes, 6, 198
Lesotho, 82–85

Malaysia
   application of erosion modeling, 257–258
   erosion risk assessment, 208–211
   soil erosion problems, 204–208
Mayan civilization, 16
Mesopotamia, 12, 25
Meyer-Wischmeier model, 223, 227–232, 238, 251
   objectives, 227
   results, 228–231
   structure, 228
Mosquito gauze experiment, 91, 185, 207
Mudflows, 77, 80
Mulch, 91, 193–195, 199, 211

Negev sediment yield model, 238

Overgrazing, 7, 10–11, 99, 151, 198, 299
Overland flow modeling, 243–244

Peat erosion, 164–165
Plant cover, 186, 189, 194, 196–198, 206–208, 211–212
   effect on rainfall properties, 91, 177, 207
   ground cover, 91, 207–208, 211
   procumbent cover, 91, 193–194, 198
Political aspects, 270
Pollution, 5, 157, 224
Population pressure on land, 262, 271
Publicity, 96

Rainfall simulation, 50, 71, 205–206, 214
Rainstorms, 170–171, 204
Recreation, 165
Reserved land, 271, 281–283, 290
Reservoir surveys, 77–80, 141, 162–163
Rill erosion, 77, 80, 165, 234
River valley projects, 140–144
Roman Empire, 26
Rotations, 6

Sahara Desert, 12
Sahel, 4–5
Sand reclamation, 263–264, 294–295, 297–299
Sand stabilization, 108, 146, 295–297
Satellite imagery, 50, 83, 85
Scale-dependence, 49
Sediment
   concentration, 63, 105–106
   delivery ratios, 48, 107
   load sampling, 62–63, 80

rating curves, 63, 105, 225, 233, 239
sinks, 48, 71, 77, 79
sources, 105
transport, 62–63, 237–238, 243
Sedimentation reservoirs, 109, 112, 116
Sheet erosion, 77, 80, 128, 165, 234
Shelterbelts, 108, 176
Shifting cultivation, 26, 139–140, 257–258
Soil conditioners, 212–213
Soil conservation
approaches, 93
discreditation, 288
principles, 108
role of the state, 93, 277
time scales, 276–277
Soil loss tolerance, 5, 38, 48, 168, 213
Soil profiles, interpretation for erosion
survey, 151, 156
Soil stability, 13, 22–23, 29
Soviet Union, 39–43
Splash erosion, 77, 228, 252
Strip cropping, 37, 41
Stubble, 37, 41

Tanzania, 77–81, 197–199. *See also* Uluguru
Mountains
Terraces, 37, 98, 264, 269
bench terracing, 7, 109–110, 133, 265,
286–288, 290
in early civilizations, 6
ladder terracing, 284, 287
narrow-based terracing, 130–132
Tied ridging, 199
Tillage, 90–92, 98, 177–178
Transmigration Plan, 275

Tree roots, use in erosion measurement, 48,
66–68
Tropical rain forest, 89, 93, 207

Uluguru Mountains, 80, 263–264, 280–291
history of conservation policy, 281–289
Land Usage Scheme, 285–288
United Kingdom, 89–90, 160–179
historical and archival evidence, 161–163
erosion risk assessment, 170–174
rates of erosion, 166–169
United States of America
colonial settlement, 17
conservation needs, 97–98
conversion of idle land to crops, 28, 90, 96
erosion, 17–19, 24, 33–39, 89, 91–92, 95–98
Soil Conservation Service, 37, 49, 95
Universal Soil Loss Equation, 153–156, 213,
222–223, 226, 227, 233, 241, 251
Upland sediment production, 233–236

Vegetation, effects of removal, 11, 13, 198,
211–212, 289
Vegetation surveys, 264

Watershed treatment, 141–144
Windbreaks, 108, 146, 151, 295–296. *See also*
Shelterbelts
Wind erosion, 13, 18, 32, 35–36, 40, 42, 77,
145, 152, 165, 263, 293
Wind erosion control, 176
Wuding Valley, 92, 99–116

Yellow River. *See* Huang He

# About the Editor

R. P. C. MORGAN, B.A., M.A., Ph.D., F. R. G. S., was educated at the Universities of Southampton, London, and Malaya. He was Assistant Lecturer in Geography at the University of Malaya from 1968 to 1971; in 1971 he joined the faculty at Silsoe College in England where he is now Reader in Applied Geomorphology. He has worked in Austria, Ecuador, Malaysia, Mexico, and the United States, and made professional visits to Belgium, Iceland, The Netherlands, and Thailand. He is author of *Soil Erosion* (Longman, 1979), editor of *Soil Conservation: Problems and Prospects* (Wiley, 1981) and coeditor of *Soil Erosion* (Wiley, 1980; Spanish transl., *Erosión de Suelos,* Limusa, 1984; Russian transl., *Erosiya Pochv,* Kolos, 1984).